Correlations and Clustering Phenomena in Subatomic Physics

NATO ASI Series

Advanced Science Institutes Series

A series presenting the results of activities sponsored by the NATO Science Committee, which aims at the dissemination of advanced scientific and technological knowledge, with a view to strengthening links between scientific communities.

The series is published by an international board of publishers in conjunction with the NATO Scientific Affairs Division

A	**Life Sciences**	Plenum Publishing Corporation
B	**Physics**	New York and London
C	**Mathematical**	Kluwer Academic Publishers
	and Physical Sciences	Dordrecht, Boston, and London
D	**Behavioral and Social Sciences**	
E	**Applied Sciences**	
F	**Computer and Systems Sciences**	Springer-Verlag
G	**Ecological Sciences**	Berlin, Heidelberg, New York, London,
H	**Cell Biology**	Paris, Tokyo, Hong Kong, and Barcelona
I	**Global Environmental Change**	

PARTNERSHIP SUB-SERIES

1. **Disarmament Technologies**	Kluwer Academic Publishers
2. **Environment**	Springer-Verlag
3. **High Technology**	Kluwer Academic Publishers
4. **Science and Technology Policy**	Kluwer Academic Publishers
5. **Computer Networking**	Kluwer Academic Publishers

The Partnership Sub-Series incorporates activities undertaken in collaboration with NATO's Cooperation Partners, the countries of the CIS and Central and Eastern Europe, in Priority Areas of concern to those countries.

Series B: Physics

Correlations and Clustering Phenomena in Subatomic Physics

Edited by

M. N. Harakeh

Nuclear Physics Accelerator Institute
Groningen, The Netherlands

J. H. Koch

National Institute for Nuclear Physics and High-Energy Physics, and
University of Amsterdam
Amsterdam, The Netherlands

and

O. Scholten

Nuclear Physics Accelerator Institute
Groningen, The Netherlands

Plenum Press
New York and London
Published in cooperation with NATO Scientific Affairs Division

Proceedings of a NATO Advanced Study Institute on
Correlations and Clustering Phenomena in Subatomic Physics,
held August 5 – 16, 1996,
in Dronten, The Netherlands

NATO-PCO-DATA BASE

The electronic index to the NATO ASI Series provides full bibliographical references (with keywords and/or abstracts) to about 50,000 contributions from international scientists published in all sections of the NATO ASI Series. Access to the NATO-PCO-DATA BASE is possible in two ways:

—via online FILE 128 (NATO-PCO-DATA BASE) hosted by ESRIN, Via Galileo Galilei, I-00044 Frascati, Italy

—via CD-ROM "NATO Science and Technology Disk" with user-friendly retrieval software in English, French, and German (©WTV GmbH and DATAWARE Technologies, Inc. 1989). The CD-ROM also contains the AGARD Aerospace Database.

The CD-ROM can be ordered through any member of the Board of Publishers or through NATO-PCO, Overijse, Belgium.

              Library of Congress Cataloging-in-Publication Data

---

Correlations and clustering phenomena in subatomic physics / edited by
  M.N. Harakeh, J.H. Koch, and O. Scholten.
      p.   cm. -- (NATO ASI series. Series B, Physics ; v. 359)
    "Proceedings of a NATO Advanced Study Institute on Correlations
  and Clustering Phenomena in Subatomic Physics, held August 5-16,
  1996, in Dronten, the Netherlands"--T.p. verso.
    Includes bibliographical references and index.
    ISBN 0-306-45612-5
    1. Angular correlations (Nuclear physics)--Congresses.
  2. Clustering phenomena (Nuclear physics)--Congresses.   I. Harakeh,
  M. N. II. Koch, J. H. III. Scholten, O. (Olaf) IV. NATO Advanced
  Study Institute on Correlations and Clustering Phenomena in
  Subatomic Physics (1996 : Dronten, Netherlands) V. Series.
  QC794.6.A5C67   1997
  539.7--dc21   97-17641
                                                                  CIP

Additional material to this book can be downloaded from http://extra.springer.com.

ISBN-13: 978-1-4684-1368-7   e-ISBN-13: 978-1-4684-1366-3

DOI: 10.1007/978-1-4684-1366-3

© 1997 Plenum Press, New York
Softcover reprint of the hardcover 1st edition 1997
A Division of Plenum Publishing Corporation
233 Spring Street, New York, N. Y. 10013

http://www.plenum.com

10 9 8 7 6 5 4 3 2 1

PREFACE

In many areas of physics, such as astrophysics, solid-state physics, nuclear physics and particle physics, a major outstanding problem is a better understanding of correlation phenomena. While in most cases the average properties of a system are rather well understood, the correlations and the resulting clustering are poorly understood. They are reflections of the force mediating the interaction among the constituents and play essential roles in determining the structure of a physical system.

At the largest scales, in astrophysics, it has recently been realized that there are huge voids in space and almost all matter is concentrated on filaments, raising interesting questions concerning the origin of this clustering of matter. In nuclear physics correlation phenomena are important in all its subfields. It has been realized that so-called fluctuations in the one-particle density, which are a manifestation of nucleon-nucleon correlations, are crucial. These are important for an understanding of heavy-ion reactions. This is the subject of modern quantum transport theories. Correlations are also crucial in the description of the high momentum components as observed in quasi-elastic knock-out reactions.

At even smaller scales, correlations play a role among the constituents of elementary particles. An example is the nucleon, primarily a three-quark cluster. However, in deep-inelastic lepton scattering other correlations, also involving the gluons, are seen to be responsible for the spin of the nucleon. Through ultra-relativistic heavy-ion reactions it is investigated under what conditions these correlations vanish and the system behaves like a cauldron of hot quark soup. In effective lagrangian models the effect of the nuclear medium on the quark correlations can be investigated in a more phenomenological approach. To address this problem by starting directly from the fundamental QCD lagrangian, one has to formulate the problem on a space-time lattice.

Our understanding of correlations and clustering in systems of interacting particles has progressed rapidly in the last decade. This is especially true in subatomic systems where, in addition to modern refined experimental techniques, theoretical frameworks and procedures have also been developed to allow an accurate and detailed study of these phenomena.

The Dronten Summer School, a NATO Advanced Study Institute, therefore provided courses by experts in these research activities to present the above-mentioned developments to advanced graduate students and young postdocs. The lectures focussed on a broad variety of aspects of correlations in subatomic physics, ranging from nucleon correlations in nuclear reactions to quark correlations in high-energy hadronic reactions. The lectures included the necessary physics background and discussed the similarities and differences in the different fields of physics.

We wish to thank all speakers for their interesting lectures, which were clear and well geared to the level of the school, and for spending extra time on problem and discussion sessions. Furthermore, the organizers wish to thank Marijke Oskam-Tamboezer for her help in preparing and running the summer school, as well as Grietje van der Tuin for organizational assistance.

Last but not least, we acknowledge the generous financial support from the Science Committee of the North Atlantic Treaty Organization which made this school possible.

M. N. Harakeh
J. H. Koch
O. Scholten

CONTENTS

QUANTUM MONTE CARLO STUDIES OF NUCLEAR GROUND STATES

V. R. Pandharipande

Department of Physics,
University of Illinois at Urbana-Champaign,
1110 West Green Street
Urbana, IL 61801-3080

ABSTRACT

In the many-body theory, nuclei are regarded as bound states of nucleons interacting via two- and three-body interactions. The recent, realistic models of these interactions, and the quantum (variational and Green's function) Monte Carlo methods used to calculate ground states of light ($A \leq 8$) nuclei from a nuclear Hamiltonian containing them are discussed. The nature of correlations among the nucleons in nuclei is studied from the calculated ground states. The correlations seem to be dominated by the pion exchange tensor and the repulsive core parts of the interaction between two nucleons. They generate neutron-proton pair distributions with interesting, femtometer sized toroidal and dumbbell shapes that also describe the short-range structure of the deuteron.

INTRODUCTION

The first glimpse of nuclear structure was provided by the liquid-drop model developed by von Weizsacker[1] and Bethe[2] in 1935. It regarded nuclei as drops of charged, incompressible, liquid nuclear matter and explained nuclear binding energies as sums of volume, surface and Coulomb terms. This model could also explain the observed fission radioactivity of heavy ($A > 230$) nuclei as due to the instability of a large charged liquid drop, which occurs in classical mechanics when the drop's Coulomb energy exceeds twice it's surface energy[3]. Including neutron-proton asymmetry terms, proportional to $(N - Z)^2/A^2$, the liquid-drop model could qualitatively explain the region covered by stable nuclei in the $N - Z$ plane, nuclear sizes, binding energies etc., and refined versions of it are presently being used to study some of the aspects of nuclear structure[4].

The shell model, proposed simultaneously by Mayer[5] and by Haxel, Jensen and Suess[6] in 1949, was the next major step in our understanding of nuclear structure. It provided an explanation of the spins and parities of odd-mass nuclei and their low-energy excitations, of the regions of deformed nuclei, and of the extra stability of nuclei having magic numbers of neutrons on protons. In the naive shell model, nuclear forces are assumed to generate an average potential well whose eigenstates are occupied by the nucleons in the nucleus respecting Pauli exclusion principle. The success of the shell model surprised[7] most physicists who, untill then, believed that collisions among nucleons would strongly limit independent-particle motion inside nuclei. However, it was soon realized that Pauli exclusion suppresses collisions among low-energy nucleons in nuclei and permits single-particle motion in shell-model orbitals. When two-nucleons come close to each other the strong interaction can momentarily transfer them from their orbitals to empty, high-energy orbitals. However, the uncertainty relationship $\Delta E \Delta t \sim \hbar$ limits the time Δt for which the interacting particles are out of their shell-model orbitals. It has been estimated that typically nucleons are in their shell-model orbitals with $\sim 80\%$ probability[8]. The shell model is presently used in a variety of nuclear structure studies.

The liquid-drop and the shell models are macroscopic in nature. They are most useful in their limited domain of applicability. Nuclear many-body theory, initiated by Brueckner[9], Bethe and Goldstone[10] in 1954, is expected to have a much wider domain of applicability. It assumes that the effects of all subnucleonic degrees of freedom associated with mesons and nucleon resonances, and eventually with quarks and gluons can be absorbed into nuclear forces, and hopes to describe all low-energy nuclear phenomena, below pion production threshold, with the Hamiltonian:

$$ H = \sum_{i \leq A} -\frac{\hbar^2}{2m} \nabla_i^2 + \sum_{i < j \leq A} v_{ij} + \sum_{i < j < k \leq A} V_{ijk}. \tag{1} $$

Here A denotes the number of nucleons in the system, and current indications are that terms representing more than three-body interactions in the nuclear Hamiltonian are small.

Nuclear interactions v_{ij} and V_{ijk} have intricate spin-isospin dependence, and therefore the nuclear Schrodinger equation is difficult to solve. Several methods are being used to study nuclear ground states. Approximate methods, useful to study large nuclei like ^{16}O, ^{40}Ca, ... and nuclear matter, have evolved out of Brueckner's approach[11] and the variational principle[12, 13, 14]. The ground states of A=3 and 4 nuclei (^3H, ^3He and ^4He) can now be exactly calculated with the Faddeev-Yakubovsky[15], quantum Monte Carlo (QMC)[16] and hyperspherical basis[17] methods. Recently the QMC method was extended to study A=6 and 7 isotopes of He, Li and Be[18, 19, 20] and studies of A=8 nuclei are underway. In these lectures we will consider only the QMC method. The present models of nuclear forces and deuteron are discussed in Sect. II, QMC is described in Sect. III, and Sect. IV is devoted to some of the aspects of nuclear structure that have been studied with QMC.

THE NUCLEAR HAMILTONIAN

The Two-Nucleon Interaction

Since pions have an exceptionally-small mass the long-range part of the two-nucleon interaction v_{ij} is given by the one-pion exchange potential v_{ij}^π derived in many texts including Sakurai's[21]:

$$v_{ij}^\pi = \frac{1}{3} \frac{f_{\pi NN}^2}{4\pi} X_{ij}^\pi (\tau_i \cdot \tau_j), \tag{2}$$

$$X_{ij}^\pi = Y_\pi(r_{ij}) \sigma_i \cdot \sigma_j + T_\pi(r_{ij}) S_{ij}, \tag{3}$$

$$Y_\pi(r) = \frac{1}{m_\pi r} e^{-m_\pi r} (1 - e^{-br^2}), \tag{4}$$

$$T_\pi(r) = \left(1 + \frac{3}{m_\pi r} + \frac{3}{m_\pi^2 r^2}\right) Y_\pi(r)(1 - e^{-br^2}), \tag{5}$$

$$S_{ij} = 3\sigma_i \cdot \hat{r}_{ij} \sigma_j \cdot \hat{r}_{ij} - \sigma_i \cdot \sigma_j. \tag{6}$$

This form of v^π is determined by the 0^- spin-parity and unit isospin of the pion. At $r \sim 1/m_\pi \sim 1.4$ fm, the tensor term is an order of magnitude larger than the Yukawa term. The $v^\pi(r)$ is modified at small r by the finite size of the nucleons and pions. The cutoff $(1 - e^{-br^2})$ represents this modification in the Urbana-Argonne models[22]; b = 2.1 fm^{-2} in the latest Argonne v_{18} model [23]. Other models use different cutoff's.

Shorter-range parts of the NN interaction are not that well established. The main problem here is that mesons heavier than the pion have masses in excess of 500 MeV, therefore the range of their interaction, given by the inverse of meson mass, is smaller than the nucleon size. In all models we can consider:

$$v_{ij} = v_{ij}^\pi + v_{ij}^R, \tag{7}$$

where v_{ij}^R denotes all the other terms besides v_{ij}^π. In addition to the long-range part given by eq.(2), pion exchange interaction has a short-range part. This part has zero range for point pions and nucleons; it acquires a small range due to the size of nucleons and pions, and we include it in the v^R. The v^R is primarily determined by fitting the NN scattering data. Recent models[24] such as Argonne v_{18}, Nijmegen 93, Reid 93 etc. fit the available data essentially perfectly with χ^2 close to unit. It is necessary to include at least 14 isoscalar terms in the interaction:

$$v_{ij} = \sum_{p=1,14} v^p(r_{ij}) O_{ij}^p. \tag{8}$$

In the Urbana-Argonne models the operators are choosen as:

$$O_{ij}^{p=1,14} = (1, \sigma_i \cdot \sigma_j, S_{ij}, L \cdot S, L^2, L^2 \sigma_i \cdot \sigma_j, (L \cdot S)^2) \otimes (1, \tau_i \cdot \tau_j), \tag{9}$$

in most other models the operator ∇^2 is used in place of L^2. The static part of v_{ij} containing the six terms with operators:

$$O_{ij}^{p=1,6} = (1, \sigma_i \cdot \sigma_j, S_{ij}) \otimes (1, \tau_i \cdot \tau_j), \tag{10}$$

dominates, however, the non-static part is not negligible and gives rise to spin-orbit splitting in nuclei. In addition to the above isoscalar 14 terms, accurate models[23, 24] of v_{ij} also contain small isovector and isotensor terms which describe the breaking of isospin symmetry. We will not discuss these for the sake of brevity.

The Deuteron

The interaction in isospin T=0, spin S=1 two-nucleon states induces most correlations in nuclei, and the only two-nucleon bound state, the deuteron, has T,S=0,1.

3

Therefore we will discuss this interaction in more detail. The $v_{0,1}$ (subscripts 0,1 denote T,S) has the form:

$$v_{0,1} = v_{0,1}^c(r) + v_{0,1}^t(r)S_{ij} + v_{0,1}^{ls}(r)L \cdot S + v_{0,1}^{lsq}(r)(L \cdot S)^2 + v_{0,1}^{lq}(r)L^2, \tag{11}$$

where the superscripts c,t,ls,lsq and lq stand for central, tensor, spin-orbit, quadratic spin-orbit and L^2. The $v_{T,S}^x(r)$, x=c,t,... are trivially related to the $v^{p=1,14}(r)$ in the v_{ij}; for example:

$$v_{0,1}^c = v^{(1)}(r) - 3v^{(2)}(r) + v^{(3)}(r) - 3v^{(4)}(r), \tag{12}$$

$$v_{0,1}^t = v^{(5)}(r) - 3v^{(6)}(r). \tag{13}$$

The $v^{p=1,6}(r)$ are associated with the six operators given in eq.(10); by convention terms with even values of p have $\tau_i \cdot \tau_j$ factor. The dominant static part:

$$v_{0,1}^{static}(\mathbf{r}) = v_{0,1}^c(r) + v_{0,1}^t(r)S_{ij}, \tag{14}$$

is not spherically symmetric due to the $\mathbf{r}$ in the tensor operator S_{ij}, and it depends upon the two-nucleon spin projection M_S:

$$M_S = \sigma_{z,i} + \sigma_{z,j}. \tag{15}$$

It can be easily varified that:

$$\langle M_S = 0|v_{0,1}^{static}(\mathbf{r})|M_S = 0\rangle = v_{0,1}^c(r) - 4v_{0,1}^t(r)P_2(cos\theta), \tag{16}$$

$$\langle M_S = \pm1|v_{0,1}^{static}(\mathbf{r})|M_S = \pm1\rangle = v_{0,1}^c(r) + 2v_{0,1}^t(r)P_2(cos\theta), \tag{17}$$

where θ is the angle between r and the z-axis. Figure 1 shows the expectation value of $v_{0,1}^{static}(\mathbf{r})$ in the $M_S = 0$ state for $\theta = 0$ and $\theta = \pi/2$ calculated from Reid[25], Paris [26], Urbana[22] and Argonne[23] v_{18} models of NN interaction. The long-range interaction, dominated by the tensor part of v^π, is very anisotropic. In the $M_S = 0$ state it is very attractive when $\theta = \pi/2$, i.e. when the antiparallel spins are side by side, and repulsive when $\theta = 0$, i.e. when the antiparallel spins are in a line. It resembles the interaction between two dipole magnets. At small r there is a repulsive core in all directions. A part of the model dependence of the static part of the potential seen in Figure 1 is due to the differences in the non-static parts of the four models considered here.

The anisotropies of $v_{0,1}$ imply that the deuteron is also anisotropic. Experimentally deuteron in known to have an electric quadrupole moment, and thus not spherically symmetric. The deuteron wave function has the form:

$$\Psi_d^M = \left[R_0(r)\mathcal{Y}_{011}^M + R_2(r)\mathcal{Y}_{211}^M\right]\frac{1}{\sqrt{2}}(pn - np), \tag{18}$$

discussed in Blatt and Weisskopf's book[27]. M is the projection of deuteron's total angular momentum on the z-axis, $R_0(r)$ and $R_2(r)$ are the L=0 and 2 (S and D) radial wave functions, the subscripts of the spin-angle $\mathcal{Y}$-functions denote values of L, S and the total angular momentum J, and the last factor of Ψ_d^M represents the T=0 isospin state. This wave function obeys the Schrodinger equation:

$$\left(-\frac{\hbar^2}{m}\nabla^2 + v_{0,1}\right)\Psi_d^M = E_d\Psi_d^M \tag{19}$$

4

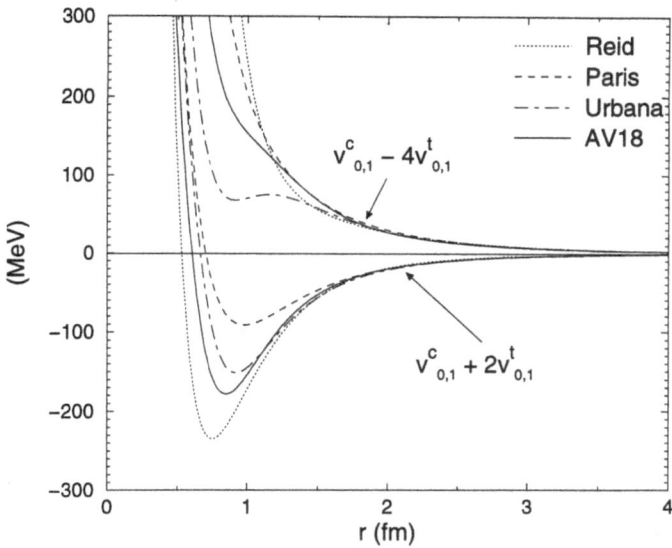

Figure 1. The upper four curves give the $v_{0,1}^{static}$ for $M_S = 0$, $\theta = 0$ in the four models considered. The $v_{0,1}^{static}$ for $M_S = \pm 1$, $\theta = 0$ equals that for $M_S = 0$, $\theta = \pi/2$ given by the lower four curves.

As discussed in texts[27], this Schrodinger equation is equivalent to two coupled equations for $u = rR_0$ and $w = rR_2$. In Urbana-Argonne models they are given by:

$$E_d u = -\frac{\hbar^2}{m}u'' + v_{0,1}^c u + \sqrt{8}v_{0,1}^t w, \qquad (20)$$

$$E_d w = -\frac{\hbar^2}{m}\left(w'' - \frac{6}{r^2}w\right) + \left(v_{0,1}^c - 2v_{0,1}^t - 3v_{0,1}^{ls} + 9v_{0,1}^{lsq} + 6v_{0,1}^{lq}\right)w + \sqrt{8}v_{0,1}^t u \quad (21)$$

They can be easily solved by standard numerical methods[28].

The $R_0(r)$ and $R_2(r)$ calculated from the four interaction models considered in Figure 1 are shown in Figure 2. They have smaller model dependence than the static parts shown in Figure 1, and contain all the information on deuteron structure. However, more insight can be obtained by studying the deuteron density distribution given by[29]:

$$\rho_d^{M=0}(\mathbf{r}') = \frac{4}{\pi}[C_0(2r') - 4C_2(2r')P_2(\cos\theta)], \qquad (22)$$

$$\rho_d^{M=\pm 1}(\mathbf{r}') = \frac{4}{\pi}[C_0(2r') + 2C_2(2r')P_2(\cos\theta)], \qquad (23)$$

$$C_0(r) = R_0^2(r) + R_2^2(r), \qquad (24)$$

$$C_2(r) = \sqrt{2}R_0(r)R_2(r) - \frac{1}{2}R_2^2(r). \qquad (25)$$

Here $\mathbf{r}' = \mathbf{r}/2$ is the vector from deuteron center of mass, $(\mathbf{r}_1 + \mathbf{r}_2)/2$, and the $\rho_d^M(\mathbf{r}')$ is normalized such that:

$$\int \rho_d^M(\mathbf{r}')d^3r' = 2, \qquad (26)$$

since deuteron has two nucleons. The deuteron density has azimuthal symmetry, thus it can be obtained from the $\rho_d^M(x', z')$ shown in Figs. 3 and 4 for the Argonne v_{18},

5

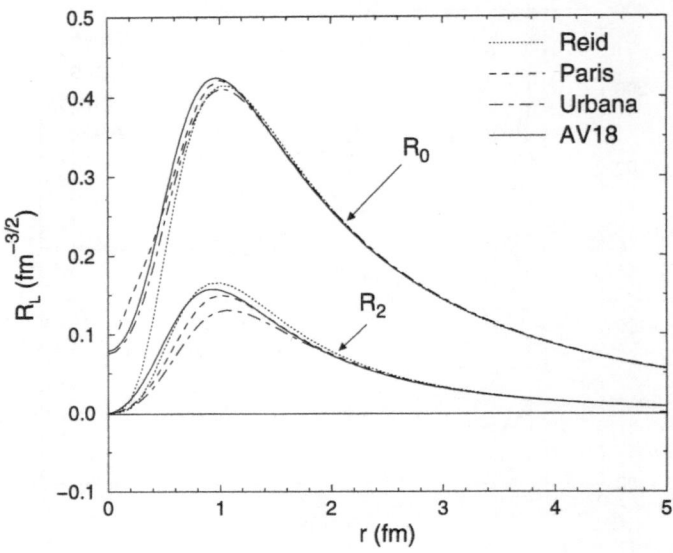

Figure 2. Deuteron radial functions calculated from four models of nuclear forces.

by rotations about z'-axis. The $\rho_d^M(r', \theta)$ obtained from the four models is shown in Figure 5. We note that:

1. The peak density of deuteron is $\sim 0.35 fm^{-3}$, i.e. about twice the nuclear-matter density.

2. The density in M=0 state is large when $\theta = \pi/2$, and it is small along the z-axis ($\theta = 0$). This follows from the static potential in the $M_S = 0$ state being attractive for $\theta = \pi/2$ and repulsive for $\theta = 0$ (Figure 1). Note that the dominant componant of the deuteron wave function in the $M = 0$ state has $M_S = 0$. The maximal tensor correlation corresponds to $\rho_d^{M=0}(r', \theta = 0) = 0$; the smallness of the calculated $\rho_d^{M=0}(r', \theta = 0)$ indicates that deuterons have near maximal tensor correlation. Equidensity surfaces of $\rho_d^{M=0}$ having densities larger than $\sim 0.05 fm^{-3}$ have toroidal shapes as illustrated in Figure 6. due to the smallness of density along z-axis.

3. The densities in the $M = \pm 1$ states of the deuteron are equal and largest along the z'-axis. The static interaction in these states is most attractive when $\theta = 0$. Equidensity surfaces of $\rho_d^{M=\pm 1}$ having densities larger than $\sim 0.2 fm^{-3}$ have dumbbell shapes shown in Figure 6. As discussed in Ref. 29, we may regard the deuteron in $M = \pm 1$ states as the $M = 0$ toroidal-shaped state rotating about an axis perpendicular to the torus symmetry axis. Most earlier literature considers only this dumbbell or cigar shape of the deuteron in the $M = \pm 1$ states. The more compact toroidal shape of the deuteron was only recently noticed[29].

4. The observed electron-deuteron elastic-scattering form factors A(q), B(q) and $T_{20}(q)$ support the above described structure of the deuteron[29].

5. A toroidal structure for the ground state of deuteron was predicted many years ago[30, 31] using classical Skyrme field theory[32] related to QCD in the limit of infinite colours. In this classical theory one obtains a toroidal ground state of $\sim 1 fm$ in size with $\sim 150 MeV$ binding energy. In the classical limit the deuteron binding energy is given by the minimum of the static potential. Figure 1 shows that realistic models of nuclear forces predict it to be 100 to 200 MeV. This agreement between classical Skyrme-model deuteron binding energy and the minimum values of the realistic potentials is

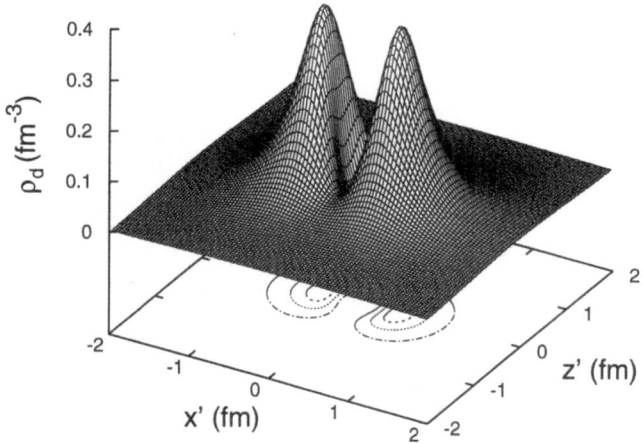

Figure 3. The $\rho_d^{M=0}(x', z')$ calculated from Argonne v_{18} interaction.

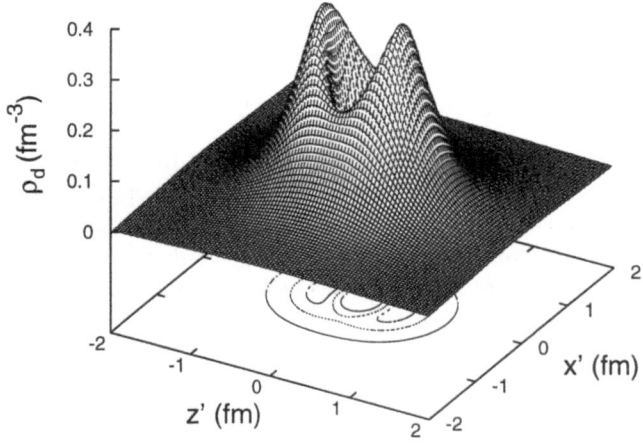

Figure 4. The $\rho_d^{M=\pm1}(x', z')$ calculated from Argonne v_{18} interaction.

7

intriguing. There have been recent attempts to include quantum corrections in Skyrme field theory[33]; they reduce the binding energy to near experimental value.

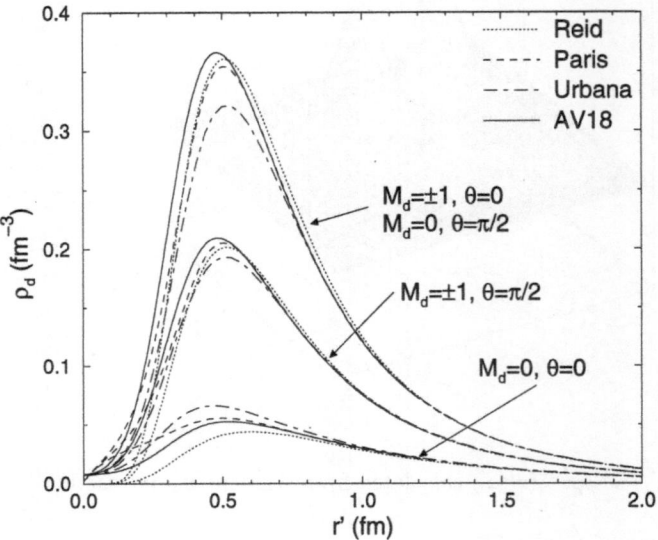

Figure 5. The $\rho_d^M(r', \theta)$ for $M = \pm 1$, 0 and $\theta = 0$, $\pi/2$

6. In nuclear many-body theory, nucleons are treated as interacting point particles. The densities shown in Figs. 3-6 are point nucleon densities. The distribution of matter inside a nucleon is not well understood. That of electromagnetic charge and current is approximately given by the dipole form factor in momentum space. For the sake of illustration, and at the risk of oversimplification, we assume that the electromagnetic (charge and current) distributions of a nucleon are given by the exponential:

$$\rho_{em}(r) = \rho_0 e^{-r/a}. \tag{27}$$

This form and the value $a = 0.23 fm$ is obtained by inverting the dipole form factor. Since this procedure is exactly valid only in the non-relativistic region, the above $\rho_{em}(r)$ may have significant relativistic corrections at $r < 1/m \sim 0.2 fm$. Figure 3 shows that at the peak density in the deuteron, the two nucleons in the deuteron are only 1 fm apart. In Figure 7 we show the ρ_{em} of two nucleons one fm apart, along the line joining them. They seem to have a significant overlap.

The Three-nucleon Interaction

The binding energy of ^{3}H calculated from realistic two-nucleon interactions alone[24] is typically -7.6 against the observed value of -8.48 MeV. The equilibrium density of nuclear-matter predicted by Hamiltonians containing only two-nucleon interactions is too large[34]. Therefore it is necessary to add at least three-nucleon interactions to the nuclear Hamiltonian to obtain agreement with experiment. The long-range part of V_{ijk} is given by the two-pion exchange interaction studied by Fujita and Miyazawa[35]. It has the form:

$$V_{ijk}^{2\pi} = \sum_{cyc} A_{2\pi} \left(\left\{ X_{ij}^{\pi}, X_{jk}^{\pi} \right\} \left\{ \tau_i \cdot \tau_j, \tau_j \cdot \tau_i \right\} + \frac{1}{4} [X_{ij}^{\pi}, X_{jk}^{\pi}][\tau_i \cdot \tau_j, \tau_j \cdot \tau_i] \right), \tag{28}$$

Figure 6. Equidensity surfaces of the deuteron in $M = \pm 1$ state (left) and $M = 0$ state (right) for $\rho_d = 0.24 fm^{-3}$ calculated from the Argonne v_{18} model.

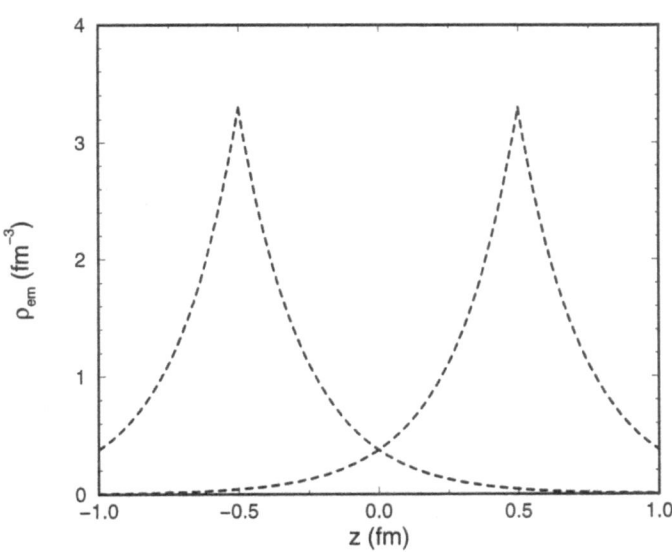

Figure 7. The dashed lines show the $\rho_{em}(z)$ of nucleons located at $z = \pm 0.5 fm$.

and is attractive. In order to obtain the empirical equilibrium density of nuclear matter, it has to be supplimented by a repulsive term which is phenomenologically assumed to have the form :

$$V_{ijk}^R = \sum_{cyc} U_0 T_\pi^2(r_{ij}) T_\pi^2(r_{ik}) \tag{29}$$

in Urbana-Argonne models. The parameters $A_{2\pi}$ and U_0 are adjusted to reproduce the binding energy of ^{3}H and the equilibrium density of nuclear matter[18]. The resulting value of $A_{2\pi}$ is generally close to that of the Fujita-Miyazawa interaction.

The three nucleon interaction is much weaker than the two-nucleon interaction, however, it enhances the tensor correlations slightly and therefore cannot be treated as a first-order perturbation in accurate calculations. Hamiltonians containing two- and three-nucleon interactions accurately reproduce the binding energy of ^{4}He, thus it seems that four-nucleon interactions are not very important in the nuclear Hamiltonian.

QUANTUM MONTE CARLO METHODS

The variational (VMC) and Green's function (GFMC) Monte Carlo methods were first used by McMillan[36] and Kalos[37] respectively to study simple Bose systems. Here we define simple systems as those in which the interaction potentials depend only on interparticle distances. Coulomb systems, atomic helium liquids etc. are examples of simple systems; due to the strong dependence of nuclear forces on the spin-isospin orientations of the interacting nucleons, nuclei are not simple systems. The Monte Carlo methods were subsequently developed to study simple Fermi systems [38, 39] and nuclei[40, 41].

Monte Carlo Integration

The Monte Carlo method offers means to calculate multidimensional integrals. Let $R = \mathbf{r}_1, \mathbf{r}_2 ... \mathbf{r}_A$ be a 3A dimensional vector that specifies the spatial configuration of the system, i.e. the positions of all particles 1 to A. Let $F(R)$ be a function of R. We choose a suitable weight function $W(R)$ which is real-positive and normalizable so that it can be regarded as a probability distribution, and write the required integral of $F(R)$ as:

$$\int F(R)dR = \int \frac{F(R)}{W(R)} W(R)dR. \tag{30}$$

Let $[R_i], i = 1, N_C$ be an ensemble of N_C configurations distributed with the weight $W(R)$. This implies that, when N_C is large, the number of configurations with R_i within a volume dR centered at R is proportional to $W(R)dR$. The ensemble average is defined as:

$$\overline{F/W} = \frac{1}{N_C} \sum_{i=1,N_C} F(R_i)/W(R_i), \tag{31}$$

and

$$\int F(R)dR = \lim_{N_C \to \infty} \overline{F/W} \int W(R)dR. \tag{32}$$

In practice, N_C is finite and the Monte Carlo method gives the value of the integral with a standard deviation δ:

$$\int F(R)dR = (\overline{F/W} \pm \delta) \int W(R)dR, \tag{33}$$

$$\delta = \sigma/\sqrt{N_C}, \tag{34}$$

10

$$\sigma^2 = \frac{1}{N_C} \sum_{i=1,N_C} [F(R_i)/W(R_i) - \overline{F/W}]^2. \tag{35}$$

Equations (33-35) assume that the values of $F(R_i)/W(R_i)$ have a normal distribution. This is generally not the case, and block averages[42] must be performed to obtain a normal distribution and calculate δ.

This method is most useful when the variance σ^2 of $F(R)/W(R)$ is small; then results with tolerably small standard deviation δ can be easily obtained. In quantum Monte Carlo calculations it is most convenient to use:

$$W(R) = \Psi^\dagger(R)\Psi(R), \tag{36}$$

where $\Psi(R)$ is the wave function. The $W(R)$ is then the probability to find the system in the configuration R, and the ensemble $[R_i]$ constitutes N_C snapshots of the system. The expectation value of an operator $\hat{O}$ is then given by:

$$\langle \hat{O} \rangle = \int \Psi^\dagger(R)\hat{O}\Psi(R)dR \Big/ \int \Psi^\dagger(R)\Psi(R)dR, \tag{37}$$

$$= \overline{\Psi^\dagger \hat{O} \Psi / \Psi^\dagger \Psi} \pm \delta. \tag{38}$$

Metropolis Sampling

The configurations $[R_i]$ are obtained by "sampling" the weight function $W(R)$. When $W(R)$ is a simple function one can use some of the specialized methods compiled by Kalos and Whitlock[43] to sample it. In many cases, however, $W(R)$ is not a simple function. It is then very convenient to use the Metropolis method[44] to obtain $[R_i]$.

The Metropolis algorithm is based on the principle of detailed balance. Consider $[R_i]$ as a set of points representing states of a dynamical system in equilibrium, and define:

$$\rho_{conf.} = \text{density of points } [R_i]. \tag{39}$$

If these points are allowed to move with transition probability $T(R \rightarrow R')$ to go from configuration R to R', then:

$$\rho_{conf.}(R')T(R' \rightarrow R) = \rho_{conf.}(R)T(R \rightarrow R'), \tag{40}$$

is the condition for dynamical equilibrium. When sampling the weight function $W(R)$ we want to have

$$\rho_{conf.} \propto W(R), \tag{41}$$

thus $T(R \rightarrow R')$ must satisfy:

$$\frac{T(R' \rightarrow R)}{T(R \rightarrow R')} = \frac{\rho_{conf.}(R)}{\rho_{conf.}(R')} = \frac{W(R)}{W(R')}. \tag{42}$$

The following $T(R \rightarrow R')$ satisfies the above and is convenient to use:

$$T(R' \rightarrow R) = 1 \text{ if } W(R) > W(R'), \tag{43}$$

$$T(R' \rightarrow R) = W(R')/W(R) \text{ otherwise.} \tag{44}$$

The Metropolis algorithm is implemented by sequentially obtaining configurations $[R_i]$ in a "Metropolis walk." Let R_i be the ith configuration. To obtain the R_{i+1}, make a random step $R' = R_i + \Delta R$ and accept it with the probability $\min[1, W(R')/W(R_i)]$. If the step is accepted $R_{i+1} = R'$ and if it is rejected, then $R_{i+1} = R_i$. More details of the algorithm can be found in Refs. 42, 43. The consecutive samples of a walk have significant autocorrelation; generally every nth step is retained in the ensemble $[R_i]$, and n is determined by studying the autocorrelation function[42].

11

VMC for Simple Systems

Consider a system of A spin 1/2 Fermions interacting with each other with a spin-independent potential $v_{ij} = v(r_{ij})$. It's states can be labeled with the number A_u of spin-up particles; $A_d = A - A_u$ is the number of spin-down particles. The simple interaction does not flip spins, therefore A_u and A_d are good quantum numbers. Moreover, we can assign particles 1 to A_u to have spin-up, $A_u + 1$ to A to have spin-down; the interactions cannot change these assignments. In this case the wave function of the system is just a complex number $\Psi(R)$. The $\Psi^*(R)\Psi(R)$ gives the probability to find spin-up particles at $\mathbf{r}_1, ...\mathbf{r}_{A_u}$ and spin-down particles at $\mathbf{r}_{A_u+1}, ...\mathbf{r}_A$. The energy expectation value:

$$\langle H \rangle = \int \Psi^*(R)H\Psi(R)dR \, / \int \Psi^*(R)\Psi(R)dR, \qquad (45)$$

can then be easily calculated for any chosen $\Psi(R)$ by the Monte Carlo method using $W(R) = \Psi^*(R)\Psi(R)$.

In variational Monte Carlo (VMC) method the ground-state wave function $\Psi_0(R)$ is approximated by a variational wave function $\Psi_V(R)$ with many variational parameters, which are determined by minimizing $\langle H \rangle$. Many researchers have contributed to the development of the forms of Ψ_V suitable to describe various systems such as atoms, quantum-liquid drops, etc., and methods to minimize the energy by varying the parameters of Ψ_V. For the sake of brevity, we will not discuss them in detail. A discussion of several forms of Ψ_V and minimization methods can be found in the Proceedings of the Elba Workshop on Monte Carlo Methods in Theoretical Physics[45].

A common choice of Ψ_V for simple Fermi systems contains a product of two- and three-body correlation functions $f(r_{ij})$ and $F(r_{ij}, r_{jk}, r_{ki})$ and antisymmetric Slater determinants $\Phi_u(\mathbf{r}_1...\mathbf{r}_{A_u})$ an $\Phi_d(\mathbf{r}_{A_u+1}...\mathbf{r}_A)$ for the spin-up and down particles:

$$\Psi_V(R) = \left[\prod_{i<j<k} F(r_{ij}, r_{jk}, r_{ki}) \right] \left[\prod_{i<j} f(r_{ij}) \right] \Phi_u(\mathbf{r}_1...\mathbf{r}_{A_u})\Phi_d(\mathbf{r}_{A_u+1}...\mathbf{r}_A). \qquad (46)$$

This Ψ_V reduces to the Hartree-Fock form when the correlation functions $f = F = 1$. When the three-body correlation $F = 1$, but the pair correlation $f \neq 1$, one obtains the Jastrow form. More refined Ψ_V have additional back-flow correlations[46].

When Ψ_V is close to the exact Ψ_0, the configuration energy $E(R)$ defined as:

$$E(R) = \Psi_V^*(R)H\Psi_V(R)/\Psi_V^*(R)\Psi_V(R), \qquad (47)$$

is close to the exact E_0 for all R. In this case, the variational energy:

$$E_V = \frac{1}{N_C} \sum_{i=1,N_C} E(R_i), \qquad (48)$$

is close to E_0, and the variance:

$$\sigma^2 = \frac{1}{N_C} \sum_{i=1,N_C} [E(R_i) - E_V]^2, \qquad (49)$$

and therefore the standard deviation δ are both small. Thus, good Ψ_V's lead to VMC calculations with small statistical errors. Perfect $\Psi_V = \Psi_0$ would give zero variance, and $E_V = E_0$ with no statistical error. Many researchers[45] find it more convenient to minimize the combination $aE_V + b\sigma^2$ with suitably chosen positive constants a and b.

VMC for Nuclei

Nucleons in nuclei do not have conserved spin-isospin projection. The nucleon i can change from being a neutron (proton) to proton (neutron) by emitting a virtual $\pi^-(\pi^+)$-meson carrying one-pion exchange interaction. The v_{ij}^π has a $\tau_i \cdot \tau_j = 2 - P_{ij}^\tau$ factor, where P_{ij}^τ exchanges the isospins of nucleons i and j. It thus can change a state where nucleons i and j are respectively p and n to that in which i is n and j is a p. The $\sigma_i \cdot \sigma_j$ component of the interaction exchanges spins, and the $\sigma_i \cdot \hat{r}_{ij}\sigma_j \cdot \hat{r}_{ij}$ term in the tensor force can flip the spin of either i or j or both i and j. Therefore simple scalar functions cannot be used to describe nuclear wave functions. Even the simple deuteron wave function (18) is not a scalar function of $\mathbf{r} = \mathbf{r}_1 - \mathbf{r}_2$.

One can regard the nuclear wave function as a vector function of R

$$\Psi(R) = \sum_\alpha \psi_\alpha(R)|\alpha\rangle; \qquad (50)$$

the $\Psi_\alpha^*(R)\Psi_\alpha(R)$ gives the probability of finding the nucleons in the configuration R and spin-isospin state $|\alpha\rangle$. The tensor forces mix all possible 2^4 spin states, while charge conservation implies that only the $A!/[(A-Z)!Z!]$ states which label Z out of the A nucleons as protons, are allowed in $\Psi(R)$. Thus the total number of spin-isospin states $|\alpha\rangle$ of a nucleus is given by:

$$\alpha_{total} = 2^A A!/[(A-Z)!Z!]. \qquad (51)$$

For example, $\alpha_{total} = 24$ in ^{3}H; and the states $|\alpha\rangle$ that can contribute to any eigenstate $\Psi(R)$ of ^{3}H are:

$$(uuu, uud, udu, udd, duu, dud, ddu, ddd) \otimes (pnn, npn, nnp) \qquad (52)$$

(note u,d are used here for spin direction, and not for quarks). In this representation the Hamiltonian is $\alpha_{total} \times \alpha_{total}$ matrix operator that depends upon R. The kinetic energy is a diagonal operator:

$$\Psi^\dagger(R)\nabla_i^2\Psi(R) = \sum_\alpha \psi_\alpha^*(R)\nabla_i^2\psi_\alpha(R), \qquad (53)$$

whereas the interactions have both diagonal and non-diagonal contributions. For example, the static part gives:

$$\Psi^\dagger(R)v_{ij}^{static}\Psi(R) = \sum_{p=1,6} \sum_{\alpha,\alpha'} \psi_\alpha^*(R)v^p(r_{ij})\psi_{\alpha'}(R)\langle\alpha|O_{ij}^p(R)|\alpha'\rangle. \qquad (54)$$

Details of the calculation of momentum-dependent interactions with operators O^p having $p = 7$ to 14, as well as three-body interactions can be found in Refs. 40 and 47.

We note that the vector length α_{total} increases very rapidly with the size of the nucleus. It is 24, 96, 1280 and 17,920 for ^{3}H, ^{4}He, ^{6}Li and ^{8}Be respectively. Therefore VMC calculations have only been carried out for nuclei with $A \leq 7$ while those for $A = 8$ are in progress. It is obvious that VMC will not be possible for $A > 10$ nuclei in the foreseeable future. New methods, such as cluster Monte Carlo[13] are being developed for larger nuclei.

It is possible to reduce α_{totle} by a factor of the order of two, by using eigenstates of total isospin T and T_z. Recent calculations [18, 19, 20] of $A = 6$ and 7 nuclei use such reduced isospin states. The total spin S is not a good quantum number in nuclei; for example, ^{4}He ground state has a large superposition of $S = 0$, $L = 0$ and $S = 2$, $L = 2$ states with $J^\pi = 0^+$. Therefore the number of spin states cannot be reduced below their maximum value 2^A.

The simplest variational wave functions for nuclei have the form:

$$\Psi_V = \left(S \prod_{i<j} F_{ij} \right) \Phi, \tag{55}$$

$$F_{ij} = \sum_{p=1,6} f^p(r_{ij}) O_{ij}^p. \tag{56}$$

The Φ is antisymmetric and the product of the pair correlation operators F_{ij} is symmetrized because F_{ij} and F_{ik} do not commute. There are $A(A-1)/2$ F_{ij} operators, and their product has $[A(A-1)/2]!$ permutations which need to be summed in the $S \prod F$. It is impractical to sum over these large number of permutations explicitly for each configuration. However, it is possible to do it implicitly in the Monte Carlo calculation. The variational energy is given by:

$$E_V = \frac{\sum_{N_L,N_R} \int \Phi^\dagger(R)(\Pi_{N_L} F_{ij}^\dagger) H (\Pi_{N_R} F_{ij}) \Phi(R) dR}{\sum_{N_L,N_R} \int \Phi^\dagger(R)(\Pi_{N_L} F_{ij}^\dagger)(\Pi_{N_R} F_{ij}) \Phi(R) dR} \tag{57}$$

where $\Pi_N F_{ij}$ denotes a product of the F_{ij} operators in a specific order labeled by N. Both N_L and N_R can have values from 1 to $[A(A-1)/2]!$ spanning all possible orders. The numerator and denominator of E_V are evaluated by the Monte Carlo method with the weight function:

$$W(N_L, N_R, R) = |\Phi^\dagger(R)(\Pi_{N_L} F_{ij}^\dagger)(\Pi_{N_R} F_{ij}) \Phi(R)| \tag{58}$$

which is sampled to obtain an ensemble of configurations labeled with $N_L(i)$, $N_R(i)$ and R_i.

Recent VMC calculations of nuclei[48, 12, 20] use more refined variational wave functions containing spin-orbit correlations, three-body correlation functions and operators. The best wave functions[48] give energies that are above the exact E_0 of ^{3}H and ^{4}He by only $\sim$ 2%. The error in the optimum Ψ_V, defined as $\langle \Delta | \Delta \rangle$, where $\Delta = \Psi_0 - \Psi_V$ (both Ψ_0 and Ψ_V have unit norm), is only $\sim$0.0008 for ^{3}H and < 0.005 for ^{4}He as estimated in Refs. 48 and 20. However, the present forms of Ψ_V are not that accurate for $A = 6$ and 7 nuclei. Their E_V's are above the exact E_0's by $\sim$ 10%, and $\langle \Delta | \Delta \rangle$ is estimated to be $\sim$0.05 in Ref. 20.

The GFMC Method

This method aims to project out the lowest-energy state having the same quantum numbers J, M, T, T_Z of a trial state Ψ_T, from Ψ_T, using:

$$\Psi(\tau) = e^{-(H-E_0)\tau} \Psi_T = \Psi_0 \ (\lim \tau \to \infty) \tag{59}$$

The τ is called imaginary time or inverse temperature or just time for brevity, and $exp[-(H - E_0)\tau]$ propagates states in (imaginary) time. It is possible to calculate the Green's function for the short-time propagator:

$$G_{\alpha,\beta}(R, R') = \langle R, \alpha | e^{-(H-E_0)\Delta\tau} | R', \beta \rangle. \tag{60}$$

It is a matrix in spin-isospin state, however, for the sake of brevity we will omit the α, β subscripts. The $\Psi(\tau)$ is then obtained as:

$$\Psi(\tau = n\Delta\tau, R_n) = [e^{-(H-E_0)\Delta\tau}]^n \Psi_T, \tag{61}$$

$$= \int G(R_n, R_{n-1})...G(R_1, R_0) \Psi_T(R_0) \, dR_{n-1}...dR_1 dR_0. \tag{62}$$

14

The mixed expectation values are defined as:

$$\langle \hat{O}(\tau) \rangle_{mixed} = \langle \Psi_T | \hat{O} | \Psi(\tau) \rangle / \langle \Psi_T | \Psi(\tau) \rangle, \tag{63}$$

$$= \frac{\int \Psi_T^\dagger(R_n) \hat{O} G(R_n, R_{n-1}) ... G(R_1, R_0) \Psi_T(R_0) \, dP_n}{\int \Psi_T^\dagger(R_n) G(R_n, R_{n-1}) ... G(R_1, R_0) \Psi_T(R_0) \, dP_n}, \tag{64}$$

where P_n denotes the "path" of points $R_0, R_1 ... R_n$, and we have to integrate over all paths.

Generally, we want expectation values calculated with the exact Ψ_0:

$$\langle \hat{O} \rangle = \langle \Psi_0 | \hat{O} | \Psi_0 \rangle / \langle \Psi_0 | \Psi_0 \rangle = \langle \hat{O}(\tau) \rangle \lim(\tau \to \infty), \tag{65}$$

$$\langle \hat{O}(\tau) \rangle = \langle \Psi(\tau) | \hat{O} | \Psi(\tau) \rangle / \langle \Psi(\tau) | \Psi(\tau) \rangle. \tag{66}$$

When Ψ_T is close to Ψ_0, i.e.

$$\delta\Psi(\tau) = \Psi(\tau) - \Psi_T, \tag{67}$$

is small we can use:

$$\langle \hat{O}(\tau) \rangle = 2\langle \hat{O}(\tau) \rangle_{mixed} - \langle \hat{O} \rangle_T + \text{ terms of order } |\delta\Psi(\tau)|^2, \tag{68}$$

$$\langle \hat{O} \rangle_T = \langle \Psi_T | \hat{O} | \Psi_T \rangle / \langle \Psi_T | \Psi_T \rangle, \tag{69}$$

and neglect terms of order $|\delta\Psi(\tau)|^2$ and higher.

There is, however, an important exception. Since H commutes with the evolution operator, the mixed estimate of the energy is given by:

$$\langle H(\tau) \rangle_{mixed} = \langle \Psi_T | H e^{-(H-E_0)\tau} | \Psi_T \rangle / \langle \Psi_T | e^{-(H-E_0)\tau} | \Psi_T \rangle \tag{70}$$

$$= \langle \Psi_T | e^{-(H-E_0)\tau/2} H e^{-(H-E_0)\tau/2} | \Psi_T \rangle / \langle \Psi_T | e^{-(H-E_0)\tau/2} e^{-(H-E_0)\tau/2} | \Psi_T \rangle \tag{71}$$

$$= \langle H(\tau/2) \rangle. \tag{72}$$

Therefore $\langle H(\tau) \rangle_{mixed}$ equals the exact E_0 in the limit $\tau \to \infty$. Moreover, since $\langle H(\tau) \rangle_{mixed}$ is an expectation value of H, by the variational theorem it is $> E_0$.

The three major steps in GFMC are:

(i) Choose Ψ_T;

(ii) Choose the short-time propagator $exp[-(H - E_0)\Delta\tau]$;

(iii) Choose an appropriate weight function to sample the paths P_n in the integrals (64).

(ii) and (iii) are discussed in the following subsections; here we consider (i) and compare VMC and GFMC.

One should obviously choose the best available approximation to the exact Ψ_0 as the trial function. Often the optimum variational wave function Ψ_V is the best available, hence we consider $\Psi_T = \Psi_V$. In VMC the error in the calculated energy is of order $|\Psi_0 - \Psi_V|^2$, while that in GFMC is zero in principle. The errors in other observables, $\langle \hat{O} \rangle$ are of order $|\Psi_V - \Psi_0|$ in VMC, and $|\Psi_V - \Psi_0|^2$ in GFMC. Thus GFMC offers a very significant improvement over VMC. Finally we note that more elaborate schemes can be developed to calculate $\langle \hat{O} \rangle$ more accurately, and secondly, it is often computationally more economical to choose a computationally simpler Ψ_T that is close to the optimum Ψ_V. The optimum Ψ_V generally has small, but computationally-intensive terms which can either be approximated or omitted in the Ψ_T without significant loss of accuracy.

The Short-Time Propagator

Consider simple systems in which:

$$H = T + V(R), \tag{73}$$

15

$$T = \sum_i -\frac{\hbar^2}{2m}\nabla^2, \tag{74}$$

$$V(R) = \sum_{i<j} v(r_{ij}) + \sum_{i<j<k} V(r_{ij}, r_{jk}, r_{ki}), \tag{75}$$

the T does not commute with $V(R)$, but the pair and triplet potentials commute with each other. Feynman's symmetric form provides the simplest approximation for the short-time propagator. It is given by:

$$e^{-(T+V)\Delta\tau} = e^{-V\Delta\tau/2}e^{-T\Delta\tau}e^{-V\Delta\tau/2} + \text{ terms of order } (\Delta\tau)^3. \tag{76}$$

In order to propagate up to time τ, we need $\tau/\Delta\tau$ time steps. Therefore the total error in using this approximation is of order $(\Delta\tau)^2\tau$. The $\Delta\tau$ is chosen small enough so that this error is smaller than the standard deviation δ. In this sense the Feynman form is not an approximation.

The free-particle propagator:

$$G_0(R, R') = \langle R|e^{-T\Delta\tau}|R'\rangle, \tag{77}$$

$$= \left[\frac{1}{8\pi}\sqrt{\frac{2\pi m}{\Delta\tau\hbar^2}}\right]^{3A} \exp\left[\frac{(R - R')^2}{2\hbar^2\Delta\tau/m}\right], \tag{78}$$

can be easily calculated with plane-wave eigenstates of T. In simple systems $|R\rangle$ are eigenstates of $V(R)$:

$$e^{-V(R)\Delta\tau/2}|R\rangle = |R\rangle e^{-V(R)\Delta\tau/2}. \tag{79}$$

Therefore:

$$G(R, R') = \langle R|e^{-(H-E_0)\Delta\tau}|R'\rangle, \tag{80}$$

$$= e^{E_0\Delta\tau}e^{-V(R)\Delta\tau/2}G_0(R, R')e^{-V(R')\Delta\tau/2}. \tag{81}$$

Note that this $G(R, R')$ is real positive normalizable, and can be used as a weight function for sampling.

GFMC calculations of nuclei[49] are carried out with a simpler Hamiltonian H':

$$H' = T + \sum_{i<j} v_8'(ij) + \sum_{i<j<k} V_{ijk}, \tag{82}$$

which contains an approximate v_8' interaction that has only eight terms with operators $[1, \sigma_i \cdot \sigma_j, S_{ij}, L \cdot S] \otimes [1, \tau_i \cdot \tau_j]$, and is identical to the full isoscalar v_{14} interaction in the $^1S_0, ^3S_1 -^3 D_1, ^3P_J, J = 0, 1, 2$ and $^1P_1$ partial waves. The small difference $(v_{14} - v_8')$ is treated as a first order perturbation. This is due to the difficulties in treating interactions with L^2 (or ∇^2) operators in the short-range propagator [49]. The v_8' interaction has the six terms of v^{static} and the spin-orbit parts. Here, for simplicity, we consider only the v^{static}; treatment of spin-orbit terms can be found in Ref. 49. Earlier calculations use the propagator:

$$G(R, R') = e^{E_0\Delta\tau}I_3(R)I_2(R)G_0(R, R')I_2(R')I_3(R'), \tag{83}$$

$$I_2(R) = S\prod_{i<j}\exp(-v_{ij}^{static}(R)\Delta\tau/2), \tag{84}$$

$$I_3(R) = \left[1 - \frac{\Delta\tau}{2}\left(\sum_{i<j<k} V_{ijk}^{2\pi}(R)\right)\right]\exp\left[-\sum_{i<j<k} V_{ijk}^{R}(R)\frac{\Delta\tau}{2}\right]. \tag{85}$$

16

Note that I_3 and $\exp(-v_{ij}^{static}(R)\Delta\tau/2)$ are operators, i.e. $\alpha_{total} \times \alpha_{total}$ matrix functions of R. It is simple to exponentiate v_{ij}^{static}, but not $V_{ijk}^{2\pi}$; therefore only the leading term of $\exp(-\sum V_{ijk}^{2\pi}\Delta\tau/2)$ is retained in the $G(R, R')$. This propagator therefore has errors of order $(\Delta\tau)^2$ containing the smaller $V_{ijk}^{2\pi}$, and of order $(\Delta\tau)^3$ containing the larger v_{ij}^{static}. These are rendered negligible by using a very small $\Delta\tau = 0.1$ GeV^{-1}.

The latest calculations[19, 20] use exact two-body propagators:

$$g_{ij}(\mathbf{r}_{ij}, \mathbf{r}'_{ij}) = \langle \mathbf{r}_{ij}| \exp\left[\left(-\frac{\hbar^2}{m}\nabla_{ij}^2 + v_8'(ij)\right)\Delta\tau\right] |\mathbf{r}'_{ij}\rangle, \tag{86}$$

$$g_{0,ij}(\mathbf{r}_{ij}, \mathbf{r}'_{ij}) = \langle \mathbf{r}_{ij}| \exp\left[-\frac{\hbar^2}{m}\nabla_{ij}^2 \Delta\tau\right] |\mathbf{r}'_{ij}\rangle, \tag{87}$$

in the short-time propagator:

$$G(R, R') = e^{E_0\Delta\tau} G_0(R, R') I_3(R) \left[S \prod_{i<j} \frac{g_{ij}(\mathbf{r}_{ij}, \mathbf{r}'_{ij})}{g_{0,ij}(\mathbf{r}_{ij}, \mathbf{r}'_{ij})}\right] I_3(R'). \tag{88}$$

The above propagator has smaller errors than that given by (83), and can be used with $\Delta\tau = 0.5$ GeV^{-1}. Since the number of time steps needed is proportional to $1/\Delta\tau$; a propagator that allows larger $\Delta\tau$ is more efficient.

Sampling the paths

Let $W(P) = W(R_0, R_1...R_n)$ be a real positive normalizable weight function. The mixed expectation value given by Eq. (64) is expressed as:

$$\langle \hat{O}(\tau) \rangle_{mixed} = N/D, \tag{89}$$

$$N = \int N(P)W(P)dP, \tag{90}$$

$$D = \int D(P)W(P)dP, \tag{91}$$

$$N(P) = \Psi_T^\dagger(R_n)\hat{O}G(R_n, R_{n-1})...G(R_1, R_0)\Psi_T(R_0)/W(P), \tag{92}$$

$$D(P) = \Psi_T^\dagger(R_n)G(R_n, R_{n-1})...G(R_1, R_0)\Psi_T(R_0)/W(P). \tag{93}$$

Integrals over dP are carried out stochastically using an ensemble $[P]$ of paths distributed with weight $W(P)$. We then obtain:

$$\langle \hat{O}(\tau) \rangle_{mixed} = \left[\sum_{[P]} N(P) \Big/ \sum_{[P]} D(P)\right] \pm \delta. \tag{94}$$

The weight function $W(P)$ is to be chosen such that δ is acceptably small. The naive choice:

$$W(P) = G(R_n, R_{n-1})...G(R_1...R_0)\Psi_T^2(R_0), \tag{95}$$

is easy to implement in simple systems whose $G(R, R')$ is real positive. We sample $\Psi_T^2(R_0)$ by the Metropolis method to obtain the ensemble $[R_0]$. For each element of $[R_0]$ the value of R_1 is obtained by sampling $G(R_1, R_0)$. The ensemble $[P]$ is obtained by continuing this process. Unfortunately this naive choice does not work. It gives for simple systems having $\Psi^\dagger = \Psi^*$:

$$N(P) = \Psi_T^*(R_n)\hat{O}\Psi_T(R_0) / \Psi_T^*(R_0)\Psi_T(R_0), \tag{96}$$

$$D(P) = \Psi_T^*(R_n)\Psi_T(R_0) / \Psi_T^*(R_0)\Psi_T(R_0). \tag{97}$$

17

The weight (95) does not depend upon $\Psi_T(R_n)$; most paths obtained by sampling that weight end up in the vast region where $\Psi_T \sim 0$. Only a few paths have significant values for $N(P)$ and $D(P)$, leading to unacceptably large variance and δ.

Kalos introduced importance sampled Green's function:

$$\tilde{G}(R, R') = |\Psi_T(R)|G(R, R')\frac{1}{|\Psi_T(R')|}, \tag{98}$$

for simple systems to avoid this difficulty. The weight function is chosen as:

$$W(P) = \tilde{G}(R_n, R_{n-1})...\tilde{G}(R_1, R_0)\Psi_T^2(R_0), \tag{99}$$

and the paths are obtained by first sampling Ψ_T^2 to obtain the ensamble $[R_0]$, then sampling $\tilde{G}(R_1, R_0)$ for each R_0 to obtain $[R_1]$ and so on. Special techniques including configuration weights and branching[16] are used to sample this weight function. With it we obtain:

$$N(P) = \Psi_T^*(R_n)\hat{O}|\Psi_T(R_0)|\,/|\Psi_T^*(R_n)|\Psi_T(R_0), \tag{100}$$
$$D(P) = \Psi_T^*(R_n)|\Psi_T(R_0)|\,/|\Psi_T^*(R_n)|\Psi_T(R_0), \tag{101}$$

which have much smaller variance.

Consider first a simple Bose system whose ground-state wave function, and thus Ψ_T is real positive. In this case $|\Psi_T(R)| = \Psi_T(R)$, and

$$N(P) = \frac{\Psi_T(R_n)\hat{O}}{\Psi_T(R_n)}, \tag{102}$$
$$D(P) = 1, \tag{103}$$

The variance of $D(P)$ is zero, while that of $N(P)$ is tolerable for many interesting operators. Note that we must operate by $\hat{O}$ on left side Ψ_T since the product of G's on the right of $\hat{O}$ has been cancelled against that in the $W(P)$.

Simple Fermi systems offer a more difficult problem. Real wave functions can be used to describe ground states of several interesting Fermi systems[46], however the many-Fermion wave function $\Psi_T(R)$ necessarily has domains in R-space where it is positive, and it is negative elsewhere. Due to antisymmetry, if $\Psi_T(R)$ is positive, then $\Psi_T(R')$ for R' obtained by exchanging particles 1 and 2,

$$\mathbf{r}'_1 = \mathbf{r}_2, \;\; \mathbf{r}'_1 = \mathbf{r}_2, \;\; \mathbf{r}'_i = \mathbf{r}_i \;\; \text{for all } i \neq 1, 2; \tag{104}$$

must be negative. In this case, we obtain:

$$N(P) = \frac{\Psi_T(R_n)\hat{O}}{\Psi_T(R_n)}\text{sign}(\Psi_T(R_n)\Psi_T(R_0)), \tag{105}$$
$$D(P) = sign(\Psi_T(R_n)\Psi_T(R_0)). \tag{106}$$

At small τ the path length is small, and the paths remain mostly inside a domain. The $\text{sign}(\Psi_T(R_n)\Psi_T(R_0))$ is then $+1$, and the variance is small. However, as τ increases, configurations cross domain boundaries, then the $\text{sign}(\Psi_T(R_n)\Psi_T(R_0))$ becomes -1. This increases the variance and decreases the average value of $D(P)$. Generally the variance is tolerable till $\overline{D(P)} > 0.6$, i.e. till about 20% of the configurations have crossed domain boundaries. This, so called "Fermion sign problem" limits the value of τ up to which mixed expectation values can be calculated with the required accuracy.

Approximate methods to extrapolate mixed expectation values to $\tau \to \infty$ have been used [18, 19].

In nuclei $G(R, R')$ is a matrix and $\Psi_T(R)$ is a vector function of R. Thus these cannot be directly used in the weight function $W(P)$. Nuclear calculations have been carried out using the weight function developed by Carlson[49]. It contains a product of scalar Green's functions:

$$G_S(R, R') = \exp(\Delta \tau E_0) \exp\left(-\sum_{i<j} v_S(r_{ij}) \Delta \tau/2\right) G_0(R, R') \exp\left(-\sum_{i<j} v_S(r'_{ij}) \Delta \tau/2\right), \tag{107}$$

with a potential $v_S(r_{ij})$ that equals the average of v_{ij} in $L = 0$ partial waves. It is meant to provide a scalar approximation to the full matrix $G(R, R')$. A vector $\bar{\Psi}_i(P_i)$ is defined as:

$$\bar{\Psi}_i(P_i) = \frac{G(R_i, R_{i-1})...G(R_1, R_0)\Psi_T(R_0)}{G_S(R_i, R_{i-1})...G_S(R_1, R_0)}; \tag{108}$$

it depends upon the path $P_i = R_0, R_1 ..., R_i$ up to R_i, and represents the full propagated wave function. The scalar imporance function $I(P_i)$ is defined as:

$$I(P_i) = |\sum_{\alpha} \psi_{T,\alpha}(R_i)\bar{\psi}_{i,\alpha}(P_i)| + \epsilon \sum_{\alpha} |\psi_{T,\alpha}(R_i)\bar{\psi}_{i,\alpha}(P_i)|, \tag{109}$$

where $\psi_{T,\alpha}$ and $\bar{\psi}_{i,\alpha}$ are the componants of Ψ_T and $\bar{\Psi}_i$ for spin-isospin state $|\alpha\rangle$. The second term, with a small $\epsilon \sim 0.01$, is added to provide sampling of regions where $\Psi_T^{\dagger}(R_i)\bar{\Psi}_i(P_i)$ may be small due to cancellation between the components $\psi_{T,\alpha}(R_i)\bar{\psi}_{i,\alpha}(P_i)$ that contribute to it. The weight-function $W(P_n)$:

$$W(P_n) = \left[\prod_{i=1,n} I(P_i) G_S(R_i, R_{i-1}) \frac{1}{I(P_{i-1})}\right] I(P_0), \tag{110}$$

is used to sample the paths. The starting points $[R_0]$ of the ensemble $[P]$ are sampled from:

$$I(P_0) = (1 + \epsilon)\Psi_T^{\dagger}(R_0)\Psi_T(R_0), \tag{111}$$

or equivalently from $\Psi_T^{\dagger}(R_0)\Psi_T(R_0)$, and the subsequent $[R_i]$ are obtained by sampling:

$$I(P_i) G_S(R_i, R_{i-1})/I(P_{i-1}) \tag{112}$$

for each of the R_{i-1}. An approximate scalar importance function is used to expedite the computation[20].

The $N(P)$ and $D(P)$ obtained with this $W(P)$ are given by:

$$N(P) = \Psi_T^{\dagger}(R_n)\hat{O}\bar{\Psi}_n(P_n)/I(P_n), \tag{113}$$
$$D(P) = \Psi_T^{\dagger}(R_n)\bar{\Psi}_n(P_n)/I(P_n). \tag{114}$$

Ignoring ϵ, the variance of $D(P)$ is mostly due to the Fermion sign problem, while that of $N(P)$ is tolerable. Useful values of $\langle \hat{O}(\tau) \rangle_{mixed}$ can be calculated until $\overline{D(P)} > 0.6$.

RESULTS

The transient energy $E(\tau)$,

$$E(\tau) = \langle H'(\tau) \rangle_{mixed} + \langle \delta H(\tau) \rangle, \tag{115}$$

19

where $\delta H = H - H'$ (Eq. 82), is evaluated every $0.004\ MeV^{-1}$ to monitor the convergence of the calculations[18]. It's values, averaged over more than 10,000 configurations, are shown in Figure 8. This function contains significant information. Let Ψ_T, normalized such that $\langle \Psi_0 | \Psi_T \rangle = 1$, contain admixtures of states $|E'\rangle$, with energy $E' = E - E_0$, with strength $\alpha(E')$:

$$|\Psi_T\rangle = |\Psi_0\rangle + \int \sqrt{\alpha(E')}|E'\rangle dE'. \tag{116}$$

Assuming that the perturbative treatment of δH is valid, we obtain :

$$E(\tau) - E_0 = \frac{\int E' \alpha(E') \exp(-E'\tau) dE'}{1 + \int \alpha(E') \exp(-E'\tau) dE'}, \tag{117}$$

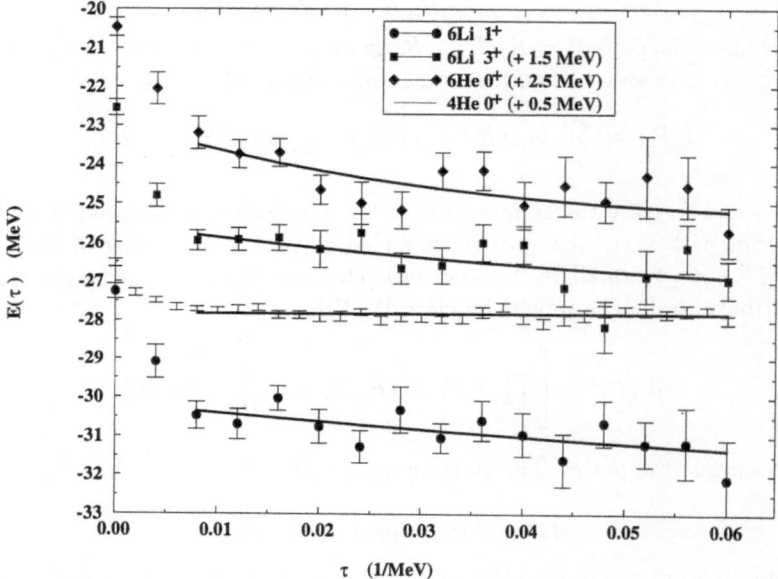

Figure 8. The values of $E(\tau)$ calculated with the Argonne v_{18} and Urbana IX two- and three-nucleon interactions for several states. Note that the energies of some of the the states are shifted for clarity.

As can be seen from Figure 8, there is a significant decrease in the $E(\tau)$ in the first $0.01\ MeV^{-1}$. Eq. (117) shows that this decrease must come from the removal of the high energy $(E' > 200\ MeV)$ contaminations in Ψ_T by the $\exp[-E'\tau]$ factor. It has been estimated[20] that the total strength of high-energy $(E' > 200\ MeV)$ contaminations in the available variational wave functions for nuclei with $A \leq 7$ is less than 1%.

The $E(\tau)$ of ^{3}He and ^{4}He (only the latter is shown in Figure 8) has little τ-dependence for $\tau > 0.01 MeV^{-1}$. Thus we can assume that the $E(\tau > 0.02 MeV^{-1})$ and $\Psi(\tau > 0.02 MeV^{-1})$ of these nuclei equal the exact E_0 and Ψ_0 within statistical errors. In contrast, the $E(\tau)$ of $A = 6$ (shown in Figure 8) and 7 nuclei has a small τ-dependence at $\tau > 0.01 MeV^{-1}$. Due to the Fermion sign problem, absent in ^{3}H, ^{3}He and ^{4}He, the statistical errors in the $E(\tau)$ of $A = 6, 7$ nuclei increase with τ, and limit the calculations to $\tau < 0.06 MeV^{-1}$. In Refs. 18 and 20 the $E(\tau)$ in the region $0.01 < \tau < 0.06 MeV^{-1}$ is fit by approximating the $\alpha(E')$ by one or two δ-functions. Such fits yield the extrapolated $\tau \to \infty$ value of E_0 listed in Table 1. This

20

Table 1. The Calculated Energies in MeV

$^A Z(J^\pi; T)$	$\overline{E(H')}$	$\overline{E}$	E_0	$E_0(expt)$
$^3\text{H}(\frac{1}{2}^+;\frac{1}{2})$	–8.70(1)	–8.47(2)	–8.47(1)	–8.48
$^4\text{He}(0^+;0)$	–30.70(5)	–28.3(1)	–28.3(1)	–28.3
$^6\text{He}(0^+;0)$	–30.6(2)	–28.0(4)	–28.7(8)	–29.3
$^6\text{Li}(1^+;0)$	–34.9(3)	–31.3(5)	–31.8(8)	–32.0
$^6\text{Li}(3^+;0)$	–32.5(4)	–28.7(5)	–28.9(7)	–29.9
$^6\text{Li}(2^+;0)$	–29.9(1)	–26.2(3)	–26.1(8)	–27.7
$^7\text{Li}(\frac{3}{2}^-;\frac{1}{2})$	–41.0(4)	–37.1(5)	–37.1(9)	–39.2
$^7\text{Li}(\frac{1}{2}^-;\frac{1}{2})$	–41.4(4)	–37.3(5)	–37.3(9)	–38.7
$^7\text{Li}(\frac{7}{2}^-;\frac{1}{2})$	–35.5(4)	–30.7(5)	–31.8(8)	–34.5

table also gives $\overline{E(H')}$, which is the average of $\langle H'(\tau)\rangle_{mixed}$ over the $\tau = 0.035$ to 0.06 MeV^{-1} region. The $\overline{E(H')}$ is an upperbound to the lowest eigenvalue E'_0 of H' with the listed quantum numbers , and $\overline{E}$, the avarage of $E(\tau)$ over the same τ-range, is an upperbound for E_0 provided the perturbative treatment of δH is valid. Note that the required extrapolation for $\tau > 0.06 MeV^{-1}$ does not appear to be too significant; the $\overline{E}$ and extrapolated E_0 are within statistical errors. The experimental values of E_0 are also listed in Table 1.

The main terms contributing to E_0 are listed in Table 2. Note that the variances of the kinetic and two-body interaction energies are much larger than that of $\overline{E}$. This is because Ψ_T is an approximate eigenfunction of H, therefore $N(P)$ (Eq. 113) has a smaller variance when $\hat{O} = H$. We also note that the one-pion exchange interaction gives a large contribution accounting for most of $\langle \sum v_{ij}\rangle$. The rest of the two-body interaction denoted by v_{ij}^R in Eq. (7), contains a repulsive core and an intermediate-range attraction. These two parts of v_{ij}^R give large contributions of apposite signs; however, due to a large cancellation, their sum is much smaller than $\langle v_{ij}^\pi \rangle$. Also the difference between calculated and experimental energies, denoted by δE_0 in table 2, is quite small, even when compared to $\langle V_{ijk}\rangle$. It may be possible to reduce δE_0 by using more realistic models of three nucleon interaction. The present V_{ijk}^R (Eq. 29) contains a single spin-isospin-independent term whose strength U_0 is fitted to the binding energy of ^{4}He. In reality we should expect the V_{ijk}^R to have spin-isospin dependence.

The quantum Monte Carlo methods are well suited to study the structure of nuclear bound states. Since the present variational wave functions have only a few-percent admixture of excited states, they give fairly accurate results for one- and two-nucleon density-distribution functions. For example, these functions for ^{6}He and ^{6}Li, calculated with GFMC are very similar to those obtained from the variational Ψ_V[19].

As mentioned earlier, the configurations generated in a Metropolitan walk sampling $\Psi_V^\dagger(R)\Psi_V(R)$ can be regarded as snapshots of the bound state. Each configuration $R_n, n = 1, N_C$ gives A nucleon coordinates $i = 1, A$:

$$\mathbf{r}'_{n,i} = \mathbf{r}_{n,i} - \frac{1}{A}\sum_j \mathbf{r}_{n,j}, \tag{118}$$

measured from the configuration center of mass. These $A \times N_C$ values of $\mathbf{r}'_{n,i}$ are

Table 2. Energy Contributions in MeV

$^AZ(J^\pi;T)$	$\langle T\rangle$	$\langle v_{ij}\rangle$	$\langle V_{ijk}\rangle$	δE_0	$\langle v_{ij}^\pi\rangle$	$\langle V_{ijk}^{2\pi}\rangle$
^{2}H$(1^+;0)$	19.9	-22.1	0.0	0.0	-21.2	0.0
^{3}H$(\frac{1}{2}^+;\frac{1}{2})$	50.0(8)	-57.6(8)	-1.20(7)	0.01(2)	-44.1(5)	-2.20(6)
^{4}He$(0^+;0)$	115.(2)	$-138.$(2)	-6.6(4)	0.0(1)	$-106.$(1)	-12.6(3)
^{6}He$(0^+;0)$	145.(4)	$-171.$(4)	-7.6(8)	0.6(8)	$-121.$(2)	-14.8(6)
^{6}Li$(1^+;0)$	157.(4)	$-186.$(3)	-7.6(9)	0.2(8)	$-145.$(3)	-15.2(7)
^{6}Li$(3^+;0)$	159.(4)	$-186.$(4)	-7.7(9)	1.0(7)	$-143.$(3)	-15.6(7)
^{6}Li$(2^+;0)$	151.(3)	$-174.$(3)	-7.3(6)	1.6(8)	$-138.$(2)	-14.5(5)
^{7}Li$(\frac{3}{2}^-;\frac{1}{2})$	198.(8)	$-233.$(8)	-9.8(12)	2.1(9)	$-173.$(3)	-19.5(11)
^{7}Li$(\frac{1}{2}^-;\frac{1}{2})$	202.(4)	$-237.$(4)	-10.1(10)	1.4(9)	$-177.$(3)	-20.0(8)
^{7}Li$(\frac{7}{2}^-;\frac{1}{2})$	198.(5)	$-227.$(4)	-9.8(10)	2.7(8)	$-169.$(4)	-20.1(8)

distributed according to the nucleon density $\rho(\mathbf{r}')$ of the nucleus in it's center of mass frame. The probability that the nucleon i is a spin up proton in the configuration number n is given by:

$$P_{n,i} = \frac{1}{4}\frac{\Psi^\dagger(R_n)(1+\tau_Z(i))(1+\sigma_Z(i))\Psi(R_n)}{\Psi^\dagger(R_n)\Psi(R_n)}, \tag{119}$$

and the $\mathbf{r}'_{n,i}$ weighted with $P_{n,i}$ give the density distribution of spin-up protons in the nucleus. The distribution of spin up/down neutrons and protons in the nucleus can thus be studied using appropriate projection operators. In a spin-isospin zero nucleus such as ^{4}He the four (n,p)$\otimes$(up,down) densities are identical, while in a nucleus such as ^{7}Li having non-zero spin and isospin they are different. The one-body densities give large contributions to elastic scattering form factors of nuclei[50].

In each configuration we have $A(A-1)$ two-nucleon separations $\mathbf{r}_{n,ij} = \mathbf{r}_{n,i} - \mathbf{r}_{n,j}$. The $A(A-1) \times N_C$ values of $\mathbf{r}_{n,ij}$ are distributed according to the two-nucleon density $\rho_2(\mathbf{r})$, which gives the probability of finding a nucleon at a distance $\mathbf{r}$ from another nucleon in the nucleus. More details of the two-body density can be obtained by using pair projection operators:

$$P_{S=0} = \frac{1}{4}(1 - \sigma_i \cdot \sigma_j), \tag{120}$$

$$P_{S=1} = \frac{1}{4}(3 + \sigma_i \cdot \sigma_j), \tag{121}$$

for the two-nucleon spin and similar projection operators for two nucleon isospin. We can also project out two-nucleon states with spin projection M_S using:

$$P_{M_S=\pm1} = \frac{1}{4}(1 \pm \sigma_Z(i))(1 \pm \sigma_Z(j)), \tag{122}$$

$$P_{M_S=0} = \frac{1}{4}[(1 + \sigma_Z(i))(1 - \sigma_Z(j)) + (1 - \sigma_Z(i))(1 + \sigma_Z(j))]. \tag{123}$$

The two-body density distributions are useful to study short-range structures in nuclei[29], sums of response functions[51] and other quantities related to the expectation value of two-nucleon operators.

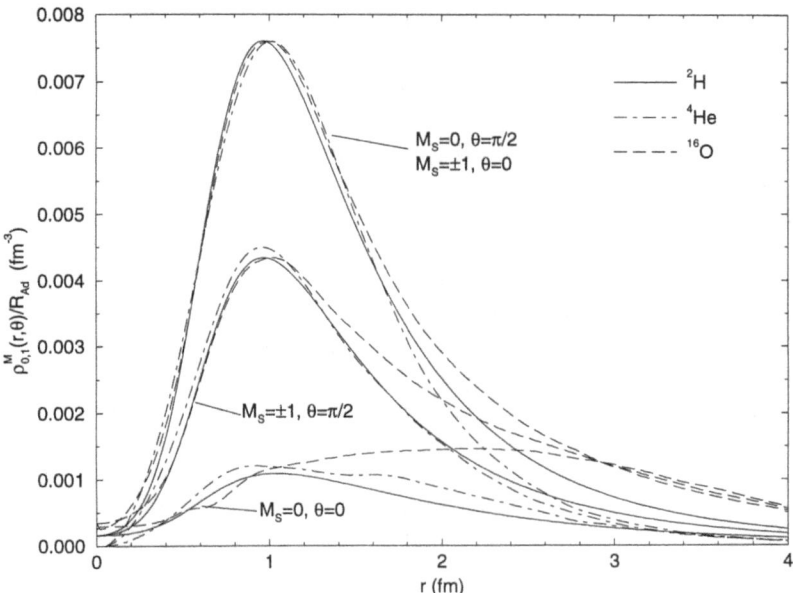

Figure 9. The $\rho_{0,1}^A(M_S, r, \theta)/R_{Ad}$ for $M_S = \pm 1$, 0 and $\theta = 0$, $\pi/2$ in ^{2}H, ^{4}He and ^{16}O. The values of R_{Ad} are 1, 4.7 and 18.8 for ^{2}H, ^{4}He and ^{16}O respectively.

The two-body densities in states with two-body spin-isospin T, S, and spin projection M_S, denoted by $\rho_{T,S}(M_S, \mathbf{r})$, have been studied[29] in nuclei with $A \leq 7$ and ^{16}O. Their dependence on the nuclear spin projection M_J is averaged out as it is in case of unpolarized targets. The shape of the densities in $T, S = 0,1$ and 1,0 states appears to be fairly independent of A at small r. Naturally the magnitude of $\rho_{T,S}$ increases with the size of the nucleus, however in $T, S = 1,0$ and 0,1 states this increase at small r can be absorbed in a single scale factor. For example, the approximation:

$$\rho_{0,1}^A(M_S, \mathbf{r}) \sim R_{Ad}\, \rho_{0,1}^d(M_S, \mathbf{r}), \qquad (124)$$

is valid at small r, particularly in the region where this density is large (Figure 9). Following the suggestion of Levinger and Bethe[52] the constant R_{Ad}, relating the magnitude of $\rho_{0,1}$ in the nucleus A to that in the deuteron, can be used to estimate the rate of absorption of intermediate-energy photons and pions by nuclei[29]. Figure 9 also indicates that the equidensity surfaces of $\rho_{0,1}^A(M_S, \mathbf{r})$ have toriadal shapes in $M_S = 0$, and dumbbell shapes in $M_S = \pm 1$ states, similar to those of the deuteron. These shapes are produced by near maximal tensor correlations at $r < 2 fm$.

The two-cluster distribution functions such as d,p in ^{3}He, dd and tp in ^{4}He, and αd in ^{6}Li etc., provide additional information on nuclear structure[29]. The two-cluster density in configuration space is defined as:

$$\rho_{ab}^{M_a, M_b, M_J}(\mathbf{r}_{ab}) = |\langle A\Psi_a^{M_a}\Psi_b^{M_b}, \mathbf{r}_{ab}|\Psi^{M_J}\rangle|^2, \qquad (125)$$

where Ψ^{M_J}, $\Psi_a^{M_a}$ and $\Psi_b^{M_b}$ are the wave functions of the initial nucleus and the clusters a and b in states with angular-momentum projections M_J, M_a and M_b. A denotes antisymmetrization between nucleons of clusters a and b, and $\mathbf{r}_{ab}$ is the distance between the centers of mass of the clusters. It gives the probability of finding the clusters a and b at distance $\mathbf{r}_{ab}$ in the nucleus. This probability often has a strong dependence on spin projections M_a, M_b and M_J, and can be easily calculated with the Monte Carlo method[47, 29]. The cross sections of $A(e, e'a)b$ reactions are related to ρ_{ab} in the plane-wave impulse approximation [50].

We will consider the tp and dd distribution functions in ^{4}He for illustration. The spin-parity of ^{4}He is 0^+, therefore it has non-vanishing tp overlap only when the spins of the triton and proton are coupled to zero and the relative angular momentum L_{tp} between them is also zero. The $\rho_{tp}(r_{tp})$ is thus spherically symmetric and has trivial dependence on M_t and M_p. The ρ_{tp} in momentum space, is called the tp momentum distribution $n_{tp}(k)$ in ^{4}He. It is compared with the momentum distribution $n_p(k)$ of protons in ^{4}He in Figure 10. Since there are two protons in ^{4}He, the integral of $n_p(k)$ equals two, however that of $n_{tp}(k)$ is only 1.65. At small k, $n_{tp}(k) \sim n_p(k)$, however at large k, $n_{tp}(k) < n_p(k)$. These results are interpreted as follows: The $n_{tp}(k)$ describes the motion of the proton in an average potential created by the triton. It represents the wave function of the lowest $1S1/2$ orbital in the nuclear-shell model. The difference, $n_p(k) - n_{tp}(k)$, is due to correlations between the proton and the nucleons in the triton. From the integral of $n_{tp}(k)$, it appears that $\sim 82\%$ of the time protons in ^{4}He act as independent particles bound in an average potential, as assumed in the shell model. The observed[54] ^{4}He(e,e'p) reaction rates seem to support this interpretation.

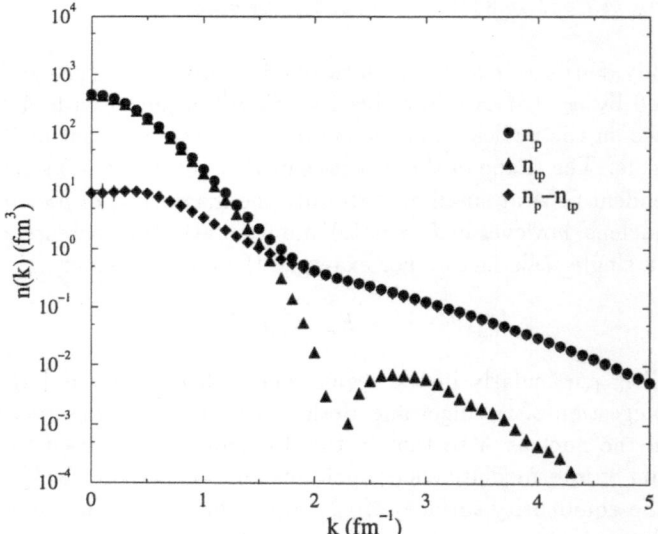

Figure 10. The calculated values of the proton-momentum distribution $n_p(k), n_{tp}(k)$ and their difference, from ref. 53.

In contrast the two-deuteron distribution function $\rho_{dd}(M_{d1}, M_{d2}, \mathbf{r}_{dd})$ in ^{4}He (Figure 11) has intricate dependence on the deuteron spins. Two states, having relative dd orbital angular momentum $L_{dd} = 1$ and 2 and the sum of deuteron spins $S_{dd} = 0$ and 2 contribute to the dd-distribution. It is anisotropic largely because of the anisotropy of the deuteron. The two toroidal-shaped deuterons, each having $M_d = 0$, like to be coaxial (on top of each other), rather than side by side, as evident from $\rho_{dd}(M_{d1} = M_{d2} = 0, r_{dd}, \theta)$ being larger for $\theta = 0$ than for $\theta = \pi/2$, while the two dumbbell shaped deuteron prefer to be side by side ($\theta = \pi/2$) over being in a line ($\theta = 0$). Several other two-cluster distributions are discussed in Ref. 29.

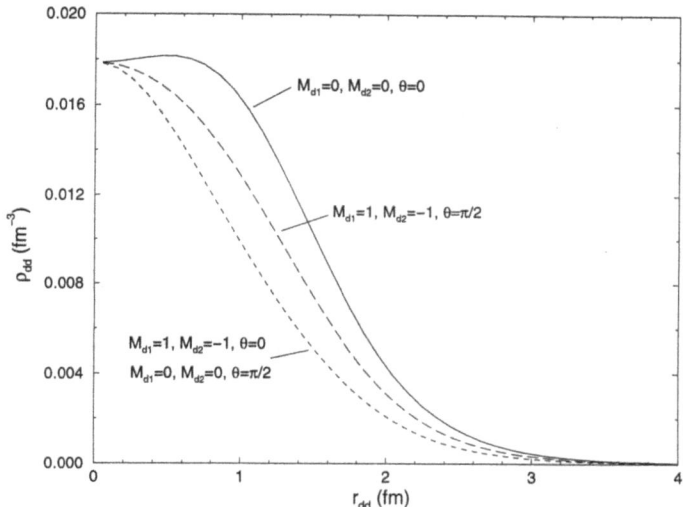

Figure 11. The dd distribution functions in ^{4}He for $M_{d1}, M_{d2} = 0, 0$ and $1, -1$, for $\theta = 0, \pi/2$.

CONCLUSIONS

In these lectures we discussed calculations of nuclear-bound states from realistic models of nuclear forces using quantum Monte Carlo methods. Such calculations make detailed predictions of nuclear structure, particularly at short distance scales. Some, but not all of the predictions have at least qualitatively been verified experimentally, and attempts to verify the other predictions will hopefully be made. Eventually, one hopes to understand these short-range structures using the more elementary quark-gluon degrees of freedom.

The research discussed here has been carried out in collaboration with A. Arriaga, J. Carlson, J. Forest, S. Pieper, B. Pudliner, R. Schiavilla, and R. Wiringa. I would like to thank them for many discussions, figures, tables and instruction in Latex. I would also like to thank the Mathematics and Computer Sciences division of Argonne National Laboratory, the Cornell Theory Center and Pittsburg Supercomputing Center for grants of computer time, and the US National Science Foundation for supporting my research via grant PHY 94-21309.

REFERENCES

1. C.F. v. Weizsacker, Z. Phys. 96:431 (1935).
2. H.A. Bethe and R.F. Bacher, Rev. Mod. Phys. 8:82 (1936).
3. A. Bohr and B.R. Mottelson. Nuclear Structure, W.A. Benjamin (1975).
4. P.J.Siemens and A.S.Jensen, Elements of Nuclei, Addison-Wesley (1987).
5. M.G. Mayer, Phys. Rev. 75:1969 (1949).

6. O. Haxel, J.H.D. Jenson, and H.E. Suess, Phys. Rev. 75:1766 (1949).

7. H.A. Weidenmuller, Nucl. Phys. A507:5c (1990).

8. V.R. Pandharipande, in Structure of Hadrons and Hadronic Matter, O. Scholten and J.H. Koch, eds., World Scientific, p. 1 (1991).

9. K.A. Brueckner, C.A. Levinson, and H.M. Mahmoud, Phys. Rev. 95:217 (1954).

10. J. Goldstone, Proc. Roy. Soc. (London) A239:267 (1957).

11. H. Kummel, K.H. Luhrmann, and J.G. Zabolitzky, Phys. Rept. C36:1 (1978).

12. R.B. Wiringa, Phys. Rev. C43:1585 (1991).

13. S.C. Pieper, R.B. Wiringa, and V.R. Pandharipande, Phys. Rev. C46:1741 (1992).

14. R.B. Wiringa, V. Fiks, and A. Fabrocini, Phys. Rev. C38:1010 (1988).

15. W. Glockle and H. Kamada, Phys. Rev. Lett. 71:971 (1993).

16. J. Carlson in Structure of Hadrons and Hadronic Matter, O. Schulten and J.H. Koch, eds., World Scientific, p. 43 (1991).

17. A. Kievsky, M. Viviani, and S. Rosati, Nucl. Phys. A551:241 (1993); A577:511 (1994).

18. B.S. Pudliner, V.R. Pandharipande, J. Carlson, and R.B. Wiringa, Phys. Rev. Lett. 74:4396 (1995).

19. B.S. Pudliner, Ph.D. Thesis, University of Illinois at Urbana-Champaign (1996).

20. B.S. Pudliner et al., to be published.

21. J.J. Sakurai. Advanced Quantum Mechanics, Addison-Wesley (1973).

22. I.E. Lagaris and V.R. Pandharipande, Nucl. Phys. A359:331 (1981).

23. R.B. Wiringa, V.G.J. Stokes, and R. Schiavilla, Phys. Rev. C51:38 (1995).

24. J.L. Friar, G.L. Payne, V.G.J. Stokes, and J.J. de Swart, Phys. Lett. B311:4 (1993).

25. R.V. Reid, Jr., Ann. Phys. (N.Y.) 50:411 (1968).

26. M. Lacombe et al., Phys. Rev. C21:861 (1980).

27. J.M. Blatt and V.F. Weisskopf. Theoretical Nuclear Physics, John Wiley (1952).

28. S.E.Koonin, Computational Physics, Benjamin/Cummings (1986).

29. J.L. Forest et al., Phys. Rev. C54:646 (1996).

30. V.B. Kopeliovich and B.E. Stern, JETP Lett. 45:203 (1987).

31. J.J.M. Verbaarschot, T.S. Walhout, J. Wambach, and H.W. Wyld, Nucl. Phys. A468:520 (1987).

32. V.G. Makhankov, Y.P. Rybakov, and V.I. Sanyuk. The Skyrme Model, Springer Verlag (1993).

33. R.A.Leese, N.S.Manton, and B.J.Schroers, Nucl. Phys. B442:228 (1995).

34. V.R. Pandharipande, Nucl. Phys. A553:191c (1993).

35. I. Fujita and H. Miyazawa, Prog. Theo. Phys. 17:360 (1957).

36. W.L. McMillan, Phys. Rev. A442:138 (1965).

37. M.H. Kalos, D. Levesque, and L. Verlet, Phys. Rev. A9:2178 (1974).

38. D. Ceperley, G.V. Chester, and M.H. Kalow, Phys. Rev. B16:3081 (1977).

39. M.A. Lee, K.E. Schmidt, M.H. Kalos, and G.V. Chester, Phys. Rev. Lett. 46:728 (1981).

40. J. Lomnitz-Adler, V.R. Pandharipande, and R.A. Smith, Nucl. Phys. A361:399 (1981).

41. J. Carlson, Phys. Rev. C38:1879 (1988).

42. D.S. Lewart and V.R. Pandharipande, in Monte Carlo Methods in Theoretical Physics, S. Caracciolo and A. Fabrocini, eds., ETS Editrice, Pisa (1992).

43. M.H. Kalos and P.A. Whitlock. Monte Carlo Methods, Wiley (1986).

44. N. Metropolis et al., J. Chem. Phys. 21:1087 (1953).

45. S. Caracciolo and A. Fabrocini, eds. Monte Carlo Methods in Theoretical Physics, ETS Editrice, Pisa (1992).

46. V.R. Pandharipande, S.C. Pieper, and R.B. Wiringa, Phys. Rev. B34:4571 (1986).

47. R. Schiavilla, V.R. Pandharipande, and R.B. Wiringa, Nucl. Phys. A449:219 (1986).

48. A. Arriaga, V.R. Pandharipande, and R.B. Wiringa, Phys. Rev. C52:2362 (1995).

49. J. Carlson and R. Schiavilla, in Few Body Systems Supplement 7, B.L.G. Bakker and R. van Dantzig, eds., Springer-Verlag, p. 349 (1994).

50. B. Frois and I. Sick, eds. Modern Topics in Electron Scattering, World Scientific (1991).

51. V.R. Pandharipande et al., Phys. Rev. C49:789 (1994).

52. J.S. Levinger, Nuclear Photo-disintegration, Oxford University Press (1960).

53. R.B. Wiringa, Private communication (1995).

54. Van Leeuwe, Ph.D. Thesis, Univ. Amsterdam, (1995).

CLUSTERING IN NUCLEAR STRUCTURE AND COLLISIONS

H. Horiuchi

Department of Physics
Kyoto University
Kyoto 606-01
Japan

INTRODUCTION

Clustering aspects appear abundantly in many problems in both nuclear structure and nuclear collision problems, implying that the formation of clusters is a fundamental property of nuclear matter together with the formation of mean field. Nuclear matter has a saturation property and due to this property nuclei can easily assemble and disassemble with only small amounts of energy input or output for such processes. This is a basic reason of the omnipresence of clustering aspects in nuclear phenomena.

The purpose of this lecture is to discuss microscopic descriptions of the phenomena of formation and dissolution of clusters both in structure problems and in collision problems. In both problems, comparison with the formation of mean field is important.

We first discuss some typical clustering phenomena in stable-nucleus region together with their microscopic descriptions. Then we discuss some novel clustering features in exotic nuclei near neutron drip line. Finally we discuss microscopic descriptions of fragmentation processes in medium energy heavy ion collisions. We will see that the newly developed theory named Antisymmetrized Molecular Dynamics (AMD) is very powerful for all these discussions.

CLUSTERING IN STABLE NUCLEUS REGION

As typical and fundamentally important topics of clustering in regions of stable nuclei, we discuss below two problems: one is the inversion doublet phenomenon and the other is the unification of the cluster model and the composite particle optical potential model. Both topics are very important for our understanding of the clustering in nuclei, because they both show clearly the existence of the clustering structure which is different from the shell-model structure (or mean-field-model structure).

Correlations and Clustering Phenomena in Subatomic Physics
Edited by Harakeh *et al.*, Plenum Press, New York, 1997

Inversion Doublet

The existence of positive and negative parity rotational bands with $K^\pi = 0^+$ and $K^\pi = 0^-$ with an energy gap between the two bands far smaller than $1\ \hbar\omega$ provides us with a strong evidence of the cluster structure $A + B$ with two clusters A and B located at spatially different positions[1].

Let $\psi(A, \mathbf{D}_A)$ $(\psi(B, \mathbf{D}_B))$ denote the wave function of the cluster A (B) whose center of mass is located at the spatial point $\mathbf{D}_A$ $(\mathbf{D}_B)$. Then the cluster structure $A + B$ is described by

$$\Phi(A + B, \mathbf{D}) = \mathcal{A}\{\psi(A, \mathbf{D}_A)\psi(B, \mathbf{D}_B)\}, \tag{1}$$

where $\mathbf{D} = \mathbf{D}_A - \mathbf{D}_B$ and $\mathcal{A}$ is the antisymmetrizing operator. When the clusters A and B are different, the intrinsic wave function $\Phi(A + B, \mathbf{D})$ is not an eigen state of parity and we have an inversion doublet $\Phi^+(A + B, \mathbf{D})$ and $\Phi^-(A + B, \mathbf{D})$ which are given by

$$\Phi^\pm(A + B, \mathbf{D}) = (1 \pm P)\Phi(A + B, \mathbf{D}), \tag{2}$$

where P is a parity operator. The energy gap ΔE between the two states $\Phi^\pm(A+B, \mathbf{D})$ is given by

$$\Delta E = \frac{\langle \Phi^-(A + B, \mathbf{D})|\hat{H}|\Phi^-(A + B, \mathbf{D})\rangle}{\langle \Phi^-(A + B, \mathbf{D})|\Phi^-(A + B, \mathbf{D})\rangle} \\ - \frac{\langle \Phi^+(A + B, \mathbf{D})|\hat{H}|\Phi^+(A + B, \mathbf{D})\rangle}{\langle \Phi^+(A + B, \mathbf{D})|\Phi^+(A + B, \mathbf{D})\rangle}, \tag{3}$$

where $\hat{H}$ is the Hamiltonian operator with the center-of-mass kinetic energy subtracted. When the spatial clustering is prominent, namely when the inter-cluster distance $\mathbf{D}$ is large, the energy gap ΔE between two states is very small, because both $\langle \Phi(A + B, \mathbf{D})|P|\Phi(A + B, \mathbf{D})\rangle$ and $\langle \Phi(A + B, \mathbf{D})|\hat{H}P|\Phi(A + B, \mathbf{D})\rangle$ become small.

Needless to say, since both positive and negative parity intrinsic states $\psi^\pm(A + B, \mathbf{D})$ are deformed, there appear rotational bands from both intrinsic states.

Most typical inversion doublet bands observed experimentally are those of ^{12}C + α structure in ^{16}O and those of ^{16}O + α structure in ^{20}Ne[1]. They are shown in Fig. 1. Note that the gap energies ΔE in both cases are far smaller than the value of $1\ \hbar\omega$ which is about 15 MeV in this mass-number region.

In the case of the negative parity rotational band states with $K^\pi = 0^-$, these states are observed as resonance states in the α-particle elastic scattering on ^{12}C and ^{16}O and the widths of these resonance states have values almost equal to the Wigner limit values of " α + core " configuration[2]. Furthermore, both positive $(K^\pi = 0^+)$ and negative $(K^\pi = 0^-)$ parity rotational band states are strongly populated by the α-transfer reactions such as the $(^7$Li, $t)$ reaction[3, 4]. This experimental evidence strengthens the argument that the inversion doublet bands $(K^\pi = 0^\pm)$ in ^{16}O and ^{20}Ne have prominent clustering structure of ^{12}C + α and ^{16}O + α, respectively.

Unification of the Cluster Model and the Composite Particle Optical Potential Model

A characteristic feature of the cluster structure is the relative motion between clusters. The cluster system can have various quantum states according to various quantum states of the inter-cluster relative motion, namely bound, resonance, and scattering states. In the case of the alpha cluster, the scattering state of the α particle

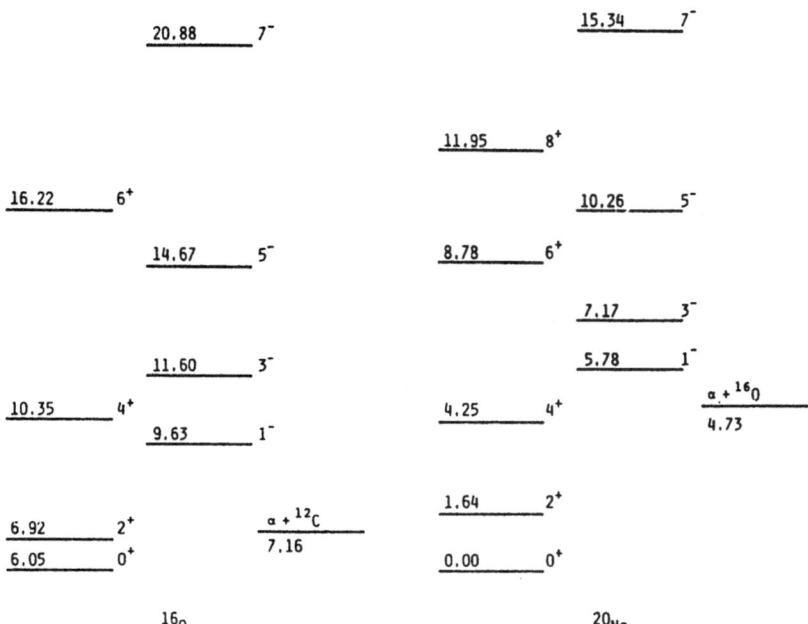

Figure 1. Inversion doublet bands in ^{16}O and ^{20}Ne. Numbers stand for excitation energies in MeV. The threshold energy of ^{12}C $+ \alpha$ in ^{16}O and that of ^{16}O $+ \alpha$ in ^{20}Ne are also shown.

shows very characteristic phenomena known as the ALAS (Anomalous Large Angle Scattering) and the nuclear rainbow at higher energy region. The α-particle optical model had a great success in describing these phenomena in elastic scattering in a wide range of incident energy up to about 40 MeV/nucleon. A key point of this success was the finding of three basic features of the α-particle optical potential by Michel and Vanderpoorten[5]. The three points are (i) deep real potential, (ii) large diffuseness of real potential, and (iii) weak imaginary potential. The overall reproduction of the data including the ALAS and the nuclear rainbow enabled us to determine the so-called unique optical potential.

An important challenge to the cluster model was to show that the cluster model, which can describe well the inversion doublet band states as the bound or resonance states of the α-core relative motion, can also describe well the scattering data of the α-core relative motion including the ALAS and the nuclear rainbow phenomena. Furthermore, it was demanded that the microscopic cluster model derives the uniquely determined optical potential of the α particle.

This unification of the cluster model and the α-particle optical potential model was achieved very satisfactorily. In the case of the $\alpha + {}^{16}$O system, it was shown that the $\alpha + {}^{16}$O RGM (Resonating Group Method) which reproduces well the inversion doublet band states in ^{20}Ne[6] reproduces also well the observed angular distribution of the $\alpha + {}^{16}$O scattering in a wide range of incident energy including the ALAS and the nuclear rainbow phenomena[7]. Furtheremore, it was shown that the equivalent local potential derived from the RGM non-local potential is very close to the real potential of the uniquely determined optical potential of $\alpha + {}^{16}$O[7]. These are shown in Fig. 2.

In the $\alpha + {}^{16}$O RGM, the system wave function Φ is given by

$$\Phi = \mathcal{A}\{\chi(\mathbf{r})\phi({}^{16}\text{O})\phi(\alpha)\}, \tag{4}$$

31

where $\phi(^{16}O)$ and $\phi(\alpha)$ stand for the internal wave functions of ^{16}O and α, respectively, and $\mathbf{r}$ is the inter-cluster relative coordinate. The inter-cluster relative wave function

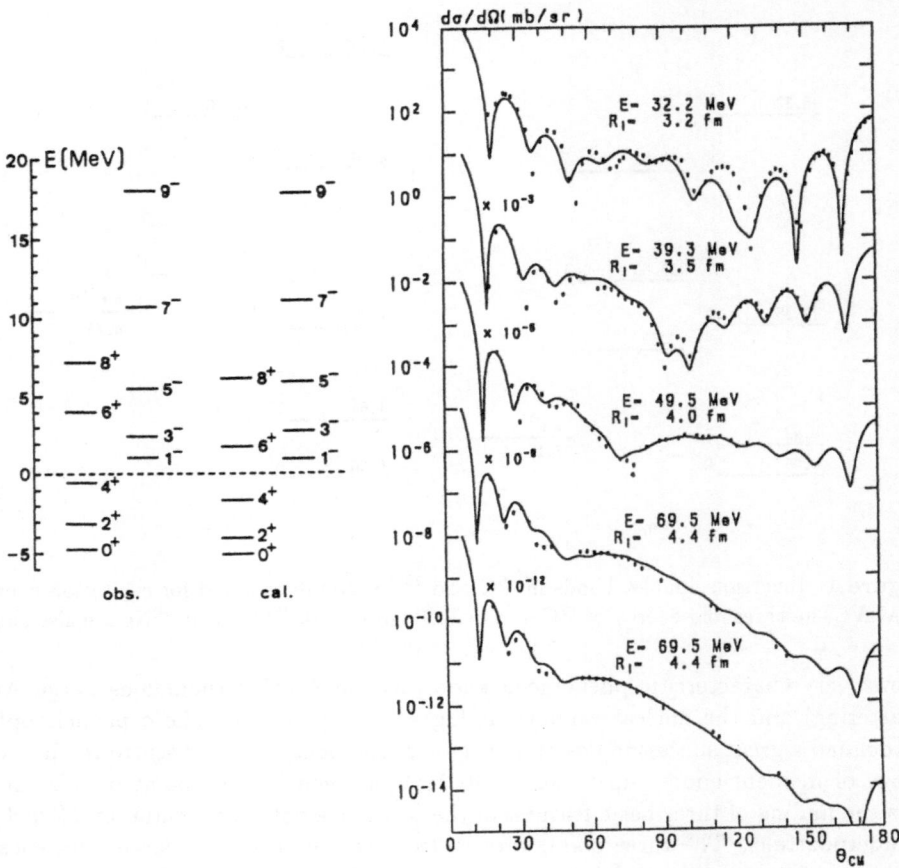

Figure 2. Comparison of the $^{16}O + \alpha$ RGM calculations with the data. Left figure shows the RGM reproduction of the inversion doublet band states of ^{20}Ne, while the right figure shows the reproduction of the differential cross sections of the $^{16}O + \alpha$ elastic scattering by the same RGM with inclusion of the imaginary potential of the unique optical potential.

$\chi(\mathbf{r})$ is obtained by solving the RGM equation of motion,

$$\langle \phi(^{16}O)\phi(\alpha)|(\hat{H} - E)|\mathcal{A}\{\chi(\mathbf{r})\phi(^{16}O)\phi(\alpha)\}\rangle = 0, \tag{5}$$

which is rewritten as

$$\int d\mathbf{a}'\{H(\mathbf{a}, \mathbf{a}') - EN(\mathbf{a}, \mathbf{a}')\}\chi(\mathbf{a}') = 0, \tag{6}$$

where

$$
\begin{aligned}
H(\mathbf{a}, \mathbf{a}') &= \langle \delta(\mathbf{r} - \mathbf{a})\phi(^{16}O)\phi(\alpha)|\hat{H}|\mathcal{A}\{\delta(\mathbf{r} - \mathbf{a}')\phi(^{16}O)\phi(\alpha)\}\rangle, \\
N(\mathbf{a}, \mathbf{a}') &= \langle \delta(\mathbf{r} - \mathbf{a})\phi(^{16}O)\phi(\alpha)|\mathcal{A}\{\delta(\mathbf{r} - \mathbf{a}')\phi(^{16}O)\phi(\alpha)\}\rangle .
\end{aligned} \tag{7}
$$

The the equivalent local potential $V^{eq}(r)$ from the RGM non-local potential is obtained as

$$V^{eq}(r) = E_r - \frac{\mathbf{p}^2(r)}{2\mu}, \tag{8}$$

where E_r is the energy of the relative motion, $E_r = E - E(^{16}\text{O}) - E(\alpha)$, and μ is the reduced mass. The local momentum $\mathbf{p(r)}$ is calculated by solving[8]

$$H^W(\mathbf{r}, \mathbf{p(r)}) - EN^W(\mathbf{r}, \mathbf{p(r)}) = 0, \tag{9}$$

where $H^W(\mathbf{r}, \mathbf{p})$ and $N^W(\mathbf{r}, \mathbf{p})$ are Wigner transforms of the RGM kernels $H(\mathbf{r}, \mathbf{r}')$ and $N(\mathbf{r}, \mathbf{r}')$, respectively. The definition of the Wigner transform $A^W(\mathbf{r}, \mathbf{p})$ of the kernel $A(\mathbf{r}, \mathbf{r}')$ is

$$A^W(\mathbf{r}, \mathbf{p}) = \int d\mathbf{s} \exp(i\frac{\mathbf{p}}{\hbar}\mathbf{s}) A(\mathbf{r} - \frac{\mathbf{s}}{2}, \mathbf{r} + \frac{\mathbf{s}}{2}). \tag{10}$$

It is well known that the real part of the unique optical potential of the α particle is close to the double-folding potential. Here, however, we should note that the double-folding potential is not a fully microscopic potential. Therefore we need a microscopic justification why the double folding potential can be close to the real part of the unique optical potential. This microscopic justification can only be obtained when we have a correct understanding of the important role of the Pauli exclusion principle. The important role of the Pauli priciple can be said to be the formation of the Pauli-forbidden region in the phase space of the inter-cluster relative motion. We explain this point below.

We give the explanation on the basis of the equivalent local potential to the RGM non-local potential[8]. It is well known that for the relative motion between clusters (A and B) there exist <u>Pauli-forbidden states</u>[8] $\chi_F(\mathbf{r})$ which satisfy

$$\mathcal{A}\{\chi_F(\mathbf{r})\phi(A)\phi(B)\} = 0, \tag{11}$$

and which are just harmonic oscillator wave functions whose number of oscillator quanta is smaller than certain value N_F characteristic to the cluster system $A + B$:

$$\chi_F(\mathbf{r}) = R_{nL}(r)Y_{Lm}(\hat{r}) \quad \text{with} \quad 2n + L \leq N_F. \tag{12}$$

The Pauli-forbidden states, which the wave function of the relative motion cannot occupy, are translated to the <u>Pauli-forbidden region</u>[8] in the language of phase space. The Pauli-forbidden region corresponding to the functional space spanned by the Pauli-forbidden states of Eq. (12) is given as

$$\frac{\mathbf{p}^2}{2\mu} + \frac{\mu\omega^2}{2}\mathbf{r}^2 \leq (N_F + 3/2)\hbar\omega. \tag{13}$$

Any physical trajectory $\mathbf{p(r)}$ cannot enter the Pauli-forbidden region in the phase space, from which we have

$$\frac{\mathbf{p}^2(\mathbf{r})}{2\mu} > (N_F + 3/2)\hbar\omega - \frac{\mu\omega^2}{2}\mathbf{r}^2. \tag{14}$$

This inequality means that the local momentum $\mathbf{p(r)}$ in small distance region ($r \approx$ small) is very large, namely the equivalent local potential $V^{eq}(r)$ for small r is very deep because of the relation $V^{eq}(r) = E_r - \mathbf{p}^2(\mathbf{r})/2\mu$ in Eq. (8). Thus we arrive at the important conclusion that **the inter-cluster potential should be very deep because of the Pauli principle.** When the potential is very deep, the local momentum is very high and the scattering is effectively a high energy scattering in the interaction region even for low incident energy. For high energy scattering, the effect of the nucleon exchange due to antisymmetrization is very small. Thus the main part of the inter-cluster potential is given by the direct potential term and the exchange potential terms give only minor modification to it. The direct potential is nothing but the double-folding potential.

The concept of the Pauli-forbidden region in the phase space plays an important role in the theory of AMD (Antisymmetrized Molecular Dynamics)[9, 11] for reactions which we discuss later in this lecture. The AMD theory is an extension of the TDCM (Time-Dependent Cluster Model)[12, 13] and the TDCM has been used extensively for the microscopic derivation of the inter-nucleus potential. The concept of the Pauli-forbidden region has played also decisively important role in the TDCM calculation of the inter-nucleus potential. We discuss the Pauli-forbidden region in the TDCM theory later when we explain the formulation of AMD.

AMD THEORY FOR FORMATION AND DISSOLUTION OF CLUSTERS

Needless to say, for the theoretical study of clustering it is highly desirable to have a theory with which we can describe formation and dissolution of clusters without assuming the existence of clusters. The Antisymmetrized Molecular Dynamics (AMD)[9, 14−18] is such a kind of theory. In this section we briefly explain the formulation of AMD and then discuss the formation and dissolution of clusters in ^{20}Ne as an example in stable-nucleus region[15]. AMD study of clustering in unstable nuclei is discussed in the next section.

Formulation of AMD for the Study of Nuclear Structure

The antisymmetrized molecular dynamics (AMD) describes the total nuclear system with a linear combination of parity-projected Slater determinants of nucleon wave packets,

$$\Phi^{\pm} \equiv \Phi^{\pm}(\{\mathbf{Z}_j, \xi_j\}) + C\Phi^{\pm}(\{\mathbf{Z}'_j, \xi'_j\}) + \cdots,$$

$$\Phi^{\pm}(\{\mathbf{Z}_j, \xi_j\}) \equiv (1 \pm P)\Phi(\{\mathbf{Z}_j, \xi_j\}),$$

$$\Phi(\{\mathbf{Z}_j, \xi_j\}) = \det\left[\phi_{\mathbf{Z}_j}(\mathbf{r}_k)\chi_j^{spin}\chi_j^{isospin}\right],$$

$$\phi_{\mathbf{Z}_j}(\mathbf{r}) \propto \exp[-\nu(\mathbf{r} - \frac{\mathbf{Z}_j}{\sqrt{\nu}})^2] \propto \exp[-\nu(\mathbf{r} - \mathbf{D}_j)^2 + i\frac{\mathbf{K}_j}{\hbar} \cdot \mathbf{r}],$$

$$\mathbf{Z}_j = \sqrt{\nu}\mathbf{D}_j + \frac{i}{2\hbar\sqrt{\nu}}\mathbf{K}_j,$$

$$\chi_j^{spin} = (1/2 + \xi_j)\chi_{\uparrow} + (1/2 - \xi_j)\chi_{\downarrow}, \tag{15}$$

with P standing for the parity operator. The parameters, $\{u_k\} \equiv (\{\mathbf{Z}_j\}, \{\xi_j\}, \{\mathbf{Z}'_j\}, \{\xi'_j\}, \cdots, C, \cdots)$ of the total wave function $\Phi^{\pm}$ are determined by

$$\frac{\langle\Phi^{\pm}|\hat{H}|\Phi^{\pm}\rangle}{\langle\Phi^{\pm}|\Phi^{\pm}\rangle} = \text{minimum}. \tag{16}$$

The AMD wave function $\Phi(\{\mathbf{Z}_j, \xi_j\})$ is just a special case of the wave function of the cluster model of Brink[19] or the time-dependent cluster model[12], where every cluster is composed of a single nucleon. It is also the same as the wave function of Feldmeier's FMD[20].

For those who are not familiar with Brink's wave function, I here show, for the case of a two-nucleon system with the same spin and isospin, the relation between Brink's wave function and the shell-model wave function:

$$\|\Phi\| = 1,$$

$$\begin{aligned}
\Phi &= n(D)\det|\phi_{-D}, \phi_{+D}| \\
&= \frac{n(D)}{2}\det|\phi_{-D} + \phi_{+D}, \phi_{-D} - \phi_{+D}| \\
&\longrightarrow \frac{1}{\sqrt{2}}\det|(0s),(0p)| \qquad \text{for} \quad D \to 0, \qquad (17)
\end{aligned}$$

where $\phi_{\pm D}$ stands for a Gaussian wave packet around a spatial point $\pm D$. By the same argument in this two-nucleon system, we can prove another example of the relation of Brink's wave function with the shell model wave function in the 16-nucleon system:

$$\begin{aligned}
\Phi &= \det|\phi_{\mathbf{Z}_j}\chi_j^{spin}\chi_j^{isospin}| \\
&\propto \det|(0s)^4(0p)^{12}| \\
&\qquad \text{for} \quad \mathbf{Z}_j \sim 0 \quad (j = 1 \sim 16) \qquad (18)
\end{aligned}$$

The determination of the parameters, $\{u_k\} \equiv (\{\mathbf{Z}_j\}, \{\xi_j\}, \{\mathbf{Z}'_j\}, \{\xi'_j\}, \cdots, C, \cdots)$ is made by the use of the frictional-cooling method: We first choose a set of arbitrary numbers $\{u_k^o\}$ as initial numbers for $\{u_k\}$. The AMD wave function with these $\{u_k^o\}$ represents a highly excited state. We cool down this initial state into the ground state by solving the following cooling equation,

$$i\hbar\frac{d}{dt}u_k = (\lambda + i\mu)\frac{\partial}{\partial u_k^*}\frac{\langle \Phi^\pm|\hat{H}|\Phi^\pm\rangle}{\langle \Phi^\pm|\Phi^\pm\rangle}.$$

One can easily prove the relation

$$\begin{aligned}
\frac{d}{dt}\frac{\langle \Phi^\pm|\hat{H}|\Phi^\pm\rangle}{\langle \Phi^\pm|\Phi^\pm\rangle} &= \frac{2\mu}{\hbar}\sum_k\left|\frac{\partial}{\partial u_k}\frac{\langle \Phi^\pm|\hat{H}|\Phi^\pm\rangle}{\langle \Phi^\pm|\Phi^\pm\rangle}\right|^2 < 0, \\
&\text{for} \quad \mu < 0, \quad \lambda = \text{arbitrary}. \qquad (19)
\end{aligned}$$

The most important character of the structure study with the AMD, combined with the frictional-cooling technique, is that it is without prejudice *i.e.* free from any model assumptions.

The AMD wave function $\Phi^\pm$ is an intrinsic state having no good angular momentum. Hence, we perform a projection of the angular momentum by numerically performing the three-dimensional integration over the Euler angles Ω;

$$\begin{aligned}
\Phi^\pm_{KJM} &= P^J_{MK}\Phi^\pm \\
&= \int d\Omega D^{J*}_{MK}(\Omega)R(\Omega)\Phi^\pm. \qquad (20)
\end{aligned}$$

Here $D^J_{MK}(\Omega)$ is the Wigner D function and $R(\Omega)$ is the rotation operator. K-mixing can be made, and actually it has been made in some cases for the calculation of energy spectra.

Formation and Dissolution of Clusters in ^{20}Ne

When we apply AMD to ^{8}Be and ^{16}O, we find that the calculated results show that ^{8}Be has an α-α dumbbell structure, while ^{16}O has a double-closed shell structure of $0s$ and $0p$ orbits: The AMD can reproduce the typical clustering structure of ^{8}Be as well as the typical shell-model structure of ^{16}O without introducing any model assumptions[9, 14].

As we saw previously, the nucleus ^{20}Ne has inversion-doublet band levels due to $\alpha + {}^{16}$O clustering. Hence, we need to check whether AMD can reproduce this $\alpha +$

^{16}O clustering without assuming any existence of clusters. Furthermore, it has been argued by many authors that this $\alpha + {}^{16}$O clustering in the ground-state rotational band becomes weaker as the spin of the band member state becomes higher and almost vanishes at the band terminal state[21]. Therefore, we also need to check whether AMD can justify these arguments on the formation and dissolution of clusters along the yrast line[15].

In the case of the study of nuclear structure along the yrast line, the energy of the system should minimized under the constraint that the expectation value of the total angular momentum should have the given value

$$W(\{\mathbf{Z}_j^*\}, \{\mathbf{Z}_j\}) = \mathbf{J} \cdot \mathbf{J} = \text{given value},$$

$$\mathbf{J} = \frac{\langle \Phi(\{\mathbf{Z}_j\}) | \hat{\mathbf{J}} | \Phi(\{\mathbf{Z}_j\}) \rangle}{\langle \Phi(\{\mathbf{Z}_j\}) | \Phi(\{\mathbf{Z}_j\}) \rangle}, \tag{21}$$

where $\hat{\mathbf{J}}$ is the total-angular-momentum operator. The equation of the constrained frictional cooling is given by

$$i\hbar \frac{d}{dt} \mathbf{Z}_j = (\lambda + i\mu) \Big[\frac{\partial}{\partial \mathbf{Z}_j^*} \frac{\langle \Phi(\{\mathbf{Z}_j\}) | \hat{H} | \Phi(\{\mathbf{Z}_j\}) \rangle}{\langle \Phi(\{\mathbf{Z}_j\}) | \Phi(\{\mathbf{Z}_j\}) \rangle} + \eta \frac{\partial}{\partial \mathbf{Z}_j^*} W(\{\mathbf{Z}_j^*\}, \{\mathbf{Z}_j\}) \Big].$$

The multiplier real function $\eta = \eta(\{\mathbf{Z}_j^*\}, \{\mathbf{Z}_j\})$ is determined so as to get the relation

$$\frac{d}{dt} W(\{\mathbf{Z}_j^*\}, \{\mathbf{Z}_j\}) = 0. \tag{22}$$

¿From Eq. (22) we get

$$\eta = -\frac{1}{2} \frac{\sum_j \Big[(1 + i\frac{\lambda}{\mu}) \frac{\partial W}{\partial \mathbf{Z}_j^*} \cdot \frac{\partial \langle \hat{H} \rangle}{\partial \mathbf{Z}_j} + (1 - i\frac{\lambda}{\mu}) \frac{\partial W}{\partial \mathbf{Z}_j} \cdot \frac{\partial \langle \hat{H} \rangle}{\partial \mathbf{Z}_j^*} \Big]}{\sum_j \frac{\partial W}{\partial \mathbf{Z}_j^*} \cdot \frac{\partial W}{\partial \mathbf{Z}_j}},$$

$$W = W(\{\mathbf{Z}_j^*\}, \{\mathbf{Z}_j\}), \quad \langle \hat{H} \rangle = \frac{\langle \Phi(\{\mathbf{Z}_j\}) | \hat{H} | \Phi(\{\mathbf{Z}_j\}) \rangle}{\langle \Phi(\{\mathbf{Z}_j\}) | \Phi(\{\mathbf{Z}_j\}) \rangle}. \tag{23}$$

Just in the case of the frictional cooling without constraints, we can easily show the following relation

$$\frac{d}{dt} \frac{\langle \Phi(\{\mathbf{Z}_j\}) | \hat{H} | \Phi(\{\mathbf{Z}_j\}) \rangle}{\langle \Phi(\{\mathbf{Z}_j\}) | \Phi(\{\mathbf{Z}_j\}) \rangle} < 0, \quad \text{for } \mu < 0, \quad \lambda = \text{arbitrary}.$$

Calculated density distributions[15] of the states of $K^\pi = 0^+$ and $K^\pi = 0^-$ bands of ^{20}Ne are shown in Fig. 3. We see clearly the formation of $\alpha + {}^{16}$O cluster structure in both bands but also the weakening of clustering with higher spin almost totally dissolved clustering at the band terminals. If we look at Fig. 3 in more detail, we find that the $\alpha + {}^{16}$O clustering is more prominent in the negative-parity band than in the positive-parity band. From the analysis of the calculated wave functions we find that the breaking of [4]-symmetry becomes rather significant near the band termination. We emphasize here again the importance of these theoretical results since they are obtained without introducing any model assumption of clustering.

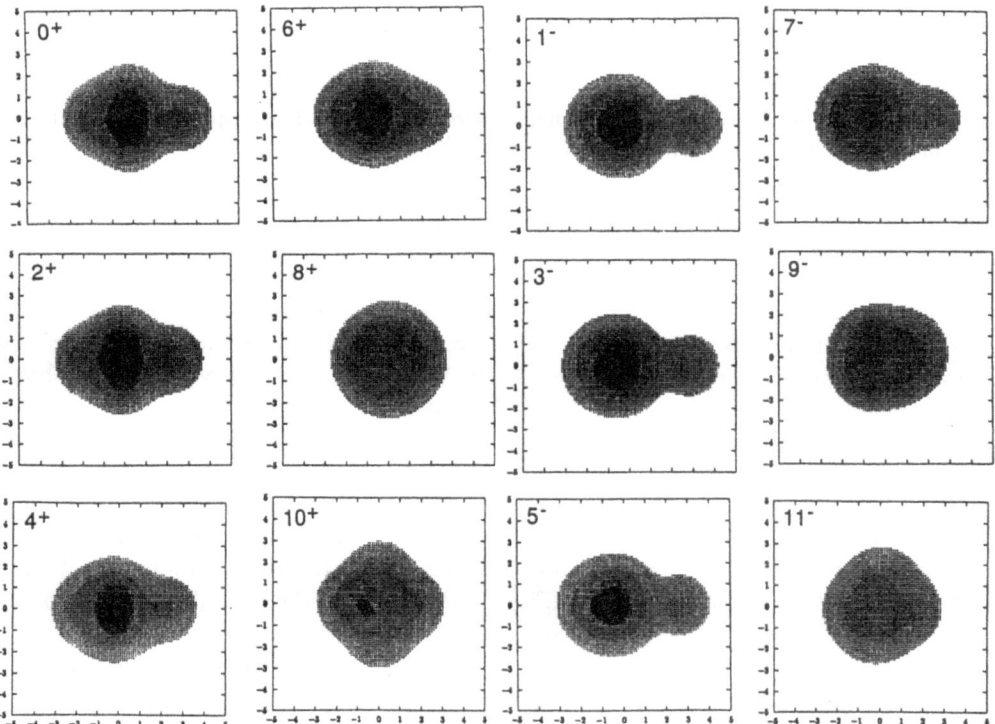

Figure 3. Calculated density distributions of states of $K^\pi = 0^+$ and $K^\pi = 0^-$ bands of ^{20}Ne. the states with 10^+ and 11^- have an entirely different structure than the states with lower spins.

FORMATION AND DISSOLUTION OF CLUSTERS IN UNSTABLE NUCLEUS REGION

The study of unstable nuclei is a novel and rapidly developing field of nuclear physics. An important question is how the clustering structure is in the region of unstable-nuclei. Let us consider Be isotopes. We know well that ^{8}Be has a prominent clustering structure of α - α. Then our question is what happens for this α - α structure when we add many neutrons to ^{8}Be until we reach the neutron drip line nucleus ^{14}Be. In order to study this problem theoretically, it is desirable that our theoretical framework can describe both clustering structure and shell-model-like structure. Namely, it is desirable to have a theoretical framework which can treat the formation and dissolution of clusters. As explained in the previous section, the Antisymmetrized Molecular Dynamics (AMD)[9, 14, 18] is such a kind of theory and below we discuss the results of the application of AMD to light unstable nuclei.

Structure Change as a Function of Neutron and Proton Numbers

We have studied the structure of Li, Be, B, and C isotopes up to the neutron drip line[16, 18]. The interaction in the Hamiltonian is as follows: The central force is the density-dependent nuclear force " MV1-case 3 "[22], where the density-dependent term is a zero-range three-body force, $v^{(3)}\delta(\mathbf{r}_1 - \mathbf{r}_2)\delta(\mathbf{r}_2 - \mathbf{r}_3)$, and the density-independent term is a two-range Gaussian interaction with exchange-mixture parameters[23]. The two-nucleon spin-orbit force G3RS[24] has the form, $u_{LS}\{\exp(-5r^2) - \exp(-2.78r^2)\}$ $P(^3O)\mathbf{L} \cdot (\mathbf{S}_1 + \mathbf{S}_2)$, with $P(^3O)$ standing for the projector onto ^3O state. The Coulomb force between protons is taken into account. The value of the oscillator constant ν is common to all nucleons and is chosen so as to get the optimum binding energy for each instrinsic state with definite parity.

In Fig. 4 the calculated density distributions of Be and B isotopes are given. Here in the case of Be isotopes the densities are not those of $\Phi^{\pm}$ but those of Φ, while in the case of B isotopes the densities of Φ^{-} as well as those of Φ are displayed. It should be noted that the energy minimization is always performed not for Φ but for the parity-projected intrinsic states $\Phi^{\pm} = (1 \pm P)\Phi$. In Fig. 4 in the case of B isotopes we show proton densities ρ_p and neutron densities ρ_n together with the total densities $\rho = \rho_p + \rho_n$. What is commonly found for Li, Be, B, and C is that the AMD wave functions with the neutron number N around 8, $N \approx 8$, have shell-model-like structure with nearly spherical shape. However, as N differs from $N \approx 8$, the AMD wave functions show large structure-change in all the isotopes of Li, Be, B, and C.

In the case of Li and Be, the structure around $N \approx Z$ is seen to have pronounced a clustering structure like $\alpha + t$ structure in ^{7}Li and $\alpha + \alpha$ structure in ^{8}Be. Needless to say, it has long been known that ^{7}Li and ^{8}Be have $\alpha + t$ and $\alpha + \alpha$ structure, respectively. On the neutron-richer side with N larger than 8, we have found that the AMD wave functions both for Be and B again show two-cluster-like structure. (Similar result was obtained by assuming α-α core in Ref.[25].) Furthermore, in the case of B isotopes near the neutron dripline, we have obtained a rather prominent neutron skin.

In the case of C isotopes, the calculated density distribution of protons which is oblate in ^{12}C does not change so much up to the neutron dripline nucleus ^{22}C, but on the other hand the calculated neutron density distribution shows a characteristic change of deformation when N increases. Near the neutron dripline AMD calculations predict not only a large oblate deformation but also a prominent neutron skin. It should be noticed that in the C isotopes AMD gives us no more a prominent clustering structure.

Comparison between Theory and Experiment

We here show that the AMD wave functions which show drastic structure-change as a function of N number reproduce well many spectroscopic data. In some cases the reproduction of data gives us direct support to the structure described by AMD wave functions.

The calculated radii reproduce the data well except for the halo nuclei such as ^{11}Li and ^{11}Be. The reproduction of data was found to be not good if we switch off the density-dependence of the nuclear force. In order to reproduce the large radii of halo nuclei, the single-particle wave functions of the halo neutrons need to be represented not by a single Gaussian wave packet but by a superposition of several Gaussian wave packets.

The binding energies are also reproduced rather well. In Fig. 5 we show a comparison between theory and experiment for the case of the B and C isotopes.

The energy spectra are also reproduced rather well, except for "halo levels" such as the $1/2^+$ ground state of ^{11}Be. Fig. 6 shows some examples of the comparison between theory and data. In order to reproduce the inversion of the lowest $1/2^+$ level and the lowest $1/2^-$ level in ^{11}Be, the single-particle wave function of a halo neutron needs to be represented by superposing several Gaussian wave packets together with a more careful choice of effective nuclear force.

The observed magnetic moments in Li, Be, B, and C isotopes are reproduced well by AMD calculations. Fig. 7 shows the comparison of magnetic moments between theory and data in Li, Be and B isotopes.

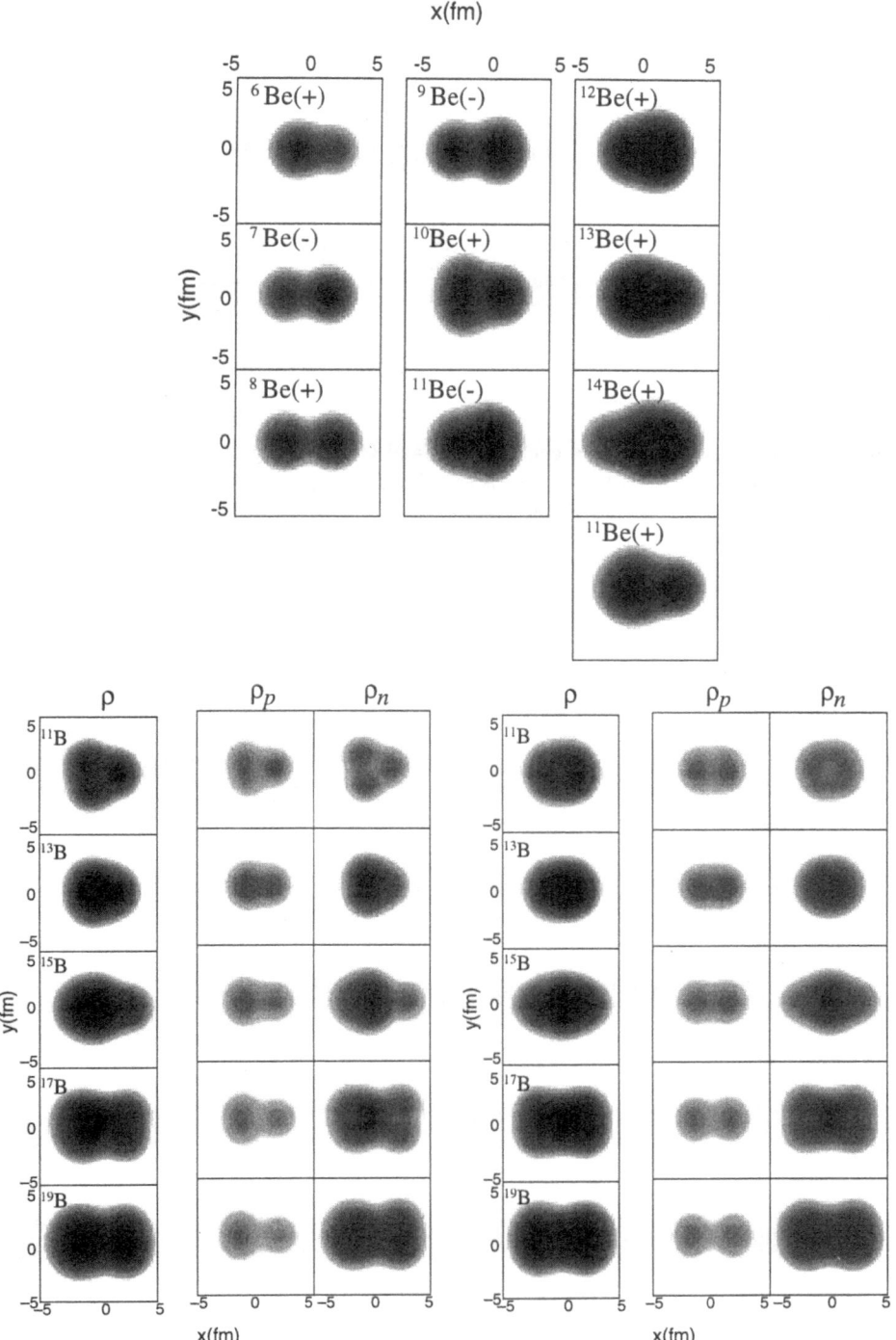

Figure 4. (a), top half: Calculated density distributions of Be isotopes. (b) lower half: Calculated density distributions of B isotopes. Left figure shows densities before parity projection, while the right figure shows the densities after parity projection. Proton ρ_p, neutron ρ_n and total $\rho = \rho_p + \rho_n$ densities are shown.

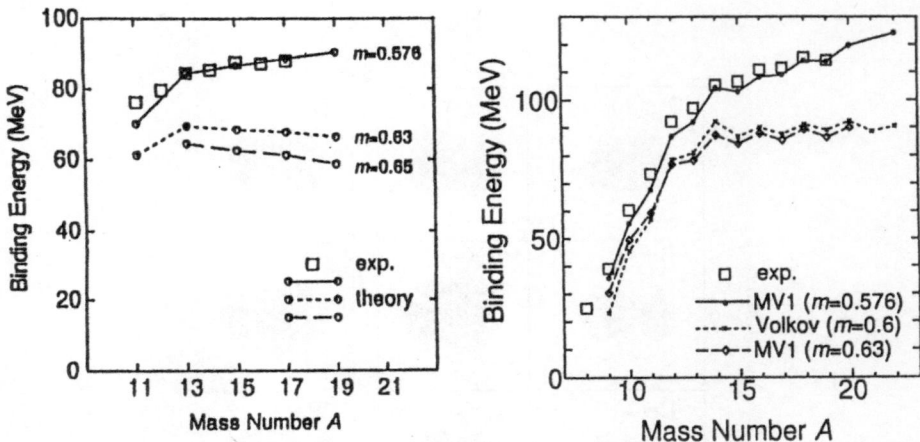

Figure 5. Comparison of binding energies of B and C isotopes between theory and experiment. In B isotopes, the VM1 force with three different choices of the Majorana parameter m is used and in C isotopes the case of the Volkov No. 1 force with Majoranaparameter $m = 0.6$ is also shown in addition to the calculation with the VM1 force.

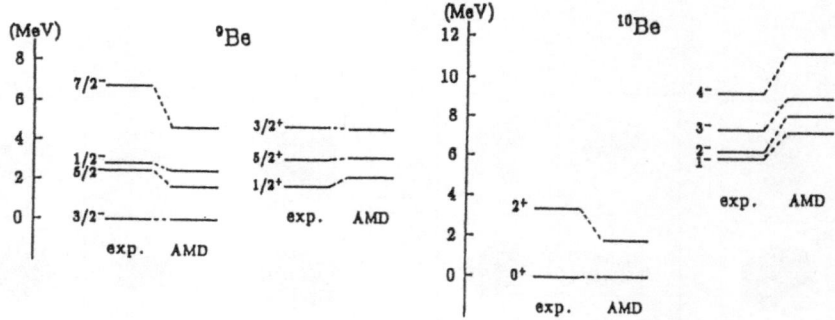

Figure 6. Energy levels of ^{9}Be and ^{10}Be.

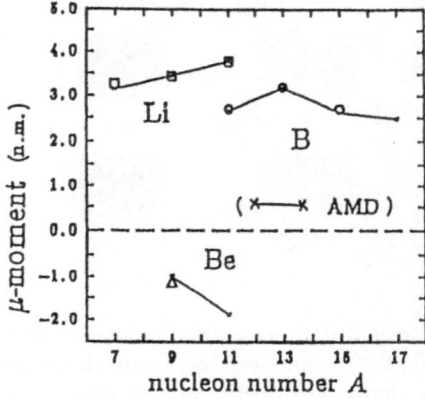

Figure 7. Magnetic moments. Crosses connected with lines are calculated values with AMD. while squares, circles and triangles represent data.

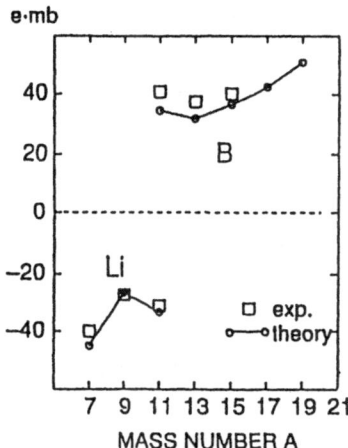

Figure 8. Electric quadrupole moments. Small circles connected with lines are AMD calculations.

Fig. 8 shows good reproduction of electric quadrupole (Q) moments in Li and B isotopes. The reproduction of E2 transition rates in Li, Be, B, and C isotopes has been found also to be rather good. Fig. 9 shows a comparison between theory and experiment

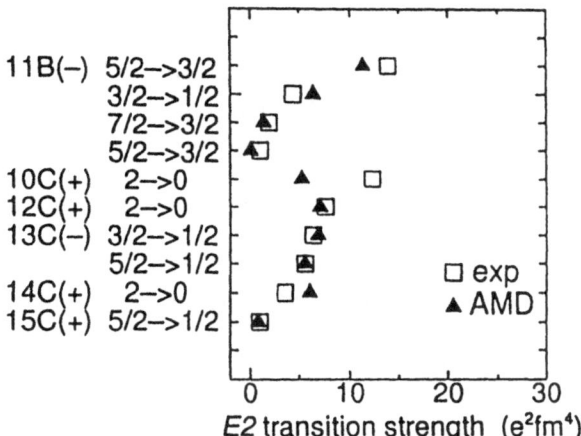

Figure 9. Comparison of B(E2,$J_1 \rightarrow J_2$) values between theory and experiment.

of $B(E2)$ values in C isotopes and in ^{11}B. Except for the case of ^{10}C($2^+ \rightarrow 0^+$), the AMD results are seen to agree well with the data. In the case of nuclei where clustering is well developed, we have to superpose several AMD determinants in order to reproduce the long tail of the inter-cluster relative wave function. Without the long tail both the E2 transition rates and Q moments are underestimated. In the case of ^{7}Li, we executed this superposition, which resulted in a good reproduction of the E2 strength from the $1/2^-$ state to the $3/2^-$ state and the Q moment as shown in Fig. 8. Without superposition the calculated E2 transition rate mentioned above is about half of the observed value.

Here, it is to be noted that no effective g-factor and no effective charge are used but only bare values of g-factors and charge are used in AMD calculations.

Novel Clustering near Neutron Dripline

We have seen that in cases of Be and B isotopes AMD calculations predict the appearance of two-cluster-like structure near the neutron dripline. This result is in accordance with the pioneering work of Ref.[25]. The structure here is basically different from the two-cluster structure in stable-nucleus region. It is because in the stable-nucleus region both clusters which form the two-cluster structure are stable clusters, such as an α cluster and a ^{16}O cluster. Stability is a necessary condition for a cluster. In the region near the neutron drip line not every nucleus can be divided into two stable clusters. It looks like the clustering structure which AMD predicts consists of at least two parts, one is the core part which has two-cluster structure and the other is the surrounding part composed of many neutrons. This is a novel clustering structure which we have not yet observed in the stable-nucleus region. We have to clarify what kind of many-body dynamics gives rise to the novel clustering structure near the neutron dripline.

FORMATION AND DISSOLUTION OF CLUSTERS IN NUCLEAR COLLISIONS

A characteristic feature of medium and high energy nuclear collisions is the production of many fragments. Since mean-field theory cannot describe the dynamical process of fragmentation, molecular-dynamics approaches have been developed. In this lecture we discuss the fragmentation problem by using the AMD theory[9, 11, 26, 29] and also discuss the characteristic feature of AMD in comparison with the mean-field theory for nuclear collisions. The AMD theory is a quantum-mechanical theory since it treats the time development of the system wave function which is exactly antisymmetrized. We start with the explanation of the formulation of AMD for the study of nuclear collisions.

Formulation of AMD for the Study of Nuclear Collisions

In AMD for collisions, the wave function of the A-nucleon total system is described by a single Slater determinant,

$$\Phi(\{\mathbf{Z}_j\}) = \frac{1}{\sqrt{A!}} \det\left[\phi_{\mathbf{Z}_j}(\mathbf{r}_k)\chi_{\alpha_j}\right],$$

where χ_{α_j} stands for the spin-isospin function and α_j represents the spin and isospin label of the j-th single-particle state, $\alpha_j = \mathrm{p}\uparrow$, $\mathrm{p}\downarrow$, $\mathrm{n}\uparrow$, or $\mathrm{n}\downarrow$. The reason why we do not use linearly superposed Slater determinants as in the case of structure studies is simply because the incorporation of the two-nucleon collision process which we discuss later is not easy for superposed Slater determinants. The time development of the coordinate parameters $\{\mathbf{Z}_j\}$ is determined by the time-dependent variational principle,

$$\delta \int_{t_1}^{t_2} dt\, \frac{\langle\Phi(\{\mathbf{Z}_j\})|(i\hbar\frac{d}{dt} - \hat{H})|\Phi(\{\mathbf{Z}_j\})\rangle}{\langle\Phi(\{\mathbf{Z}_j\})|\Phi(\{\mathbf{Z}_j\})\rangle} = 0, \tag{24}$$

which leads to the equation of motion for $\{\vec{Z}_j\}$;

$$i\hbar \sum_{j\tau} C_{k\sigma,j\tau} \frac{d}{dt} Z_{j\tau} = \frac{\partial}{\partial Z_{k\sigma}^*} \frac{\langle\Phi(\{\mathbf{Z}_j\})|\hat{H}|\Phi(\{\mathbf{Z}_j\})\rangle}{\langle\Phi(\{\mathbf{Z}_j\})|\Phi(\{\mathbf{Z}_j\})\rangle},$$

$$C_{k\sigma,j\tau} \equiv \frac{\partial^2}{\partial Z_{k\sigma}^* \partial Z_{j\tau}} \log\langle\Phi(\{\mathbf{Z}_j\})|\Phi(\{\mathbf{Z}_j\})\rangle, \qquad (25)$$

where $\sigma,\tau = x,y,z$.

In order to make AMD applicable to heavy-ion collisions, the following two problems should be further treated. One is the *initialization problem*, namely the construction of the ground-state wave functions of colliding nuclei, and the other is the *incorporation of two-nucleon collisions*. The problem of the construction of the ground-state wave function is solved by the use of the frictional-cooling method which we have already explained in the previous section. The incorporation of two-nucleon collisions is not a simple problem because the position parameters $\{\mathbf{D}_j\}$ do not represent nucleon positions and also the momentum parameters $\{\mathbf{K}_j\}$ do not represent nucleon momenta. This fact can be easily understood when we consider the following example. Let all $\mathbf{D}_j$ and $\mathbf{K}_j$ be vanishingly small in the ^{16}O system. The resulting total wave function Φ is almost equal to the harmonic-oscillator shell-model wave function $\Phi_{c.s.}$ of the doubly-closed shell configuration of $0s$ and $0p$ shells, $(0s)^4(0p)^{12}$. Nucleons of the state $\Phi_{c.s.}$ are neither all located at the space-coodinate origin nor all condensed to zero momentum state in spite of $\mathbf{D}_j \approx 0$, $\mathbf{K}_j \approx 0$ for all j.

The problem of the incorporation of two-nucleon collisions is solved by the introduction of the physical nucleon coordinate parameters, $\{\mathbf{W}_j\}$, defined as follows,

$$
\begin{aligned}
\mathbf{W}_j &= \sqrt{\nu}\mathbf{R}_j + \frac{i}{2\hbar\sqrt{\nu}}\mathbf{P}_j \equiv \sum_{k=1}^{A}\left(\sqrt{Q}\right)_{jk}\mathbf{Z}_k, \\
Q_{jk} &\equiv \frac{\partial \log\langle\Phi(\{\mathbf{Z}_j\})|\Phi(\{\mathbf{Z}_j\})\rangle}{\partial(\mathbf{Z}_j^* \cdot \mathbf{Z}_k)}.
\end{aligned} \qquad (26)
$$

In the absence of antisymmetrization we of course have $\mathbf{W}_j = \mathbf{Z}_j$. We can easily prove the following relations:

$$
\begin{aligned}
N_{\text{osc}} &= \frac{\langle\Phi(\{\mathbf{Z}_j\})|\sum_j \mathbf{a}_j^\dagger \mathbf{a}_j|\Phi(\{\mathbf{Z}_j\})\rangle}{\langle\Phi(\{\mathbf{Z}_j\})|\Phi(\{\mathbf{Z}_j\})\rangle} = \sum_{jk} \mathbf{Z}_j^* \mathbf{Z}_k Q_{jk} \\
&= \sum_j \mathbf{W}_j^* \mathbf{W}_j = \frac{1}{\hbar\omega}\sum_j \left(\frac{1}{2m}\mathbf{P}_j^2 + \frac{m\omega^2}{2}\mathbf{R}_j^2\right), \\
\frac{\langle\Phi(\{\mathbf{Z}_j\})|\hat{\mathbf{L}}|\Phi(\{\mathbf{Z}_j\})\rangle}{\langle\Phi(\{\mathbf{Z}_j\})|\Phi(\{\mathbf{Z}_j\})\rangle} &= (-i\hbar)\sum_{jk} \mathbf{Z}_j^* \times \mathbf{Z}_k Q_{jk} \\
&= (-i\hbar)\sum_j \mathbf{W}_j^* \times \mathbf{W}_j = \sum_j \mathbf{R}_j \times \mathbf{P}_j, \qquad (27)
\end{aligned}
$$

where $\mathbf{a}_j$ stands for the destruction operator of harmonic-oscillator quanta of j-th nucleon, $\mathbf{a}_j = \sqrt{\nu}\mathbf{r}_j + (i/2\hbar\sqrt{\nu})\mathbf{p}_j$, and $\hat{\mathbf{L}}$ is the total orbital angular momentum operator. Furthermore, as for the center-of-mass coordinate we can show that $\sum_j \mathbf{Z}_j = \sum_j \mathbf{W}_j$. The following relation also holds: if the $\mathbf{Z}_j$ are simultaneously displaced by $\mathbf{c}$, $\mathbf{Z}_j \to \mathbf{Z}_j + \mathbf{c}$, then the $\mathbf{W}_j$ are simultaneously displaced by the same amount $\mathbf{c}$, $\mathbf{W}_j \to \mathbf{W}_j + \mathbf{c}$. From these properties of $\{\mathbf{W}_j\}$, we regard the $\{\mathbf{W}_j\}$ as physical coordinates.

Once we have physical nucleon coordinates, we can treat two-nucleon collisions just as the two-nucleon collisions in QMD (quantum molecular dynamics)[30, 32]. When $\mathbf{R}_j$ and $\mathbf{R}_k$ come near each other, two nucleons j and k are made to scatter. Namely the initial $\mathbf{P}_j$ and $\mathbf{P}_k$ are changed into the final $\mathbf{P}_j'$ and $\mathbf{P}_k'$, keeping the initial $\mathbf{R}_j$ and $\mathbf{R}_k$ unchanged. Speaking in terms of complex coordinates $\mathbf{W}$, the initial $\mathbf{W}_j$ and $\mathbf{W}_k$ are changed into the final $\mathbf{W}_j'$ and $\mathbf{W}_k'$. In order to continue the calculation of the time development of the wave function after this two-nucleon collision, we need to transform $\{\mathbf{W}_1, \cdots, \mathbf{W}'_j, \cdots, \mathbf{W}'_k, \cdots, \mathbf{W}_A\}$ back into $\{\mathbf{Z}'_1, \mathbf{Z}'_2, \cdots, \mathbf{Z}'_A\}$. However, in general, the back-transformation from $\{\mathbf{W}_j, j = 1 \sim A\}$ to $\{\mathbf{Z}_j, j = 1 \sim A\}$ does not always exist. When the back-transformation does not exist, we consider the two-nucleon collision as Pauli-blocked.

The existence of $\{\mathbf{W}_j, j = 1 \sim A\}$ for which we have no corresponding $\{\mathbf{Z}_j, j = 1 \sim A\}$ is easily verified by using the relation $N_{\mathrm{osc}} = \sum_j \mathbf{W}_j^* \mathbf{W}_j$ of Eq. (26). This immediately shows that the coordinate origin $\{\mathbf{W}_j = 0, j = 1 \sim A\}$ cannot be constructed from any $\{\mathbf{Z}_j, j = 1 \sim A\}$, because N_{osc} cannot be zero if $A > 4$. When we choose the center-of-mass at the coordinate origin, $\sum_j \mathbf{W}_j = 0$, we have

$$(N_{\mathrm{osc}})_{\mathrm{min}} \leq \sum_j \mathbf{W}_j^* \mathbf{W}_j = \frac{1}{A} \sum_{i<j} |\mathbf{W}_i - \mathbf{W}_j|^2, \tag{28}$$

from which we get $\sqrt{\langle |\mathbf{W}_i - \mathbf{W}_j|^2 \rangle_{\mathrm{average}}} = \sqrt{2(N_{\mathrm{osc}})_{\mathrm{min}}/(A-1)} \geq \sqrt{2}$. If the relative distances $|\mathbf{W}_i - \mathbf{W}_j|$ are too small as to violate the relation of Eq. (28), such $\{\mathbf{W}_j, j = 1 \sim A\}$ cannot be transformed back to any $\{\mathbf{Z}_j, j = 1 \sim A\}$. $\{\mathbf{W}_j, j = 1 \sim A\}$ is defined to be in the Pauli-forbidden region if it cannot be back-transformed to any $\{\mathbf{Z}_j, j = 1 \sim A\}$.

The introduction of physical nucleon coordinates and the concept of the Pauli-forbidden region is due to the mimicking the theory of the canonical coordinates of the TDCM (time-dependent cluster model)[12, 13]. We explain briefly at the end of this subsection the theory of the canonical coordinates of the TDCM.

In AMD, each nucleon is represented by a wave packet and this wave packet does not split. In quantum mechanics, however, if the high-momentum component of a wave packet can escape from the nucleus, the wave packet splits and the high-momentum component goes out of the nucleus, leaving the low-momentum component inside the nucleus. In AMD, since the wave packet does not split, if the wave packet centroid cannot escape from the nucleus, the whole-wave packet does not escape the nucleus even if the high-momentum component can escape. Recently we have studied effects of this defect of the AMD theory, and have found that the modification which remedies this defect can have rather large influence on results of AMD calculations. The wave-packet splitting is made stochastically like the two-nucleon collision. A detailed explanation how we describe wave-packet splitting in AMD is given in Ref.[27].

After the violent reaction stage we stop the AMD calculation. At this time we have many excited fragments due to the violent collision. These primordial fragments decay by evaporationg particles. We use multistep statistical decay theory to describe this final stage of the reaction.

At the end of this subsection, we explain briefly the theory[13] of the canonical coordinates of the TDCM (time-dependent cluster model), because physical nucleon coordinates and the concept of the Pauli-forbidden region in the AMD are based on this theory of the TDCM. In TDCM the wave function of the total system is described by a single Slater determinant which in the case of a two-cluster system $A + B$ has the form

$$\Phi(\{\mathbf{Z}_j\}) = \mathcal{A}\{\psi(A, \mathbf{Z}_A)\psi(B, \mathbf{Z}_B)\},$$

where $\psi(A, \mathbf{Z}_A)$ ($\psi(B, \mathbf{Z}_B)$) stands for the Slater-determinant wave function of the cluster A (B) whose center of mass is located at complex vector point $\mathbf{Z}_A$ ($\mathbf{Z}_B$),

$$\psi(j, \mathbf{Z}_j) = \exp[-A_j\nu(\mathbf{X}_j - \frac{\mathbf{Z}_j}{\sqrt{A_j\nu}})^2]\phi(j)$$

$$\propto \ \exp[-A_j\nu(\mathbf{X}_j - \mathbf{D}_j)^2 + i\frac{\mathbf{K}_j}{\hbar}\cdot\mathbf{X}_j]\phi(j)$$

$$\mathbf{Z}_j \ = \ \sqrt{A_j\nu}\mathbf{D}_j + \frac{i}{2\hbar\sqrt{A_j\nu}}\mathbf{K}_j,$$

$$(j \ = \ A, B), \tag{29}$$

where $\mathbf{X}_j$ is the center-of-mass coordinate of the cluster j and A_j is the mass number of the cluster j $(j = A, B)$. The time development of $\{\mathbf{Z}_j\}$ is determined by the time-dependent variational principle

$$\delta \ \int_{t_1}^{t_2} dt\, L = 0,$$

$$
\begin{aligned}
L \ &= \ \frac{\langle\Phi(\{\mathbf{Z}_j\})|(i\hbar\frac{d}{dt} - \hat{H})|\Phi(\{\mathbf{Z}_j\})\rangle}{\langle\Phi(\{\mathbf{Z}_j\})|\Phi(\{\mathbf{Z}_j\})\rangle} \\
&= \ \frac{i\hbar}{2}\sum_{j,k}(\frac{d}{dt}\mathbf{Z}_k\cdot\mathbf{Z}_j^* - \frac{d}{dt}\mathbf{Z}_j^*\cdot\mathbf{Z}_k)\frac{\partial\log\langle\Phi(\{\mathbf{Z}_j\})|\Phi(\{\mathbf{Z}_j\})\rangle}{\partial(\mathbf{Z}_j^*\cdot\mathbf{Z}_k)} \\
&+ \frac{\langle\Phi(\{\mathbf{Z}_j\})|\hat{H}|\Phi(\{\mathbf{Z}_j\})\rangle}{\langle\Phi(\{\mathbf{Z}_j\})|\Phi(\{\mathbf{Z}_j\})\rangle} - \frac{d}{dt}\frac{i\hbar}{2}\log\langle\Phi(\{\mathbf{Z}_j\})|\Phi(\{\mathbf{Z}_j\})\rangle.
\end{aligned}
\tag{30}
$$

The equation of motion for $\{\mathbf{Z}_j\}$ is of course of the same form as Eq. (25) of AMD theory. As is clear from the appearance of the coefficients $C_{k\sigma,j\tau}$ in the equation of motion of Eq. (25), the equation of motion is not of canonical form. It means that the coordinates $\{\mathbf{Z}_j\}$ are not canonical coordinates. The canonical coordinates $\{\mathbf{W}_j\}$ should satisfy

$$
\begin{aligned}
&\sum_{j,k}(\frac{d}{dt}\mathbf{Z}_k\cdot\mathbf{Z}_j^* - \frac{d}{dt}\mathbf{Z}_j^*\cdot\mathbf{Z}_k)\frac{\partial\log\langle\Phi(\{\mathbf{Z}_j\})|\Phi(\{\mathbf{Z}_j\})\rangle}{\partial(\mathbf{Z}_j^*\cdot\mathbf{Z}_k)} \\
&= \sum_j(\frac{d}{dt}\mathbf{W}_j\cdot\mathbf{W}_j^* - \frac{d}{dt}\mathbf{W}_j^*\cdot\mathbf{W}_j) + \frac{dF}{dt}.
\end{aligned}
\tag{31}
$$

By using this relation in the Lagrangian of Eq. (30), we see easily that the equation of motion for $\{\mathbf{W}_j\}$ is of Hamilton's canonical form:

$$i\hbar\frac{d}{dt}\mathbf{W}_j = \frac{\partial}{\partial\mathbf{W}_j^*}\frac{\langle\Phi(\{\mathbf{Z}_j\})|\hat{H}|\Phi(\{\mathbf{Z}_j\})\rangle}{\langle\Phi(\{\mathbf{Z}_j\})|\Phi(\{\mathbf{Z}_j\})\rangle}.$$

In the case of the two-cluster system, the canonical coordinates $\{\mathbf{W}_j\}$ are given just by the same equation as Eq. (26) in AMD theory[13]. A very important property of the canonical coordinates $\{\mathbf{W}_j\}$ is that in the phase space of $\{\mathbf{W}_j\}$ there exists a Pauli-forbidden region, since there holds the relation

$$\mathbf{W}_A^*\cdot\mathbf{W}_A + \mathbf{W}_B^*\cdot\mathbf{W}_B \geq N_F + 1, \tag{32}$$

where N_F is the same number as in Eq. (12). The Pauli-forbidden region is the region of $\mathbf{W}_A^*\cdot\mathbf{W}_A + \mathbf{W}_B^*\cdot\mathbf{W}_B \leq N_F$.

The importance of the canonical coordinates is most dramatically seen in the calculation of the inter-cluster (or inter-nucleus) potential[8]. We first note that the total center-of-mass coordinate $\mathbf{X}_G$ $(= (A\mathbf{X}_A + B\mathbf{X}_B)/(A + B))$ is separated in the wave function $\Phi(\{\mathbf{Z}_j\})$ as

$$\Phi(\{\mathbf{Z}_j\}) = \exp\left[-(A + B)\nu(\mathbf{X}_G - \frac{\mathbf{Z}_G}{\sqrt{(A + B)\nu}})^2\right]\hat{\Phi}(\mathbf{Z}_r),$$

where

$$\mathbf{Z}_G = \sqrt{\frac{A}{A+B}}\mathbf{Z}_A + \sqrt{\frac{B}{A+B}}\mathbf{Z}_B = \sqrt{(A+B)\nu}\,\mathbf{D}_G + \frac{i}{2\hbar\sqrt{(A+B)\nu}}\mathbf{K}_G,$$

$$\mathbf{Z}_r = \sqrt{\frac{B}{A+B}}\mathbf{Z}_A - \sqrt{\frac{A}{A+B}}\mathbf{Z}_B = \sqrt{\mu_0\nu}\,\mathbf{D}_r + \frac{i}{2\hbar\sqrt{\mu_0\nu}}\mathbf{K}_r,$$

$$\mathbf{D}_G = \frac{A\mathbf{D}_A + B\mathbf{D}_B}{A+B}, \quad \mathbf{K}_G = \mathbf{K}_A - \mathbf{K}_B,$$

$$\mathbf{D}_r = \mathbf{D}_A - \mathbf{D}_B, \quad \mathbf{K}_r = \mu_0\left(\frac{\mathbf{K}_A}{A} - \frac{\mathbf{K}_B}{B}\right),$$

$$\mu_0 = \frac{AB}{A+B}. \tag{33}$$

The canonical coordinates $\mathbf{W}_G$ and $\mathbf{W}_r$ which correspond to $\mathbf{Z}_G$ and $\mathbf{Z}_r$, respectively, are given by

$$\mathbf{W}_G = \mathbf{Z}_G, \quad \mathbf{W}_r = \sqrt{\frac{\partial \log\langle\hat{\Phi}(\mathbf{Z}_r)|\hat{\Phi}(\mathbf{Z}_r)\rangle}{\partial(\mathbf{Z}_r^* \cdot \mathbf{Z}_r)}}\,\mathbf{Z}_r.$$

We always put $\mathbf{Z}_G = 0$ (which means $\mathbf{W}_G = 0$) and hence the only parameter we are concerned with is the relative-motion parameter $\mathbf{Z}_r$ or its canonical version $\mathbf{W}_r$. We should note that since $\mathbf{W}_G = 0$ the relation of Eq. (32) takes the form,

$$N_F + 1 \le \mathbf{W}_r^* \cdot \mathbf{W}_r = \frac{1}{\hbar\omega}\left(\frac{\mathbf{P}_r^2}{2\mu} + \frac{\mu\omega^2}{2}\mathbf{R}_r^2\right), \tag{34}$$

where

$$\mathbf{W}_r = \sqrt{\mu_0\nu}\,\mathbf{R}_r + \frac{i}{2\hbar\sqrt{\mu_0\nu}}\mathbf{P}_r,$$

The local momentum of the inter-cluster relative motion is calculated as usual from the energy-conservation relation

$$\frac{\langle\Phi(\{\mathbf{Z}_j\})|\hat{H}|\Phi(\{\mathbf{Z}_j\})\rangle}{\langle\Phi(\{\mathbf{Z}_j\})|\Phi(\{\mathbf{Z}_j\})\rangle} = \frac{\langle\hat{\Phi}(\mathbf{Z}_r)|\hat{H}|\hat{\Phi}(\mathbf{Z}_r)\rangle}{\langle\hat{\Phi}(\mathbf{Z}_r)|\hat{\Phi}(\mathbf{Z}_r)\rangle} = E. \tag{35}$$

If we regard the quantity $\mathbf{K}_r(\mathbf{D}_r)$ which is the solution of Eq. (35) as the local momentum of the inter-nucleus relative motion, the inter-nucleus potential $V'(D_r)$ is given by

$$V'(D_r) = E_r - \frac{\mathbf{K}_r^2(\mathbf{D}_r)}{2\mu}, \tag{36}$$

Several authors calculated $V'(D_r)$ for many pairs of nuclei. They found that the resultant potentials are very shallow and their incident-energy dependence is such that they become deeper as the incident energy becomes higher[8]. This result is of course entirely different from the unique optical potential of α particles. We now know that since $\mathbf{D}_r$ and $\mathbf{K}_r$ are not canonical coordinates, the potential $V'(D_r)$ can not be regarded as a meaningful quantity. Thus, we should rewrite the relation $\mathbf{K}_r = \mathbf{K}_r(\mathbf{D}_r)$ into $\mathbf{P}_r = \mathbf{P}_r(\mathbf{R}_r)$ and regard the quantity $\mathbf{P}_r(\mathbf{R}_r)$ as the correct local momentum, which leads us to calculate the inter-nucleus potential $V(R_r)$ by

$$V(R_r) = E_r - \frac{\mathbf{P}_r^2(\mathbf{R}_r)}{2\mu}, \tag{37}$$

The resulting potential is very deep and very similar to the potential $V^{eq}(r)$ calculated from RGM as in Eq. (8). The reason why $V(R_r)$ is deep is simple and is the same as that why $V^{eq}(r)$ is deep. Namely the physical local momentum $\mathbf{P}_r(\mathbf{R}_r)$ should satisfy Eq. (34),

$$\frac{\mathbf{P}_r^2(\mathbf{R}_r)}{2\mu} \geq (N_F + 1)\hbar\omega - \frac{\mu\omega^2}{2}\mathbf{R}_r^2,$$

which means that the local momentum $\mathbf{P}_r(\mathbf{R}_r)$ in the small distance region ($R_r \approx$ small) is very large, namely the potential $V(R_r)$ for small R_r is very deep.

Comparison of AMD Theory with Mean Field-Theories

A useful way to understand basic characteristics of AMD is to compare AMD with mean-field theories. Since the AMD wave fuction is a Slater determinant of Gaussian wave packets, one may think that AMD is an approximation of TDHF (time-dependent Hartree-Fock) theory, which is completely wrong. The most important point in comparing AMD with TDHF is the fact that in the situation where many channels are open, the TDHF wave function inevitably contains spurious correlations between channels. In TDHF, the whole system is regarded as being governed by one mean field in spite of the fact that each channel has its own different mean field. Thus in the situation where many channels are open, the TDHF wave function can not be trusted. On the other hand, in AMD (and in any other molecular-dynamics approach) different channels are treated by different events, and there can be no such spurious correlations between different channels. One important feature of the AMD theory is that the AMD is based on calculations over many events.

One may think that the use of the Gaussian wave packet for the single-particle wave function is simply a poor prescription of the AMD theory compared with the single-particle wave function of the TDHF theory which has no restriction in its functional form. This is, however, wrong again. What should be stressed here is that the use of the Gaussian wave packet for the single-particle wave function is made intentionally in the AMD theory. Let us consider a situation where we put no restriction on the functional form of the AMD single-particle wave function. Then the AMD Slater determinant becomes equivalent to the TDHF Slater determinant which has spurious channel correlations in itself. Thus, the Gaussian wave packet is adopted intentionally for the single-particle wave function in AMD so as to prevent AMD from approaching to TDHF.

Applications to Heavy-Ion Collisions

AMD has been very successful in analysing heavy-ion collision data[10, 11, 26-29]. Here, we show only two examples, which are the mass distribution of the fragments and the reproduction of the flow data of nucleons and fragments. Fig. 10 shows the good reproduction[10, 11] of the mass distribution of fragments[33]. This figure shows that the AMD can describe the shell effect in the dynamical production of fragments. It means that the dynamical production cross section of alpha particles is larger than that of mass three fragments. An important conclusion which follows from this result is that the AMD is suited for the analysis of the alpha particle flow. Fig. 11 is the comparison between AMD theory[27] and data[34] about the $Z = 1$ and $Z = 2$ flow of the ^{40}Ar $+ \ ^{27}$Al system. We see that the data are well reproduced when we use the Gogny force[35] which gives us momentum-dependent mean field and soft EOS. On the other hand, the SKG2 force[27], which gives us momentum- independent mean field and stiff

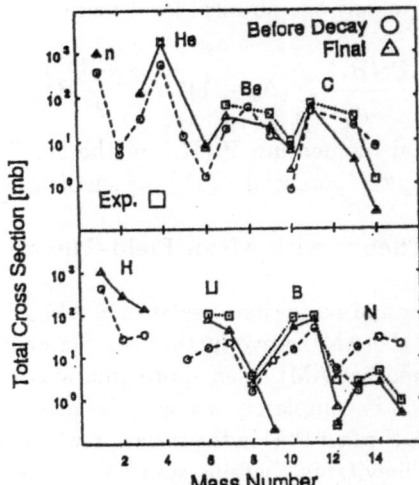

Figure 10. Production cross sections of fragments of $^{12}C + ^{12}C$ collisions at 28.7 MeV/u. Squares are data. Triangles and circles are AMD results with and without inclusion of the statistical decay process of the primordial fragments, respectively. Solid, dashed, and dotted lines connect isotopes.

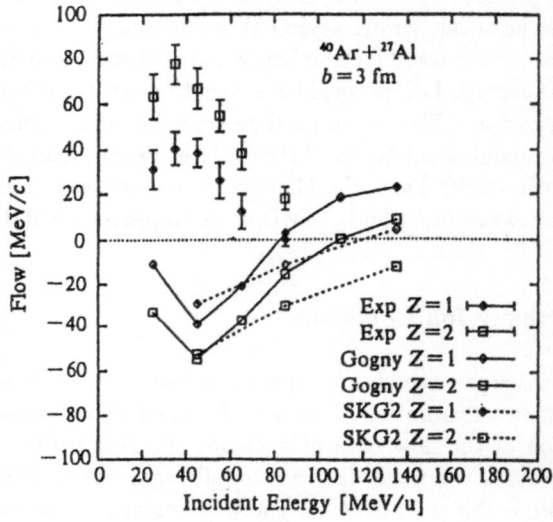

Figure 11. Comparison between AMD calculations and experiments of the flow of fragments with charge Z=1 and Z=2. The experimental data are shown on the positive flow side by changing sign artificially.

EOS, does not reproduce the flow data well. These results are not so sensitive to the magnitude of the in-medium two-nucleon cross sections.

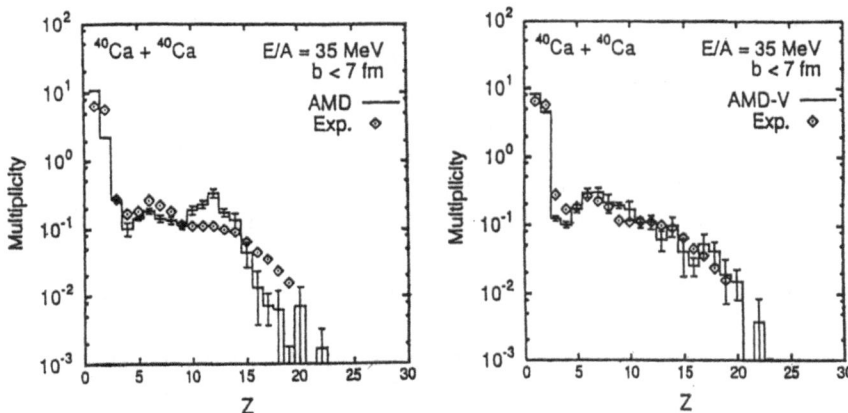

Figure 12. Charge distribution of $^{40}Ca+^{40}Ca$ at 35 MeV/u calculated with usual AMD is compared to that with modified AMD (AMD-V) which includes the wave-packet splitting process.

As explained before, we studied recently the effect of the wave-packet splitting in AMD. Fig. 12 shows the charge distributions[27] of $^{40}Ca+^{40}Ca$ at 35 MeV/u calculated-without and with wave-packet splitting both of which are compared with data[36]. We see that the modified AMD with inclusion of wave-packet splitting gives a better fit of the data. Furthermore, we found that the multiplicity distributions of protons, alphas, and IMF's are better reproduced by the modified AMD than the usual AMD.

SUMMARY

Due to the saturation property of nuclear matter, nuclei can easily assemble and disassemble with only a small amount of energy input or output. Thus, the clustering aspect is a fundamental property of nuclear matter together with the mean-field aspect. In this lecture we discussed microscopic descriptions of the phenomena of formation and dissolution of clusters both in nuclear-structure problems and in nuclear-collision problems. We saw that the Antisymmetrized Molecular Dynamics (AMD) provides us with a very powerful theoretical tool for studying a variety of clustering dynamics in the nuclear many-body system.

REFERENCES

1. H. Horiuchi and K. Ikeda, Prog. Theor. Phys. **40**, (1968) 277.
2. R.H. Davis, *Proc.III Conf. on Reactions between Complex Nuclei*, Asilomar, eds. A. Ghiorso, R.M. Diamond and H.E. Conzett, (1963) p.67.
3. *Proc. II Int. Conf. Clustering Phenomena in Nuclei*, Maryland , eds. D. A. Goldberg J. B. Marion and S. J. Wallace (1975) (ORO-4856-26, US Energy Research and Development Admininstration)
4. *Proc. III Int. Conf. Clustering Aspects of Nuclear Structure and Nuclear Reactions*, Winnipeg, eds. W. T. H. Van Oers, J. P. Svenne, J. S. C. McKee and W. R. Fark,

(1978) (*AIP Conference Proceedings* **Vol. 47**)

5. F. Michel and A. Vanderpoorten, Phys. Rev. **C 16** (1977) 142.

6. T. Matsuse, M. Kamimura, and Y. Fukushima, Prog. Theor. Phys. **53** (1975) 706.

7. T. Wada and H. Horiuchi, Phys. Rev. Lett. **58** (1987) 2190.

8. H. Horiuchi, in: *Trends in Theoretical Physics*, Vol.II, eds. P. J. Ellis and Y. C. Tang (Addison-Wesley, Redwood City, 1991), pp. 1-349.

9. H. Horiuchi, *Proc. Symp. in Honor of Akito Arima on Nuclear Physics in the 1990's*, Santa Fe (1990), eds. D. H. Feng, J. N. Ginocchio, T. Otsuka, and D. D. Strottman, Nucl. Phys. **A522** (1991) 257c.

10. A. Ono, H. Horiuchi, Toshiki Maruyama, and A. Ohnishi, Phys. Rev. Lett. **68** (1992) 2898.

11. A. Ono, H. Horiuchi, Toshiki Maruyama, and A. Ohnishi, Prog. Theor. Phys. **87** (1992) 1185.

12. S. Drożdż, J. Okolowcz, and M. Ploszajczak, Phys. Lett. **109B** (1982) 145; E. Caurier, B. Grammaticos and T. Sami, Phys. Lett. **109B** (1982) 150; W. Bauhoff, E. Caurier, B. Grammaticos and M. Ploszajczak, Phys. Rev. **C 32** (1985) 1915.

13. M. Saraceno, P. Kramer, and F. Fernandez, Nucl. Phys. **A405** (1983) 88.

14. H. Horiuchi, Toshiki Maruyama, A. Ohnishi, and S. Yamaguchi, *Proc. Int. Conf. on Nuclear and Atomic Clusters*, Turku (1991), eds. M. Brenner, T. Lönnroth, and F. B. Malik, (Springer, Berlin, 1992), p. 512; *Proc. Int. Symp. on Structure and Reactions of Unstable Nuclei*, Niigata (1991), eds. K. Ikeda and Y. Suzuki, (World Scientific, Singapore, 1992), p. 108.

15. Y. Kanada-En'yo and H. Horiuchi, Prog. Theor. Phys. **93** (1995) 115.

16. Y. Kanada-En'yo, H. Horiuchi, and A. Ono, Phys. Rev. **C 52** (1995) 628.

17. Y. Kanada-En'yo and H. Horiuchi, Phys. Rev. **C 52** (1995) 647; Phys. Rev. **C 54** (1996) R468.

18. H. Horiuchi, Y. Kanada-En'yo, and A. Ono, *Proc. Second Int. Conf. on Atomic and Nuclear Clusters*, Santorini (1993), eds. G. S. Anagnostatos and W. von Oertzen, Z. Phys. **A349** (1994) 279; H. Horiuchi and Y. Kanada-En'yo, *Proc. Int. Symp. on Physics of Unstable Nuclei*, Niigata (1994), eds. H. Horiuchi, K. Ikeda, K. Sato, Y. Suzuki, and I. Tanihata, Nucl. Phys. **A588** (1995) 121c; H. Horiuchi and Y. Kanada-En'yo, *Proc. Int. Conf. on Exotic Nuclei and Atomic Masses*, Arles (1995), eds. M. de Saint Simon and O. Sorlin (Edition Frontieres, 1996), p. 461.

19. D. M. Brink, *Proc. Int. School of Phys. "Enrico Fermi"*, **course 36** (1965), ed. C. Bloch, p. 247.

20. H. Feldmeier, Nucl. Phys. **A515** (1990) 147 ; *Proc. NATO Advanced Study Institute on the Nuclear Equation of State*, Peñiscola(1989), eds. W. Greiner and H. Stöcker (Plenum, New York, 1989), p. 375.

21. A. Arima, H. Horiuchi, K. Kubodera, and N. Takigawa, *Advances in Nuclear Physics*, Vol. 5, eds. M. Baranger and E. Vogt (Plenum, New York, 1972) p. 345.

22. T. Ando, K.Ikeda, and A. Tohsaki, Prog. Theor. Phys. **64** (1980) 1608.

23. A. B. Volkov, Nucl. Phys. **74**, 33 (1965).

24. N. Yamaguchi, T. Kasahara, S. Nagata, and Y. Akaishi, Prog. Theor. Phys. **62** (1979) 1018; R. Tamagaki, Prog. Theor. Phys. **39** (1968) 91.

25. M. Seya, M. Kohno, and S. Nagata, Prog. Theor. Phys. **65** (1981) 204; H. Furutani et al., Prog. Theor. Phys. Supple. **68** (1980) 193.

26. A. Ono, H. Horiuchi, Toshiki Maruyama, and A. Ohnishi, Phys. Rev. **C 47** (1993) 2652; A. Ono, H. Horiuchi, and Toshiki Maruyama, Phys. Rev. **C 48** (1993) p.?

27. A. Ono, H. Horiuchi, Phys. Rev. C **51** (1995) 299; Phys. Rev. C **53**, 845 (1996); Phys. Rev. C **53**, (1996) 2341; Phys. Rev. C **53** (1996) 2958.

28. H. Horiuchi, *Proc. NATO Advanced Study Institute on Hot and Dense Nuclear Matter*, Bodrum (1993), eds. W. Greiner, H. Stöcker, and A. Gallmann, NATO ASI Series B335 (Plenum, New York, 1994), p. 215; *Proc. Fifth Int. Conf. on Nucleus-Nucleus Collisions*, Taormina (1994), eds. M. Di Toro, E. Migneco, and P. Piattelli, Nucl. Phys. **A 583** (1995) 297.

29. E.I. Tanaka, A. Ono, H. Horiuchi, Tomoyuki Maruyama, and A. Engel, Phys. Rev. C **52** (1995) 316; A. Engel, E.I. Tanaka, Tomoyuki Maruyama, A. Ono, and H. Horiuchi, Phys. Rev. C **52** (1995) 3231; H. Takemoto, H. Horiuchi, A. Engel, and A. Ono, Phys. Rev. C **54** (1996) 266.

30. J. Aichelin and H. Stöcker, Phys. Lett. **176B** (1986) 14; J. Aichelin, Phys. Reports **202** (1991) 233.

31. G. Peilert, H. Stöcker, W. Greiner, A. Rosenhauer, A. Bohnet and J. Aichelin, Phys. Rev. C **39** (1989) 1402.

32. D.H. Boal and J.N. Glosli, Phys. Rev. C **38** (1988) 1870.

33. J. Czudek, L. Jarczyk, B. Kamys, A. Magiera, R. Siudak, A. Strzałkowski, B. Styczen, J. Hebenstreit, W. Oelert, P. von Rossen, H. Seyfarth, A. Budzanow-ski, and A. Szczurek, Phys. Rev. C **43** (1991) 1248.

34. J. Péter, *Proc. Int. Symp. on Heavy Ion Physics and Its Application*, Lanzhou (1990), eds. W. Q. Shen, Y. X. Luo, and J. Y. Liu, (World Scientific, Singapore, 1991) p. 191.

35. D. Gogny, *Proc. Int. Conf. on Nuclear Self-Consistent Field*, Trieste (1975), eds. G. Ripka and M. Porneuf, (North Holland, Amsterdam, 1975), p. 176, p. 209, p. 265, p.266; J. Decharge and D. Gogny, Phys. Rev. C **43** (1980) 1568.

36. K. Hagel, M. Gonin, R. Wada, J.B. Natowitz, F. Haddad, Y. Lou, M. Gui, D. Utley, B. Xiao, J. Li, G. Nebbia, D. Fabris, G. Prete, J. Ruiz, D. Drain, B. Chambon, B. Cheynis, D. Guinet, X.C. Hu, A. Demeyer, C. Pastor, A. Giorni, A. Lleres, P. Stassi, J.B. Viano, and P. Gonthier, Phys. Rev. C **50** (1994) 2017.

NUCLEON-NUCLEON CORRELATIONS IN PHOTON-INDUCED REACTIONS

Peter Grabmayr

Physikalisches Institut
der Universität Tübingen
Auf der Morgenstelle 14
D 72076 Tübingen, Germany

INTRODUCTION

First suggestions about the existence of nucleons and their motion in orbitals inside atomic nuclei appeared through the work of Heisenberg, Bartlett and Elsässer. However, at its very beginning the possibility of a nuclear-shell model was viewed with scepticism if not disregarded on various accounts: i) the liquid-drop model and the notion of a compound nucleus were quite successful; ii) the independent-particle motion inside the nucleus seemed impossible due to the short mean free path; iii) the scarce experimental data for heavy nuclei did not support the model; and iv) the strong repulsion at short distances found in the free nucleon-nucleon interaction destroyed the hopes of using this interaction directly for the calculation of matrix elements between states of independent nucleons.

The breakthrough came through the realization by Göppert-Mayer[1, 2] and independently by Haxel, Jensen and Süss[3] about the important role of the spin-orbit potential. The new model produced a level scheme obeying the magic numbers and explained many single-particle properties. The wealth of increasingly more precise data strengthened the notion of the independent-particle model. Of course, in due time the nuclear many-body theory was developed employing effective two-body interactions; the shell models in their modern versions account for effects due to pairing or deformation.

Most of the data were obtained from hadronic reactions, as e.g. elastic and inelastic scattering of neutrons, protons, deuterons or alpha particles, because these projectiles were easily accelerated at convenient intensities. One- and two-nucleon transfer reactions contributed significantly to these investigations about nuclear properties. In particular, from the one-nucleon knockout reactions the number of nucleons was extracted, which occupied specific orbitals identified with a set of quantum numbers α. Distortions to the strong interaction prevented an absolute determination of the so-called spectroscopic factors; they were adjusted globally to range within the shell-model expectations. Due to the general success of the shell model most of the criticism present at the time of its installation more than 40 years ago was more or less forgotten. As nuclear many-body theory had progressed, methods were developed to create effective interactions from the free nucleon-nucleon force. Today's understanding of nucleon properties does not assume any strong medium modification at normal nuclear densities. Thus, the remaining objection concerns the short-range part of the interaction: Remnants of the strong repulsion of the free-nucleon force are expected to rescale or modify single-particle properties as defined by the smooth mean field. The generic name 'nucleon-nucleon correlation' is widely used — often without clear distinction to its origin: spatial, spin or isospin correlations can be classified while another scheme considers their symmetry as being spherical or of tensor nature. Their typical range can be connected to the mass of exchanged mesons. Nucleon-nucleon interactions and correlations of outgoing hadrons are summarized under the term 'final-state interaction' (FSI). Nucleon-nucleon correlations of interest here are ground-state properties of the atomic nucleus.

Various attempts to locate and quantify the nucleon-nucleon correlations of short-range nature (SRC) so far have failed. Several claims of direct observation from analyses of pion- or photon-induced reactions as well as from inclusive or exclusive quasi-free electron scattering had to be revoked in view of improved models and a more detailed understanding of reaction mechanisms. Recent progress in technical and theoretical developments warrant a new attack, however. The electromagnetic probe has proved to be an ideal tool for the investigation of bulk properties as e.g. the charge radius or charge density as well as single-particle properties via the knockout reactions. In appropriate kinematics it is expected to gain new insights from the analysis of the electromagnetically induced two-nucleon-knockout reactions. After describing two series of analysis, which established the size of effects due to SRC, a review of the present status in photon-induced reactions is given.

THE DEPLETION OF SHELL-MODEL ORBITALS

In the shell-model calculations all two- and three-body interactions are summarized by a mean field in which a single nucleon with quantum numbers α moves in an independent way. The effects due to the hard core of the free nucleon-nucleon interaction are partly taken care of by effective interactions. Still, the delineation of the total charge density of ^{208}Pb in terms of single-particle densities based on Hartree-Fock-Bogolyubov calculations by Gogny[4] was only moderately successful. The predictions fail to describe the experimental charge density in the interior region suggesting a reduction of occupation numbers for normally occupied orbitals. This holds also for more sophisticated calculations accounting for correlations of RPA or BCS-type — a feature observed in other mass regions as well[5, 6]. The inclusion of long-range correlations had little effect on the interior part of the density distribution[7, 8]. The mean-field calculations by Hasse[9] e.g. predict occupation numbers from a semi-classical self-energy employing dispersion relations in the range of ~ 80 % at the Fermi surface.

Other approaches comprise the derivation of the shell-model potential from the optical potential via dispersion relations[10], the Brückner-Hartree-Fock and G-Matrix approach[11] and variational calculations starting from the free nucleon-nucleon interaction[12, 13, 14]. In particular, for the occupation numbers of levels close to the Fermi surface they predict occupancies of typically 0.75 ± 0.10.

A very stimulating experiment had been the measurement of the charge-density difference of ^{205}Tl and ^{206}Pb [15, 16, 17]. In terms of the independent-particle model these two isotopes differ by one $3s_{1/2}$-proton, which fortunately is situated in the highest orbit close to the Fermi surface. s-orbitals are distinguished from all others by their non-vanishing amplitude at the origin, which is reflected by a prominent peak in momentum space. The measured cross-section ratio of ^{205}Tl to ^{206}Pb confirms the shell-model interpretation very nicely. However, at the same time a quantitative analysis results in a $3s_{1/2}$ charge-density difference $z = 0.64(6)$[18]. The reduction from the independent-particle model value of **1** is partly due to fragmentation and partly due to SRC.

Two approaches towards analyses which are less model dependent than previous ones have been pursued since: at NIKHEF the quasi-free proton-knockout reaction (e,e′p) had been further explored[19, 20, 21] and the CERES method[22, 23, 24] had been developed at Tübingen.

For simplicity the spherical case of a closed-shell nucleus of mass A modelled in terms of Slater determinants is discussed first. The ground-states of neighbouring nuclei A and $(A-1)$ differ by just one (the last) nucleon described by the single-particle wave function ϕ_α. α includes at least the set of the most relevant quantum numbers nlj; n specifying the principal quantum number and lj the orbital and total angular momentum. Thus, the mass A system can be built from the $(A-1)$ nucleus by application of the creation operator $a_\alpha^\dagger$ which adds one particle with quantum numbers α

$$a_\alpha^\dagger |\phi_1 \phi_2 \ldots \phi_{A-1}\rangle = |\phi_1 \phi_2 \ldots \phi_{A-1} \phi_A\rangle$$

$$a_\alpha^\dagger |\Phi^{A-1}\rangle = |\Phi^{A-1} \phi_A\rangle = |\Phi^A\rangle$$

where Φ^{A-1} and Φ^A denote the respective product states. Time reversal permits the removal of nucleons from the system A via the annihilation operator a_α

$$a_\alpha |\Phi^A\rangle = a_\alpha |\Phi^{A-1} \phi_A\rangle = |\Phi^{A-1}\rangle$$

The spectroscopic factor S_α^- measures the overlap of initial state with the product of the final state and wave function of the removed nucleon

$$S_\alpha^- = |\langle \Phi^{A-1} \phi_\alpha | \Phi^A\rangle|^2$$

or equivalently in notation of second quantization

$$S_\alpha^- = |\langle \Phi^{A-1} | a_\alpha | \Phi^A\rangle|^2$$

The above assumptions of this independent-particle model make clear that for the closed-shell nuclei the spectroscopic factor S^- amounts to $(2j+1)$ because this is the number of nucleons residing in the shell-model orbit α allowed by the Pauli principle. In this model all orbits below the Fermi surface are completely filled, those above are empty. When adding one nucleon to the closed-shell nucleus A the spectroscopic factor S_α^+ becomes unity.

$$S_\alpha^+ = |\langle \Phi^{A+1} | a_\alpha^\dagger | \Phi^A\rangle|^2$$

These spectroscopic factors S_α^- can be measured e.g. by electron-induced proton-knockout reactions, whereas the hadronic one-nucleon transfer reactions allow the measurement of S_α^- and S_α^+ for both, protons and neutrons.

It is noticed immediately from the obtained excitation-energy spectra for the residual nuclei that the nuclear many-body problem is more complex. Particle-hole and particle-particle excitations due to short- or long-range forces remove the degeneracy of the simple shell-model states. As a consequence single-particle strength is fragmented and can be shifted above the Fermi surface, even to very-high energies. As a consequence the occupation of normally filled orbitals can be reduced and the strength can be distributed over many states. Both is reflected in a reduction of the spectroscopic factors. It remains the task of the experiments to find all states belonging to the quantum numbers α and determine the respective spectroscopic factors.

The connection between the spectroscopic factors and the number of nucleons residing in the respective orbital is achieved via the sum rule by French and Macfarlane[25, 26].

Since the single-particle wave functions Φ_f^{A-1} form a complete orthonormal basis, the sum over all spectroscopic factors can be rewritten using the completeness relation:

$$
\begin{aligned}
\sum_f S_{f\alpha}^- &= \sum_f |\langle \Phi_f^{A-1}|a_\alpha|\Phi_0^A\rangle|^2 \\
&= \sum_f \langle \Phi_0^A|a_\alpha^\dagger|\Phi_f^{A-1}\rangle\langle \Phi_f^{A-1}|a_\alpha|\Phi^0\rangle \\
&= \langle \Phi_0^A|a_\alpha^\dagger a_\alpha|\Phi_0^A\rangle = n_\alpha
\end{aligned}
$$

Whereas the spectroscopic factors $S_\alpha^\mp$ depend on the nuclear structure of two nuclei the summation over all final states generates the occupation number n_α, a quantity dependent on the ground state features of one nucleus only. For ease of comparison the occupation numbers are normalized to the shell-model expectation $(2j+1)$; the occupancies o_α, defined as

$$
o_\alpha = n_\alpha/(2j+1)
$$

express the fractional occupation of shell-model orbitals. It should be clear that this concept can be used for inner shells and also for nuclei other than closed-shell ones in order to determine the respective shell-model occupancies.

Immediately two problems arise for the application of this sum rule: firstly the sum has to extend to infinity with respect to the excitation energy of the residual nucleus in order to cover all states. The separation of states and the identification of their quantum numbers creates enormous experimental difficulties which limits the accessible range to a few tens of MeV. Secondly, one has to rely on the absolute value of the spectroscopic factors which depend on the precision of the data as well as on the models for the reaction mechanism.

Moreover, the experiments actually determine overlap functions. This concept was originally created for the analysis of (p,2p) reactions by Berggren[27] and further developed by Clement and Satchler[28, 29, 30]. Clement could relate the spectroscopic factors from overlap and shell-model wave functions. With the assumption of dominance of one principal quantum number in the valence region of $\sim 1\hbar\omega$ and the near equality of shell-model wave functions for neighbouring nuclei the differences diminish in particular for heavy nuclei.

Nevertheless, the above definitions of spectroscopic factors and occupation probabilities have been used extensively in the past. Regarding the analysis of transfer reactions the ground states of magic nuclei were traditionally assumed to be uncorrelated and the hole states observed in pickup reactions as being pure. The reactions were described in distorted-wave Born approximation[30, 31] with optical-model potentials accounting summarily for the secondary interactions of the projectile or ejectile. Hence, those parametrizations have been selected which yielded spectroscopic factors for magic nuclei close to the shell-model expectation. The weak mass dependence of optical-model parameters (apart from the usual $A^{1/3}$ scaling) was taken as justification for their use in neighbouring nuclei and extend the analysis into the open-shell region.

The electron-induced knockout reactions (e,e'p) are less affected by hadronic distortions, because the experiments are performed in the quasi-free regime and thus warrant high kinetic energies of the proton. The electromagnetic distortions of the incoming and/or scattered electron can be calculated; however, the treatment of the long-range Coulomb force causes some numerical problems. In combination, the two-spectrometer, high-resolution experiments[20] at NIKHEF and the model calculations of the Pavia group[32, 33] have resulted in the consistent picture that the sum of valence strengths amounts to about 60–70% of the shell-model expectation. This valence strength is identified with the respective occupation numbers and compared to model predictions. Qualitative agreement with e.g. nuclear-matter predictions corrected for surface effects was found [19, 20, 23].

Whereas the above approach relies on the absolute determination of spectroscopic factors the second one tries to find the occupation numbers via a link of several reactions. The Combined Evaluation of Relative spectroscopic factors and elastic Electron-Scattering (CERES) approach makes use of relative spectroscopic factors and proton-(charge-)density differences[22, 23]. The accuracy of charge-density distributions obtained from elastic electron scattering had been demonstrated[5, 34]. The difference between neighbouring isotones may be interpreted in a single-particle model as

$$\rho_A(r') - \rho_{A-1}(r') = \Delta_\rho(r') = \sum_\beta \Delta n_\beta^c \cdot \rho_\beta(r')$$

with the normalization

$$4\pi \int_0^\infty r'^2 \Delta_\rho(r') dr' = 1$$

The superscript c denotes that the difference in occupation numbers Δn_β^c was derived from the charge densities. A respective quantity Δn_α^r can be obtained using the French and Macfarlane sum rule[25, 26] for one-proton pickup experiments on the same pair of nuclei.

$$\Delta n_\alpha^r = \sum_f S_{f\alpha}^-(A) - \sum_f S_{f\alpha}^-(A-1)$$

Equating the two differences the occupation number of any nearby nucleus A^* can be expressed as

$$n(A^*) = \Delta n_\alpha^c \left[\frac{\sum_f S_{f\alpha}^-(A) - \sum_f S_{f\alpha}^-(A-1)}{\sum_f S_{f\alpha}^-(A^*)} \right]^{-1}$$

This general expression relies now on relative spectroscopic factors, and the sums still go to infinity. The restriction of the summation to the experimentally accessible region relies on a result by Birse and Clement [35], who showed that depletions due to Jastrow-type correlations are not sensitive to long-range effects. Therefore, it is assumed that the same fraction of strength is lost in all neighbouring nuclei involved in the CERES approach. This truncation assumption has been checked via the proton-stripping reaction.

Note that by employing relatively slight assumptions only the CERES approach gives direct access to occupation numbers themselves. In contrast, the absolute analysis of (e,e'p) reactions provides the contribution from the valence region to the occupation number.

^{208}Pb seemed to provide the best testing ground as it represents the heaviest doubly-magic nucleus, which should facilitate comparison to predictions based on nuclear-matter calculations. Furthermore, the proton shell closest to the Fermi surface is the $3s$ orbital with its pronounced structure. Since ^{207}Tl is unstable the three isotones ^{204}Hg, ^{205}Tl and ^{206}Pb have been investigated; in the independent-particle model they differ by one $3s_{1/2}$-proton. Cross sections and asymmetries of one-proton transfer reactions have been measured to obtain detailed spectroscopic information (see ref. [23]). Results from stripping reactions and the proton knockout had also been accumulated. In combination with the extracted charge-density difference z the CERES analysis yielded occupation numbers as summarized[24] in Fig. 1. The CERES approach

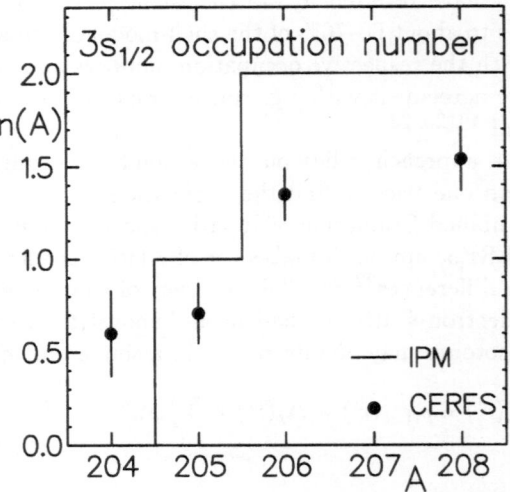

Figure 1. occupation numbers for protons in the 3s orbitals of ^{204}Hg, ^{205}Tl and ^{206,208}Pb obtained via the CERES method[23].

permits furthermore the determination of the occupancy in ^{208}Pb. The deviations from the single mean-field predictions (full line) are recognized, as the $3s$ orbital in ^{204}Hg is partially filled whereas it is expected to be empty. Similarly, the occupancy in the Pb isotopes is reduced to about 75% of the shell-model expectation. The depletion found in the CERES approach is in agreement with the result from the absolute analysis of $(e,e'p)$. The comparison to predictions suggest that the nucleon-nucleon correlations of short-range or tensor nature cause the observed shift of strength from below the Fermi surface to energies above. Both analyses can only give the amount of the depletion but they cannot give any detailed information about the underlying mechanisms.

At some time there was hope to have access to the SRC via the high-momentum part of the spectral functions for low-lying final states [21]. However, no substantial deviations from a relativistic mean-field calculation was found for the high-momentum range[36]. Furthermore, it was shown by Müther and Dickhoff[37] that only the quasi-particle strength resides there and that SRC contribute at high momentum and high excitation energies of the residual nucleus. As short distances are connected with high momenta, more information about SRC is expected from those reactions where both nucleons which had a violent interaction at small distances can be observed.

PHOTOABSORPTION

Due to the technical improvements, electromagnetically induced two-nucleon-knockout reactions are presently the most promising tools for further investigations of effects due to SRC. In accordance with experimental developments, theoretical models have progressed which is essential for proper interpretation. As virtual photons will be discussed in the contribution by H. Blok the absorption of real photons is discussed further.

Absorption of real photons had been heavily used for investigations of nuclear and nucleon properties, as e.g. the study of giant dipole resonances where the photon induces a collective motion of neutrons versus protons. Hydrodynamical models have been quite successful in describing the strength, the frequencies and the widths[38]. At higher photon energies the wavelength becomes smaller and thus becomes sensitive to the nucleon and even to quark degrees of freedom as e.g. the excitation of the Δ-resonance. In Fig. 2 the total absorption cross section per nucleon is shown for the four

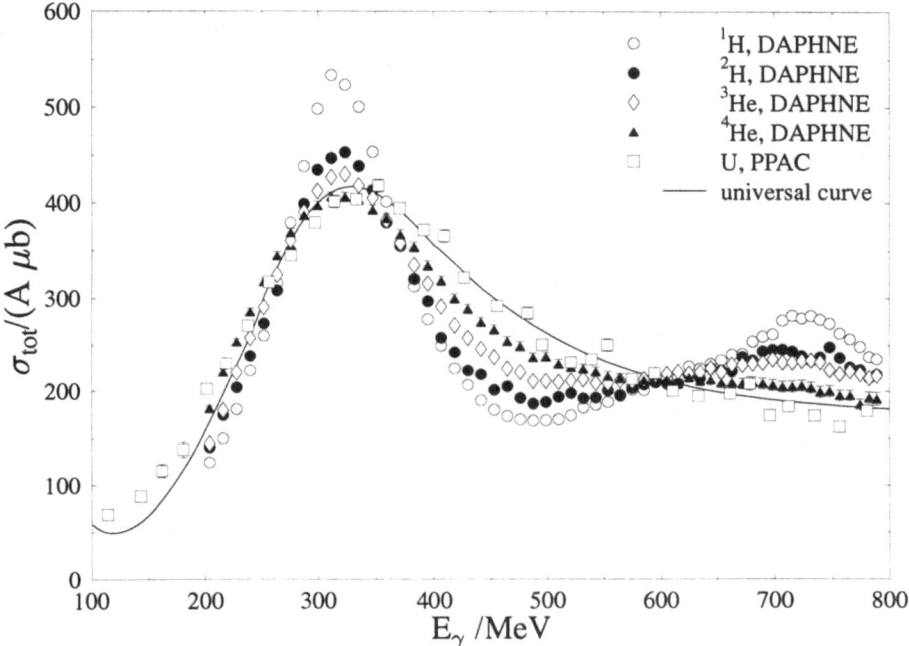

Figure 2. The total absorption cross sections for light and heavy nuclei measured at Mainz in the range of $200 < E_\gamma < 800$ MeV.

lightest nuclei[39, 40] and the average of the ^{235}U and ^{228}U results[41]. These data have been collected at MAMI B using the DAPHNE [42] and PPAC[43] detectors. The line represents the average of previous results from the literature, indicating an "universal" behaviour [44] in the photon range of about 200 MeV to 800 MeV. The proton data exhibit two structures, the Δ-resonance and around 700 MeV some strength due to several smaller resonances and the opening of the two-pion production channels. The broadening of these peaks for heavier nuclei can be taken as a clear sign for medium modifications. The most trivial one, the Fermi motion of nucleons inside nuclei, cannot explain all of the changes in the data. ^{4}He seems to have already reached the same outcome as those for heavy nuclei. Without yielding to speculations about the underlying mechanisms — vanishing of resonances versus two-pion production — the need for more exclusive experiments is evident in order to understand the underlying mechanisms.

The only process competing with photoabsorption, Compton scattering, is only a very small fraction of the total cross section and the data at $E_\gamma \sim 300$ MeV are well described by e.g. the Δ-hole model[45]. Photon-induced one-nucleon knockout was used as a tool for the study of the giant dipole resonances. At higher photon energies around 100 MeV angular distributions for emitted neutrons and protons are almost identical in shape and magnitude. This was taken as evidence for an underlying two-body absorption mechanism [46]. Even at photon energies around 300 MeV respective yields differed less than an order of magnitude.

When mono-energetic electrons of several hundred MeV impinge on a thin foil a continuous spectrum of photons, extending up to the electron energy, is produced. In these bremsstrahlung experiments the photon energy is not known and in general the flux is not very accurately determined. However, already in the pioneering experiments at Δ resonance energies the correlated emission of neutron-proton pairs had been observed[47, 48, 49, 50, 51]. The quasi-free knockout of single nucleons is strongly suppressed due to the momentum mismatch between the incoming photon and the ejected nucleon which is taking most of the photon energy. If the photon is absorbed on nucleon pairs their relative motion prior to the interaction can balance the momentum mismatch. In contrast to proton-proton or neutron-neutron pairs, neutron-proton pairs have a dipole moment which favours the photon absorption on a deuteron-like nucleon pair[46]. The additional observation of the absorption cross-section scaling with the free deuteron disintegration cross section and to the number of np-pairs had created the term Quasi-deuteron process[46, 52, 53].

This was the underlying idea of the phenomenological model by Levinger[52], the Quasi-deuteron model. A more microscopic treatment of the two-nucleon photoabsorption process was started by Gottfried[53] in 1958. However, for several reasons assumptions were needed to make calculations possible. The model applies only to closed-shell nuclei and the nucleon pair has zero separation and angular momentum $L = 0$. This led to a factorized ansatz which separates the dependencies on the pair momentum $\vec{P}$ and the relative momentum of the two nucleons $\vec{p}_r$.

$$d\sigma \sim F(\vec{P}) \cdot S_{fi}(\vec{p}_r)$$

The latter factor contains the short-range effects whereas the first term is responsible for the global features of the cross sections.

These first experiments were too crude to separate the absorption process on two nucleons, which were understood to yield hints about the SRC, from the competing process of quasi-free pion production or Δ excitation. The latter are initially one-nucleon photoabsorption processes which produce real pions. However, the pions may easily be reabsorbed by the nucleus which leads to emission of a nucleon pair in similar characteristic kinematics as the direct photon absorption on a pair. Due to technical difficulties these types of experiments had not been pursued until about 1980, when electron beams of higher duty factor allowed the tagging, i.e. the energy measurement of bremsstrahlung quanta.

Two-nucleon emission experiments using tagged photons in the Δ-resonance region had been performed in Tokyo[54] and Bonn [55, 56]. They confirmed the two main absorption mechanisms of quasi-free pion production and two-nucleon emission. Neutron-proton pairs are emitted about five times more likely than proton-proton pairs. Unfortunately the experimental resolution in several variables did not allow for detailed studies, e.g. a comparison to the cross section calculated by Weise et al.[57] for the $^{16}O(\gamma,np)$ reaction employing Jastrow type correlation functions. Apart from this work no real progress in theory of two-nucleon emission had been made at that time. However, the advent of the continuous-wave electron accelerators had a strong impact on the field. At low photon energies MAMI-A and MAXlab had reached an energy resolution of ~ 10 MeV, SAL, TAGX and LEGS concentrated their efforts on the Δ region and finally MAMI B[58] came into operation, which is the only tagging facility to cover the complete range of photon energies from 40 to 800 MeV. Furthermore, the Glasgow tagging spectrometer in Mainz excels through an intrinsic resolution of about 100 keV, which matches the beam spread. However, at present the focal plane is equipped with scintillation detectors of varying width in order to obtain a uniform photon-energy resolution of ~ 2 MeV. In addition new equipment has been constructed.

NEW THEORETICAL APPROACHES

Clearly the early models by Levinger and Gottfried were too crude on the one hand and on the other hand had not been challenged by good data. The Pavia group had gained experience in microscopic calculations for the one-nucleon-knockout reactions. They were the first to realize the potential of the new experiments being prepared. Thus, they expanded their model to describe also the electromagnetically induced two-nucleon emission, the (γ,NN) and $(e,e'\text{NN})$ reactions[59, 60]. Due to the transverse nature of the real photons the leading Feynman diagrams are meson-exchange (MEC) and isobar currents as shown in Fig. 3. First attempts had been made in a factorized

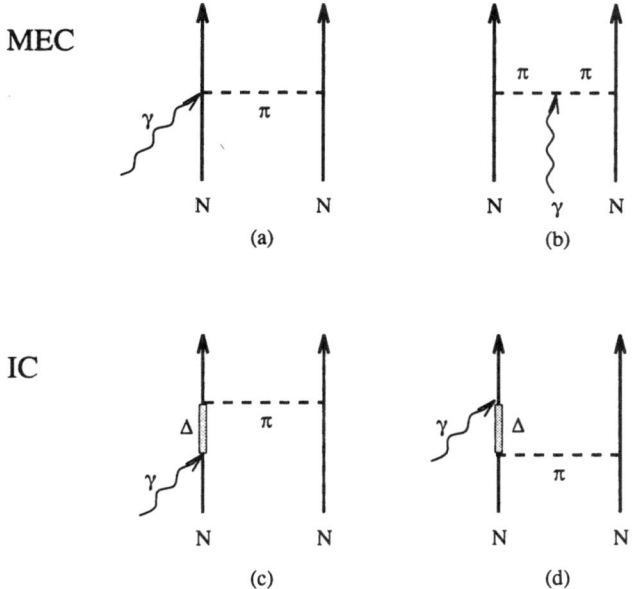

Figure 3. Leading Feynman diagrams of two-body photoabsorption: a) sea-gull, b) pion-in-flight, c) and d) isobar currents involving the Δ.

model. Boato and Gianinni[61] proved that the (γ,NN) cross sections still factorize for the leading meson-exchange terms, if the Gottfried assumptions are envoked.

The Gent group[62, 63] has demonstrated the need of inclusion of heavier meson and the use of the unfactorized model. The cross section in the laboratory frame is described by

$$d\sigma \sim 2\pi\delta(E_f - E_i)\,|T_{fi}(k)|^2 \cdot \frac{d^3p_1}{(2\pi)^3} \cdot \frac{d^3p_2}{(2\pi)^3} \cdot \frac{d^3p_R}{(2\pi)^3}$$

$$T_{fi}(k) = J^\mu(\vec{k}) = \int \psi_f^*(\vec{r}_1,\vec{r}_2)J^\mu(\vec{r},\vec{r}_1,\vec{r}_2)\psi_i(\vec{r}_1,\vec{r}_2)e^{i\vec{k}\vec{r}} \cdot d^3r \cdot d^3r_1 \cdot d^3r_2$$

$$J^\mu = J^{[1]} + J^{[2]}$$

Model dependencies arise here through the details of the assumptions for the one- and two-body currents $J^{[1]}$ and $J^{[2]}$. The spin-isospin structure of a dominant two-body current is exemplified by the seagull term

$$J_{sea}(\vec{k}_1, \vec{k}_2) = i \frac{g_\pi^2}{4M^2} (\vec{\tau}_1 \times \vec{\tau}_2)_3 \left[\frac{\vec{\sigma}_2(\vec{\sigma}_1 \cdot \vec{k}_1)}{m_\pi^2 + k_1^2} - \frac{\vec{\sigma}_1(\vec{\sigma}_2 \cdot \vec{k}_2)}{m_\pi^2 - k_2^2} \right]$$

Also, different diagrams of intermediate isobar configurations which contribute to $J^{[2]}$ had been employed. The two-nucleon initial $\psi_i(\vec{r}_1, \vec{r}_2)$ and the final state $\psi_f(\vec{r}_1, \vec{r}_2)$, where the two nucleons are in the continuum, are true many-body problems. Furthermore, the final-state wave function is affected by final-state effects of mutual interactions and those of the individual nucleon with the residual nucleon. Since this quantum mechanical problem still cannot be solved the initial-state wave function may be approximated by

$$\psi_i(\vec{r}_1, \vec{r}_2) \sim \phi_{n_1 l_1}(\vec{r}_1) \phi_{n_2 l_2}(\vec{r}_2) g(|\vec{r}_1 - \vec{r}_2|)$$

where ϕ_1 and ϕ_2 are single-particle shell-model wave functions correlated by the Jastrow-type function $g(|\vec{r}_1 - \vec{r}_2|)$. The final state is a scattering state where the final-state interaction between the nucleons and the residual nucleus is approximated by distorted waves.

Due to the complexity of the many problems several attempts are being made in order to find the leading terms. The model calculations have been shown to be sensitive to the number and parametrizations of diagrams due to interference effects. The present treatment of final-state effects causes only a reduction of cross sections up to a factor of 2. Finally, different correlation functions derived from NN potentials or variational calculations do not affect angular distributions; however, they modify the relative strengths of one- and two-body contributions. The mean-field wave function strongly determines the shape of differential cross sections[60].

A different approach was formulated by the Valencia group [64, 65, 66]. The attention is not focused on the nuclear-structure part and a few dominant diagrams like in the Pavia and Gent models, but aims at the best microscopic treatment of the photo-absorption process. The interaction of the photon with nuclear matter is calculated by evaluation of many self-energy diagrams up to third order, including π and heavy-meson exchange and isobar excitation. Then the imaginary part is identified with the photo-absorption process.

Local Fermi moments are calculated within the Thomas-Fermi model, which then are needed to adjust the results obtained for nuclear matter via the local-density approximation to real nuclei. These rather fundamental calculations of photoabsorption are followed by a semiclassical treatment of final-state interactions. All hadrons of the primary reaction are tracked through the nucleus. They may undergo any secondary reaction as nucleon or pion scattering. An important role is played by pion re-absorption. The total cross sections are well described; in addition the direct two-nucleon absorption part can be identified. Its energy dependence shows only a slight enhancement in the Δ region, and mass dependence slightly larger than A is predicted[64]. The Monte Carlo type treatment of FSI facilitates the comparison to the experiment because detector acceptances and thresholds can easily be incorporated. Additionally a tagging of the primary and secondary processes has been introduced for the assessment of the various reactions in different kinematical regions of the experiment.

Both types of calculations are very useful for the interpretation of the data. As shown below, the Valencia model will give the general features of the contributing absorption mechanisms whereas the Pavia and Gent calculations show the details of the genuine two-body absorption.

EXPERIMENTAL SETUP

Here the experimental setup of the (γ,NN) experiments performed at MAMI B will be described. Other facilities concentrated on charged channels only or restricted themselves to the few-body systems which allowed them the use of compact (often 4π) detectors. For the investigation of the (γ,NN) reaction mechanisms over a wide energy range and of nucleon-nucleon correlations an efficient and precise neutron detection is of paramount importance. In order to achieve high-energy and -momentum resolutions at neutron energies up to $\sim$ 250 MeV long flight paths are required. To keep the solid angle large enough for good statistics large detectors with position resolution are needed. The optimization process resulted in the development of 107 bars of long plastic scintillation counters which are read out by phototubes on each end. Eight of the $50 \times 200 \times 3000$ mm^3 bars were mounted onto a frame (see Fig. 4,left); up to

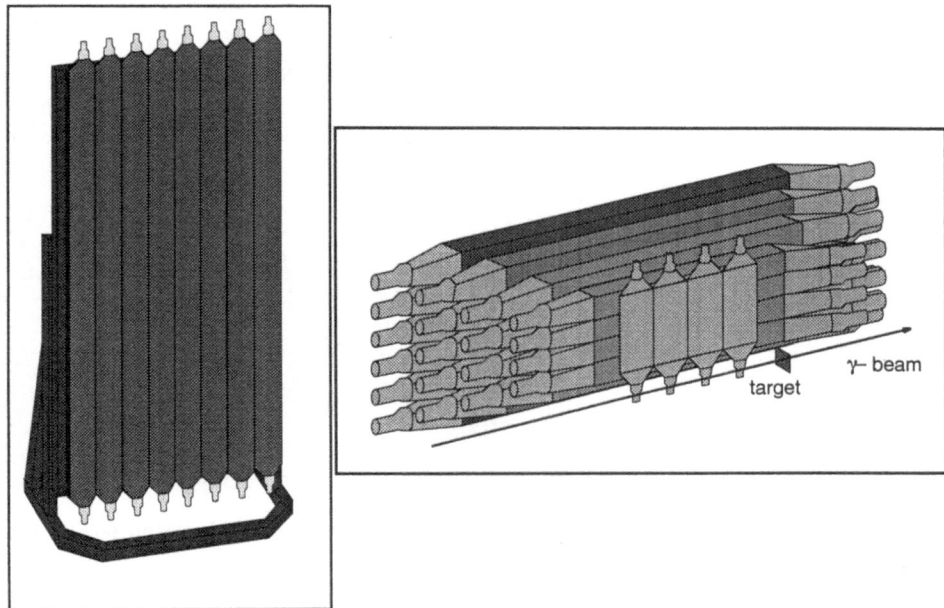

Figure 4. Schematic drawing of a TOF frame equipped with eight scintillation bars (left) and of PiP (right).

four frames can be mounted onto a stand[67]. Protons and positively-charged pions are identified in the charged-particle hodoscope PiP which subtends a solid angle of 1 sr[68]. The complete setup is depicted in Fig. 4, right part. An 855 MeV electron beam of 10 – 30 nA impinge on the $4\mu m$ thick Ni radiator foil in front of the Glasgow tagging spectrometer[69]. 352 scintillators determine the energy E' of the slowed down electron which in turn allows the determination of the photon energy E_γ as

$$E_\gamma = E_0 - E'$$

The unperturbed beam is also bent by the magnet and is dumped outside the experimental hall for the benefit of a very low background level. The collimated photon beam irradiates the target located $\sim$ 7 m downstream. As target so far self-supporting sheets of ^{6,7}Li, ^{12}C and ^{40}Ca had been used. CH$_2$ and CD$_2$ targets had been used for calibration purposes. A remotely controlled target ladder positioned the targets in general at

30° with respect to the beam in order to minimize energy losses. The support was also used for a ring of thin plastic scintillators which served as start- and veto-detectors for PiP and the TOF detectors. The setup of Fig. 5 was chosen for roughly equal neutron

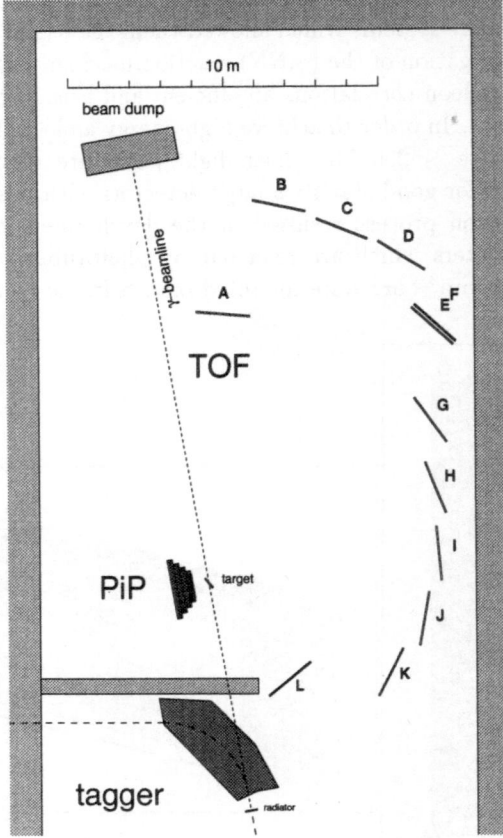

Figure 5. The layout of the (γ,NN) experiments at the tagged-photon facility of the Mainz electron accelerator MAMI-B.

resolution and large polar-angle coverage; flight paths of up to 14 m were possible. Cosmic ray tests determined a timing resolution of 0.5 μs (FWHM) which results in position resolutions of about 6 cm; the latter being determined from the time difference between the signals from each end[67, 70]. The neutron efficiency was determined from Monte-Carlo simulations[71] and amounted to about 6%. The proton energy measured in PiP was derived from the pulse-height information corrected for quenching, energy loss in intermediate absorbing material as, e.g., the wrapping of the detectors or the air between the target and the detector. Reaction losses in the scintillator become important at higher kinetic energies. The simulations based on GEANT[72] which are in good agreement with the data of Measday[73] amount to 30% at the highest energy of 250 MeV. Further corrections arise for the losses due to the photon collimation. The so-called tagging efficiency is determined at low intensities by a 100% efficient lead glass detector put into the beam: It is determined from the coincident rate between lead glass and each tagger channel, normalized to the free tagger rate. This ratio is found to be almost energy independent and amounted to about 50%. All the correction factors are applied as weights when histogramming the data in various variables.

The most interesting quantities are the missing momentum p_m, the missing energy

E_m and the relative momentum p_r. For two-nucleon emission these observables are defined as

$$\vec{p}_m = \vec{k} - \vec{p}_1 - \vec{p}_2$$
$$E_m = k - T_1 - T_2 - T_R$$
$$T_R = p_m^2/2M_R$$
$$\vec{p}_r = \frac{1}{2}(\vec{p}_1 - \vec{p}_2)$$

where $\vec{k}, \vec{p}_1$ and $\vec{p}_2$ are the photon and the nucleon momenta, respectively. The missing energy is given by the difference of kinetic energies of the incoming and outgoing particles which includes the recoil kinetic energy T_R calculated from p_m and the mass of the $(A - 2)$ residual nucleus. The missing energy shifted by the Q value can also be interpreted as excitation energy of the residual nucleus. The relative momentum $\vec{p}$ of the nucleons inside the pair is of further interest for the search of high-momentum components.

The present experiments aimed at energy resolutions better than the binding energy of nucleons in order to determine the shells from which the emission took place. For this task the kinematically overdetermined reaction $D + \gamma \rightarrow n + p$ was employed because there is no residual nucleus. In practice a CD_2 target was used for the results shown in Fig. 6. The deuterium events are clearly separated from the C events due

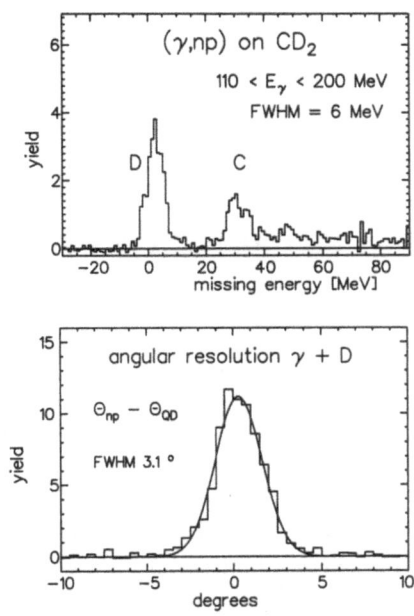

Figure 6. The missing-energy resolution is determined from the kinematically overdetermined $D + \gamma \rightarrow n + p$ reaction.

to a 6 MeV FWHM resolution in missing energy. The angular resolution is $\sim 3°$ and is dominated by PiP, which is positioned at a distance of 0.5 m only. From this a missing-momentum resolution of 35 MeV/c follows.

Finally, the precision of the data shown in the following sections was made possible on account of exceptional stability of the electron accelerator.

EXPERIMENTAL RESULTS

The discussion below concentrates on the recent data obtained at MAMI B because they cover many features observed in other experiments of restricted kinematics or with low statistics. First some general features are shown to demonstrate the understanding of the reaction mechanisms of photon absorption. The $\alpha - d$ cluster structure of ^{6}Li helps in the interpretation on basis of the data alone because of the large energy gap between the s- and the p-shell knockout. Except for the ground-state cluster wave function [74] no useful model predictions can be obtained. Thus, for their test the comparison will be made with respect to the ^{12}C data.

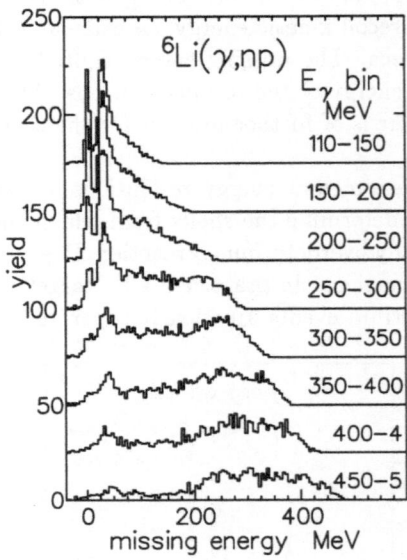

Figure 7. The missing-energy spectra for the ^{6}Li(γ,np) reaction for photon-energy bins of 50 MeV width.

The energy dependence of the various processes is shown in fig. 7. Clearly two structures are seen at low missing energy which correspond to transitions to the ground state ($E_m \sim 4$ MeV) and to the excited states ($E_m \sim 30$ MeV) of the residual ^{4}He. These are the dominant states populated at low photon energies. At intermediate energies up to $E_\gamma \approx 250$ MeV one observes in addition a structureless contribution, which increases with photon energy. Strength beyond $E_m \sim 80$ MeV cannot originate from the removal of nucleons from shell-model states any more. This yield arises either through FSI of the outgoing nucleons, more complex multi-nucleon absorption processes or through QFπ production followed by pion-induced absorption processes. At the highest photon energies the low missing-energy part exhibits vanishing strengths, which only to a small extent is due to the experimental cuts. Whereas the part of large missing energies (above $E_m \sim 200$MeV) are populated through primary processes related to Δ–production and QFπ production followed by FSI or deexcitation of the residual nucleus such that a nucleon pair had been detected. This observation was made possible through the broad range of the Glasgow tagger.

The structures mentioned above become more apparent in fig. 8 when selecting events on basis of Θ_{diff} which is the angle between the measured neutron momentum and a hypothetical one; the latter is calculated assuming the free-deuteron disintegration using the measured proton and photon energies. For a homogeneous coverage the

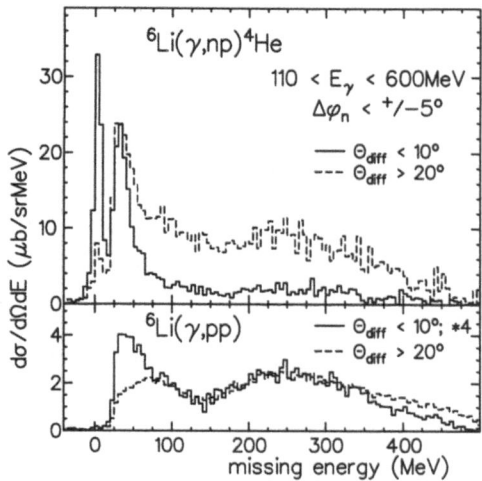

Figure 8. The missing-energy spectra of the ^{6}Li(γ,np) and ^{6}Li(γ,pp) reactions for the photon range of $114 < E_m < 600$ MeV. The histograms are accumulated for two different cuts on the angle Θ_{diff} (see text).

azimuthal range of the neutrons was limited to $\Delta\varphi_n = \pm 5°$ about the horizontal plane. Thus Θ_{diff} is a measure of the deviation from the naive QD process; it is strongly related to the missing momentum (see ref. [75]). Note the enhancement of the low-missing-energy region for low values of Θ_{diff}. In contrast, larger values of Θ_{diff} enhance the yield at higher missing energies through processes like e.g. the quasi-free pion production including FSI. Also the width of the excited ^{4}He-core might be used as an indicator of the presence of FSI. However, further studies have to prove that the width is not affected by the indirect selection of missing momenta via the angle Θ_{diff}.

The insensitivity of the (γ,pp) reaction channel towards the angle Θ_{diff} indicates that a large fraction of the yield originates through FSI. Only that small structure below $E_m \sim 80$ MeV for $\Theta_{\text{diff}} < 10°$ could contain some useful information about the genuine two-body absorption. Furthermore, fig. 8 exemplifies the relative feeding of the (γ,np) and the (γ,pp) channels. At low E_m a ratio of up to 20 can be observed, whereas above $E_m \sim 100$ MeV a ratio of 4 is found.

The energy dependence of the cross sections is demonstrated in fig. 9. The top panel in fig. 9 compares the ground-state transition of the ^{6}Li(γ,np)^{4}He reaction with that for the photodisintegration of the free deuteron. The data have been integrated over the complete solid angle of PiP and are presented as differential cross sections for $\Theta_{lab}^p = 90°$. Because the ^{6}Li can be viewed as an α-d cluster the removal of the outer np-pair ("d-cluster") should show similarities to the photoabsorption on the free deuteron. Effects of the Fermi motion are not noticeable in fig. 9 as the PiP detector covers the Fermi cone for the ground-state transition. The agreement between the two data sets is extraordinarily good, both in shape as well as in magnitude. Within the errors the data are well described by parameterizations of the $\gamma + D$ cross sections by Jenkins et al.[76]. The Monte Carlo predictions using the cluster wave functions of the Moscow group [74] also account for this excitation function. These predictions obtained with the code MORGAINE [77] are based on the QD-model as formulated by Gottfried [53]. It was shown[78] that these wave functions can account quite well for the measured-momentum distribution F(P) of the nucleon pair with respect to the residual ^{4}He.

The bottom panel of fig. 9 compares the excitation functions of the ground- and

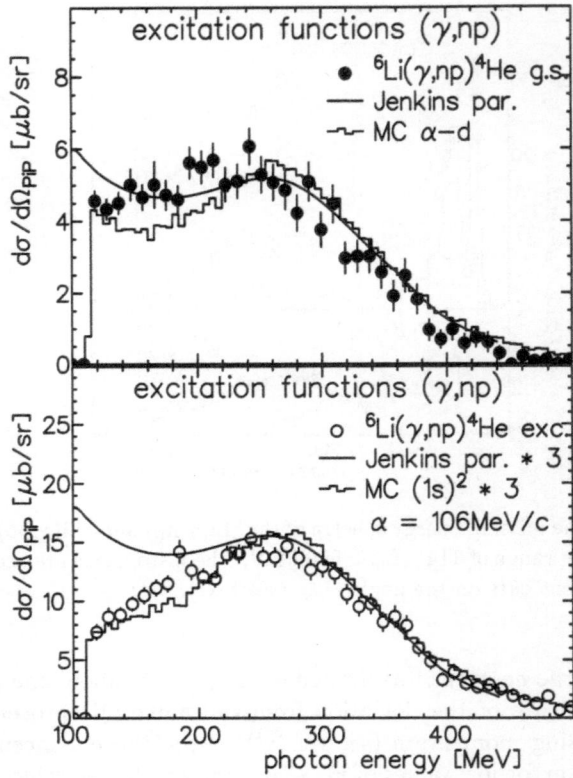

Figure 9. The excitation function of the ^{6}Li$(\gamma,np)^4$He ground (full dots) and excited-state (open circles) transitions are compared to the parameterized (full line,[76]) excitation functions of the $\gamma + D \to n + p$ reaction. In addition the predictions of the Monte Carlo simulations using an α-d cluster wave function of Kukulin [74] are given.

excited-state transitions. Note that the "excited states" in ^{4}He comprise the missing-energy range from 20 to 50 MeV. Good agreement on the high photon-energy side is obtained for this excitation function by scaling the result for the ground-state transition by a factor 3. Monte-Carlo simulations have shown that the deviation below E_γ = 180 MeV are the result of the finite detector acceptance because the np-pair is not at rest within the target nucleus, i.e. not the complete Fermi cone is covered. The loss of those events enhances the apparent resonance-like structure. However, it is concluded that both excitation functions scale with the free deuteron photodisintegration cross section.

To obtain the excitation functions integrations had been carried out over all observables except the photon energy. Thus, the expectedly small deviations from the QD behaviour due to short-range phenomena might easily be averaged out. Therefore, to enhance the sensitivity to these deviations other observables must be displayed and their respective correlations have to be investigated using higher-differential cross sections.

Next, the missing-energy spectra of the ^{12}C(γ,NN) reactions are discussed with the aim to disentangle the contributions of the various reaction mechanisms to two-nucleon emission. For this purpose we rely on the Valencia model described earlier. The results of the simulations for all events in which one proton is detected in PiP and (at least) one neutron or proton is registered in TOF are given as histograms in Fig. 10. The events are classified according to the following scheme: "2N" represents

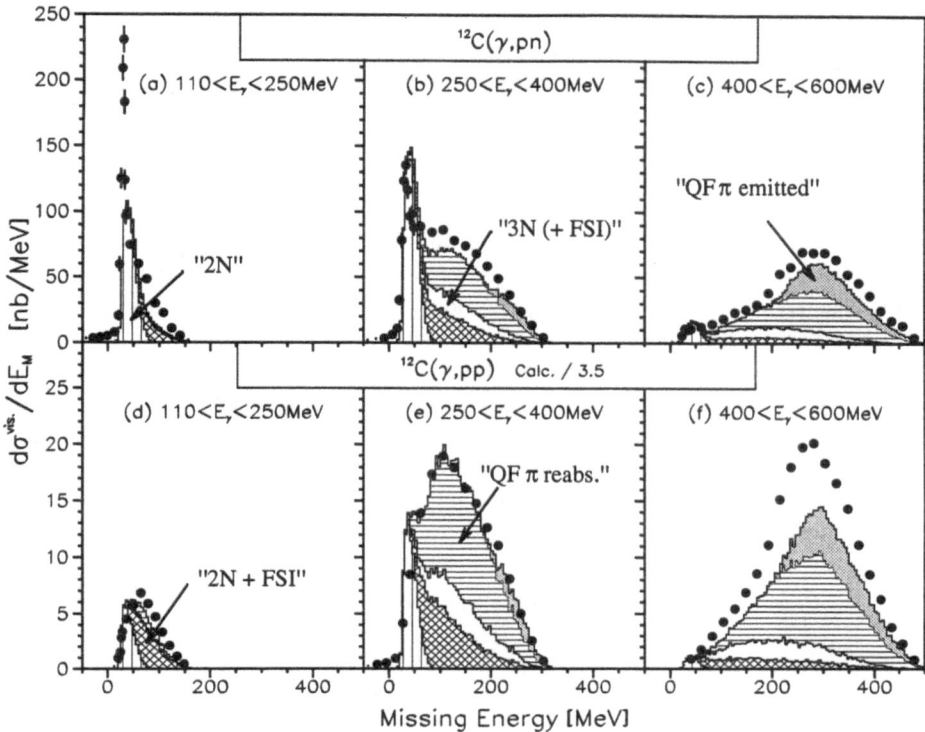

Figure 10. Comparison of missing-energy spectra for the ^{12}C(γ,NN) reaction at three different photon-energy bins.

direct photoabsorption on a nucleon pair; "2N+FSI" denotes events of this type where one or both outgoing nucleons undergo a FSI which in the model means through a hard collision with another nucleon. Another interesting category is photoabsorption on a three-nucleon cluster with or without final-state interactions: "3N(+FSI)". Finally, there are two classes of events which start with quasi-free pion production. In one case the pion is reabsorbed on its way out ("QFπ reabs."), whereas in the other case it leaves the nucleus ("QFπ emitted"). In the latter case a FSI of the nucleon or the pion following the quasi-free pion production is necessary in order to produce the required two-nucleon trigger of the experiment.

The calculated cross sections, including all event types, reproduce all the essential features of the missing-energy spectra (see Fig. 10). The narrow structure at $E_m \approx$ 40 MeV is indeed revealed as due to photoabsorption on nucleon pairs. Actually, the structure is narrower than predicted by the calculation (see Fig. 10a). This deviation is the result of the simple Fermi-gas approximation for ^{12}C. The observed (γ,pn) yield is underestimated by $\approx$ 25% for all three photon-energy bins. However, regarding the available theories the overall agreement is acceptable. The more serious problem for the Valencia model is the overestimation of the (γ,pp) cross sections by a factor of about 3.5 . This factor seems to be photon-energy dependent, since the value of 3.5 does not work in Fig. 10f. In accordance with our result Kolb et al.[79] obtained a similar normalization factor of 3.9 in their ^{12}C(γ,pp) experiment at $\approx$ 200 MeV photon energy, whereas the spectral shapes were again in good agreement. As the cross sections in the (γ,pn) channel are typically an order of magnitude larger than the (γ,pp) cross sections one might naively suspect that an overestimation of the FSI effect due to charge-exchange might have created the problem because this effect moves strength from the strong into the weak channel. Indeed, a detailed analysis of the Monte Carlo events shows that about 80% of the "2N + FSI" type events in the (γ,pp) channel originate from (γ,pn) absorption. However, the FSI-driven "cross-talk" depends on the missing energy and hence is unlikely to be an explanation of the normalization problem in view of the generally good reproduction of the spectral shapes. Also the semi-classical treatment of the hadronic interactions which neglects e.g. the antisymmetrisation must be reconsidered.

The discussion above exemplifies the advantages of a wide kinematical coverage of the (γ,NN) reactions when testing model predictions because different regions exhibit specific sensitivities to the competing reaction mechanisms. In this sense the Valencia model has been tested in such detail for the first time. It successfully describes the general features of the reactions except for the normalisation of the weak (γ,pp) channel. From this we conclude that the dominant reaction mechanisms are understood. Most importantly for the further search for effects due to short-range correlations is that the direct two-body photon channel could be localised in the low missing-energy region below 80 MeV. It had been shown by McGeorge et al. [80] that this low E_m region can be well described by folding the strength functions of s- and p-shell single-particle knockout.

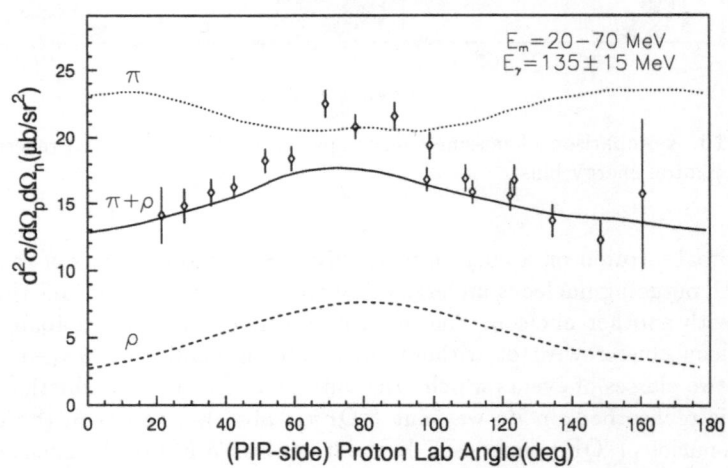

Figure 11. The angular distribution for the low missing-energy bin compared to (rescaled) model prediction by Gent.

Concentrating on the low E_m region where the shell-model structure is still of some significance we consider the Gent predictions [62, 81] which employ a more rigorous model for the nuclear-structure part than the Valencia model. The strong angular correlation of the emitted np pair had been shown which in addition is influenced strongly by the interference of different amplitudes [62]. To that purpose the yield of coincident np events is plotted in Fig. 11 versus the proton angle with respect to the beam axis. Only those events had been selected when both nucleon obeyed the two-body kinematics. As a result an angular distribution with a broad maximum around 80° is observed. The data are well described by only the one curve which includes the interference of π and ρ-mesons. Actually predictions for the $^{16}O(\gamma,np)$ reaction have been scaled down considering the number of active nucleons[82]. Definitively, more detailed calculations will be needed for the present ^{12}C data.

70

Other important guides towards the separation and identification of the genuine two-body photoabsorption processes are the missing-momentum distributions. In the

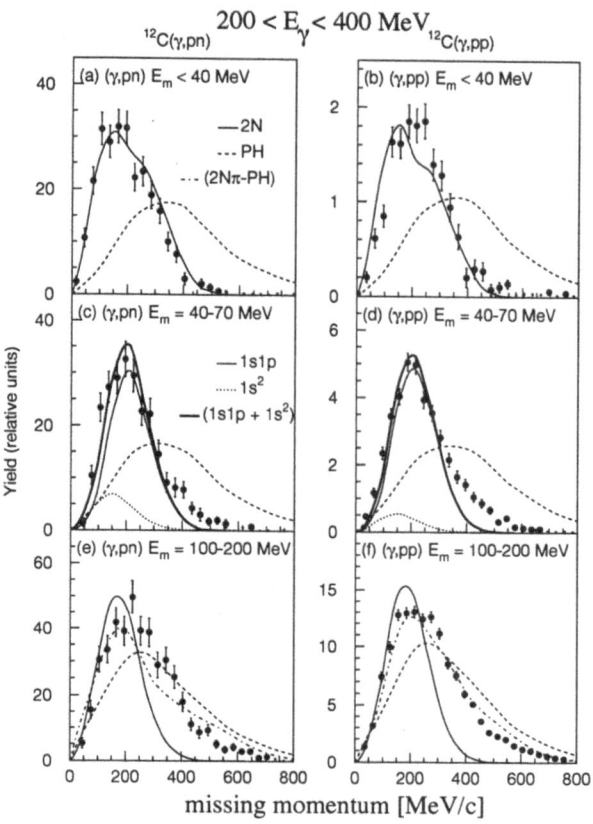

Figure 12. The missing-momentum distributions for the $^{12}C(\gamma,np)$ and the $^{12}C(\gamma,pp)$ reactions for three missing-energy bins are compared to a two-nucleon absorption model (full line). Phase-space distributions are shown for 2N emission (dashed line) and for 2N and pion emission (dot-dashed line).

plane-wave assumption the missing momentum p_m can be identified with the pair momentum inside the nucleus before the interaction. This is similar to the case of the one-proton knockout reaction (e,e'p). The underlying two-nucleon absorption can be demonstrated by a comparison of the data with simple-model predictions which employ harmonic-oscillator single-particle wave functions to construct the pair momentum [84] as shown by the full line in Fig. 12. For $E_m < 40$ MeV only p^2-shell emission is assumed while for $40 < E_m < 70$ MeV the contributions from sp and s^2 are shown also separately. Distributions (PH) according to the phase space available for two nucleons and the residual nucleus have been calculated. They are shown as dashed lines. The very-good agreement of the model predictions for the low E_m range confirm the notion of p^2 and sp shell knockout. This result is in line with the findings of ref. [80]. Also the (γ,pp) date are reasonably well reproduced. Notoriously the high-momentum tails are less well described. The disagreement at high E_m (Fig. 12 e,f) with the harmonic-oscillator model predictions is no surprise because of the large contributions from quasifree pion production followed by FSI. Due to the latter the phase-space distributions assuming 2N and pion emission (dot dashed line) describe the data satisfactorily. Also predictions based on the Valencia model citecar,car2,car3 yield fair description of the high-E_m

range.

The missing-momentum distributions of other nuclei can also be described equally well (e.g. ref. [77, 78, 88]). Apparently, in these light nuclei the distortions seem not to affect the shapes of the angular distributions drastically. A smooth behaviour is observed as noted by Gottfried. It is in this spirit to investigate also the second factor contributing to the two-nucleon-emission cross section. Therefore, the respective variable, the relative momentum between the two nucleons will be studied next. There, the short-range correlations are expected to possibly show some effects. The detailed comparison must be done later using unfactorised predictions. In a first step we will compare the results from finite nuclei to the most basic system the deuteron. In order

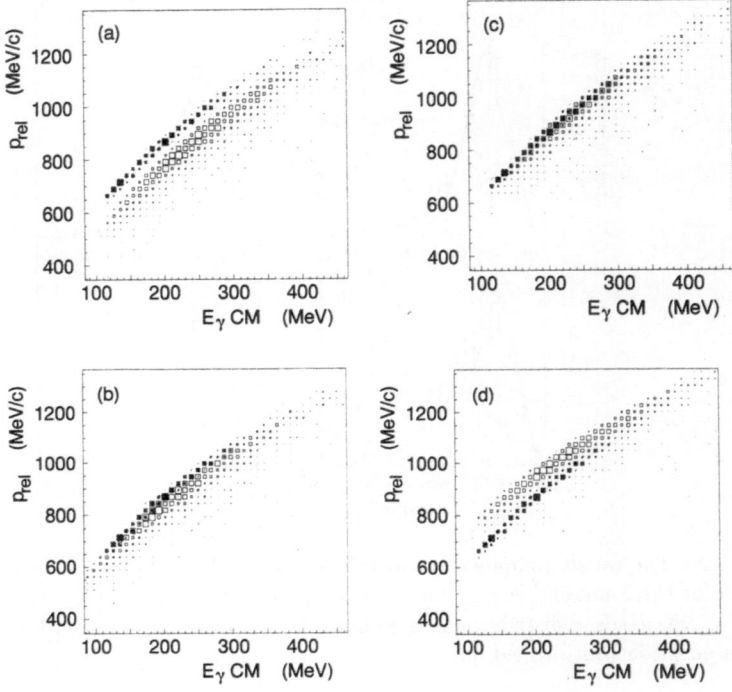

Figure 13. Plots of total relative momenta versus photon energy in the NN system. for photoabsorption on a CD_2 target. a) without Q-value correction, b) E_γ corrected by Q-value c) momentum corrected for E_m d) potential ansatz (see text for details).

to take out the artifacts due to the motion of the pair inside the nucleus a transformation into that system is performed in which the nucleon-nucleon pair is at rest. Impulse approximation is again assumed and the relative velocity between the two systems is derived from the (eventwise) measured-missing momentum. Note that with the assumption of an equal sharing of the momentum of the incoming photon between the two nucleons this relative momentum is not changed through the interaction. This assumption is valid due to the transverse nature of the photoabsorption process. It is however different than usually employed for the analysis of the (e,e'pp) reaction, where the photon is assumed to interact with only one proton. There an ambiguity exists as the experiment cannot discriminate which proton was struck. For the (γ,NN) analysis, we will use the total relative momentum p_{rel} to contrast this experimentally derived quantity from the sought after relative momentum before the interaction as defined earlier. In Fig. 13a the correlation between p_{rel} and E_γ is demonstrated for deuterium and the low E_m range of carbon. The range of 20 to 40 MeV in E_m was selected for ^{12}C from this CD_2 target. Due to energy conservation the events fall into two clearly separated ridges, the upper for the deuterium the lower for carbon with the difference given by the difference in reaction Q-values. We wish to compare the momenta of the free and the bound two-nucleon systems, however we would like to correct for the binding-energy effect. In (e,e′p) the so-called deForest prescription yields the off-shell correction. No such prescription is available for the infinitely more difficult two-nucleon case. Three possible Q-value corrections are investigated. Firstly, the photon energy is reduced (see Fig. 13b) by the Q-value assuming a modification taking place in the entrance channel. Secondly, a correction of the momenta is applied on the basis that asymptotic momenta are measured. Because the nucleons have to overcome the mean-field potential in order to escape into the continuum their momenta are modified by the depth of that potential. A depth of 40 MeV produces the effects seen in Fig. 13d). Thirdly, the missing-energy calculated for each event is taken as the actual Q-value and is added appropriately to the nucleon momenta (Fig. 13c). The best overlap of the two ridges is achieved with the last method which is therefore taken for the further comparisons.

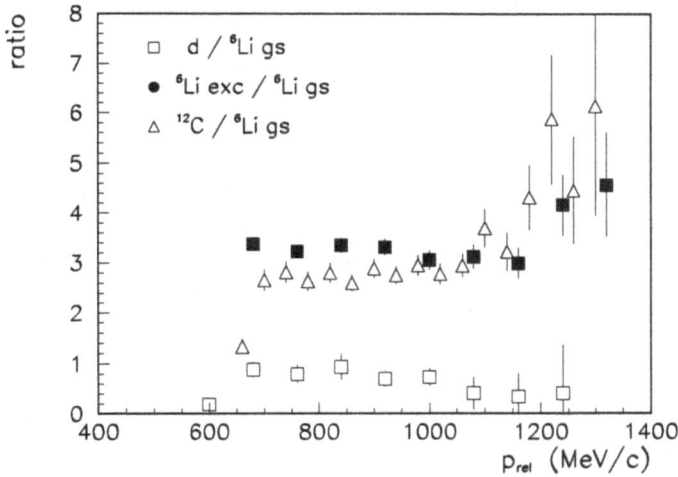

Figure 14. Ratios of relative nucleon momenta p_{rel} for low missing-energy ranges in carbon and Li (20 $<E_m<$ 50 MeV).

The (corrected) momentum distributions for deuterium, lithium and carbon are bell shaped with a maximum around 900 MeV/c.

They extend to very high momenta of 1400 MeV/c (see Fig. 13), which is still 700 MeV/c per nucleon and thus about 3 times the classical one-nucleon Fermi momentum inside the nuclear mean field. However, the (e,e′p) experiments [21] have observed high momenta in this range, but with very small probability only. In order to avoid any speculations without having a detailed calculation available the yields will be studied relative to the deuterium. This way the excess of high momenta, beyond those already present in the deuterium, can be investigated. This will give also allow comparisons to the predictions of the variational calculations which arrange nucleons in dumb-bell shaped configurations; in similar ways in the deuterium as in other light nuclei (see contribution of V. Pandharipande).

Unfortunately, the statistics of the deuterium data is insufficient for the present purpose. Therefore, we resorted to the ^{6}Li(γ,np)^{4}He ground-state transition which was shown to exhibit the same strength as the deuterium photodisintegration (see Fig. 9) as a function of photon energy. The momentum dependence of this ratio is presented in Fig. 14 by the open squares and amounts to 1 within errors at lower momenta. The respective ratios for breakup of the α cluster in ^{6}Li and in ^{12}C are given by diamonds and triangles. Below 1.1 GeV/c the ratios are flat and their values reflect the relative strengths. The rise around 1.3 GeV/c can be taken as a hint of effects due to an increase in the high-momentum components in denser matter than in the loosely bound deuterium. Further investigations will have to corroborate these findings and to rule out possible artifacts due to FSI. It can be noted, that the ratio of (γ,pp)/(γ,np) appears to rise stronger with p_{rel} for ^{12}C than for ^{6}Li.

CONCLUDING REMARKS

After introducing the experimental evidence for occupancies of shell-model orbitals in the order of 75% the need for more detailed experiments, which involve more directly the two nucleons at short relative distance, became apparent. Several model calculations had predicted this depletion to be the result of the remnants of the hard core of the nucleon-nucleon interaction in the mean field. Electromagnetically induced two-nucleon emission seems to be a promising tool to investigate the resulting short range correlations. The processes contributing to photoabsorption had been investigated and it was made clear that the genuine two-body absorption process can be identified and separated from others by kinematical selections. Modern microscopic calculations are now able to describe the data quite well. It is also noticed that the dominance of the meson-exchange currents produces features of (γ,np) reactions on nuclei which are very similar to those on the free deuteron. This holds to a certain extent also for the (γ,pp) channel. Hints of high relative-momentum components being more abundant in nuclei than in the deuterium must be further investigated to establish whether these comprise real effects due to the long searched for short-range nucleon-nucleon correlations inside nuclei.

Furthermore, the most recent (γ,NN) experiments have been improved by use of linearly polarised photons. The cross section of nucleon pairs from the classical quasi-deuteron process is expected to be larger in the direction of the electric-dipole field. Polarised photons had been produced at MAMI via coherent bremsstrahlung from a diamond crystal where the rotation of the diamond can turn the polarisation axis easily at convenient intervals. Orientations of the polarisation parallel and perpendicular to the plane of the emitted nucleon pairs were chosen. As expected, the yield for parallel polarisation exceeds that for perpendicular orientation (see very preliminary results in Fig. 15). This holds only for the shell-model range below E_m about 80 MeV. Above this energy the multi-particle process destroys any asymmetry. The very preliminary online results will be improved after thorough calibration and weighting with the photon polarisation of about 50%. Model predictions suggest a strong sensitivity of the asymmetries to SRC effects. It remains to be seen if a further measurement of the polarisation of the outgoing nucleons will exhibit some sensitivity to SRC.

Finally, the studies of real photon-induced reactions will also be fruitful for the investigations of SRC via the (e,e'NN) reactions because of the necessary separation of longitudinal and transverse components. Sofar the (e,e'pp) experiments have restricted themselves to relatively small solid angles. At present (e,e'np) experiments on ^{16}O are in the planning stage to be performed at MAMI. The tensor part of the SRC should

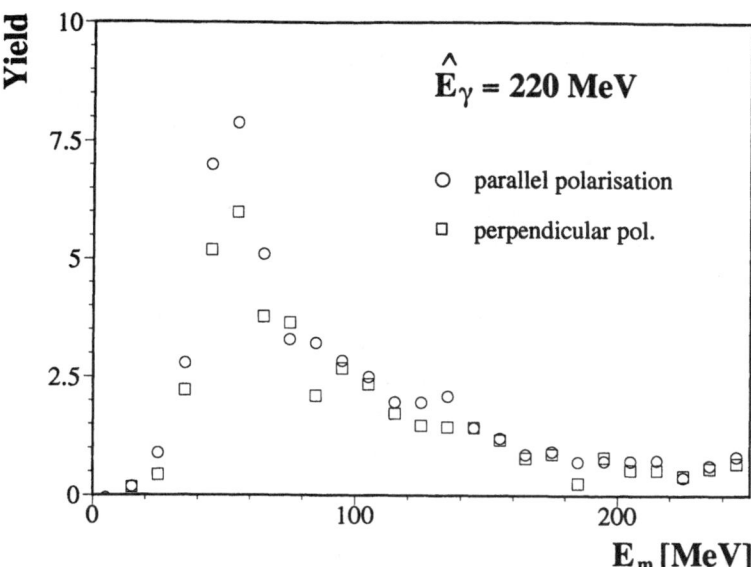

Figure 15. Missing-energy spectra from the ^{4}He(γ,np) reaction for two orientations of linearly-polarised photons. The maximum polarisation of $\approx$ 70% was reached at $\hat{E}_\gamma = 220$ MeV.

be visible there.

Acknowledgement

The author would like to thank all the experimental and theoretical colleagues of the PiP/TOF group of the Mainz A2 collaboration for numerous interesting discussions during the experimental phase and during analysis. In particular I am grateful to T. Hehl and Th. Lamparter for their endurance in the analysis. This work was supported in part by the German Federal Minister for Research and Technology (BMFT 06 Tü 656), the United Kingdom Science and Engineering Research Council, the British Council, the Deutsche Forschungsgemeinschaft (Mu 705/3, SFB201), DAAD (313-ARC-VI-92/118), the European Community (SCI.0910.C(JR)) and NATO (CRG 920171). Thanks also to M. Khalil for efficiently typing this manuscript.

REFERENCES

1. M. Goeppert-Mayer, Phys. Rev. **75** (1949) 1969
2. M. Goeppert-Mayer, Phys. Rev. **78** (1950) 16
3. O. Haxel, J.H.D. Jensen and H.E. Süss, Phys. Rev. **75** (1949) 1766
4. J. Decharge and D. Gogny, Phys. Rev. **C21** (1980) 1568
5. J.L. Friar and J.W. Negele, Theoretical and Experimental Determination of Nuclear Charge Distributions, in Advances in Nuclear Physics, Vol. 8, (Plenum Press, New York-London, 1975) p. 219
6. I. Sick, Lecture Notes in Physics **236** (1985) 137

7. A. Faessler, S. Krewald, A. Plastino and J. Speth, Z. Phys. **A276** (1976) 91

8. J. Dechargé and L. Sips, Nucl. Phys. **A407** (1983) 1

9. R.W. Hasse, B.L. Friman and D. Berdichevsky, Phys. Lett. **B181** (1986) 5

10. C. Mahaux and R. Sartor, Phys. Rev. Lett. **57** (1986) 3015

11. W.H. Dickhoff and H. Müther, Rep. Prog. Phys. **55** (1992) 1947

12. R. Schiavilla, V.R. Pandharipande, R.B. Wiringa, Nucl. Phys. **A449** (1986) 219

13. V.R. Pandharipande *et al.*, Phys. Rev. Lett. **53** (1984) 1133

14. C. Mahaux and H. Ngo, Nucl. Phys. **A431** (1984) 486

15. H. Euteneuer, J. Friedrich and N. Voegler, Nucl. Phys. **A298** (1978) 452

16. J.M. Cavedon *et al.*, Phys. Rev. Lett. **49** (1982) 978

17. B. Frois *et al.*, Nucl. Phys. **A396** (1983) 409c

18. I. Sick and P.K.A. de Witt Huberts, Comm. Nucl. Part. Phys. **40** (1991) 177

19. A.E.L. Dieperink and P.K.A. de Witt Huberts, Ann. Rev. Nucl. Part. Sci. **40** (1990) 239

20. P.K.A. de Witt Huberts, J. Phys. **G16** (1990) 507

21. I. Bobeldijk *et al.*, Phys. Rev. Lett. **73** (1994) 2684

22. G.J. Wagner, AIP Conf. Proc. **142** (1985) 220

23. P. Grabmayr, Prog. Part. Nucl. Phys. **29** (1992) 221 and references therein

24. P. Grabmayr *et al.*, Phys. Rev. **C49** (1994) 2971

25. M. H. Macfarlane and J.B. French, Revs. Mod. Phys. **32** (1960) 567

26. J.B. French and M.H. Macfarlane, Nucl. Phys. **26** (1966) 168

27. T. Berggren, Nucl. Phys. **A72** (1965) 337

28. C.F. Clement, Nucl. Phys. **A213** (1973) 469

29. C.F. Clement, Nucl. Phys. **A213** (1973) 493

30. G.R. Satchler, *Introduction to Nuclear Reactions*, (The MacMillan Press Ltd., London and Basingstoke, 1980)

31. G.R. Satchler, *Direct Nuclear Reactions*, (Oxford University Press, Oxford, 1983)

32. S. Boffi, C. Giusti and F. Pacati, **A336** (1980) 416

33. S. Boffi, C. Giusti and F. Pacati, **A386** (1982) 599

34. B. Frois and C.N. Papanicolas, Ann. Rev. Nucl. Part. Sci. **37** (1987) 133

35. M.C. Birse and C.F. Clement, Nucl. Phys. **A351** (1981) 112

36. J.M. Urias *et al.*, Phys. Rev. **C53** (1996) R1488

37. H. Müther and W.H. Dickhoff, Phys. Rev. **C49** (1994) R17

38. J. Speth, ed., Int. Review on Nuclear Physics **7**, (World Scientific, Singapore, 1991)

39. M. MacCormick *et al.*, Phys. Rev. **C**

40. J. Habermann, Diploma thesis, University Tübingen (1996), unpublished

41. Th. Frommhold *et al.*, Z. Phys. **A350** (1994) 249

42. G. Audit *et al.*, Nucl. Instr. Meth. **A301** (1991) 473

43. W. Wilke *et al.*, Nucl. Instr. Meth. **A272** (1988) 785

44. J. Ahrens, Nucl. Phys. **A446** (1985) 229c

45. J. Koch, Ann. Phys. (N.Y.) **154** (1984) 99

46. M. Gari and H. Hebach, Prog. Rep. **72** (1981) 1

47. C. Levinthal and A. Silverman, Phys. Rev. **82** (1951) 822

48. J.C. Keck, Phys. Rev. **85** (1952) 410

49. M.Q. Barton and J.H. Smith, Phys. Rev. **110** (1958) 1143

50. P.C. Stein *et al.*, Phys. Rev. **119** (1960) 348

51. I.L. Smith et. al., Nucl. Phys. **B1** (1967) 483

52. J.S. Levinger, Phys. Rev. **84** (1951) 43

53. K. Gottfried, Nucl. Phys. **5** (1958) 557

54. M. Kanazawa *et al.*, Phys. Rev. **C35** (1987) 1828

55. J. Arends *et al.*, Z. Phys. **A298** (1980) 103

56. J. Arends *et al.*, Nucl. Phys. **A526** (1991) 479

57. W. Weise, M.G. Huber and M. Danos, Z. Phys. **236** (1970) 176

58. H. Herminghaus, Proc. Lin. Accel. Conf., Albuquerque, USA, (1990); Nucl. Instr. Meth. **A187** (1981) 103

59. C. Giusti *et al.*, Nucl. Phys. **A546** (1992) 607

60. S. Boffi *et al.*, Nucl. Phys. **A564** (1993) 473

61. L. Boato and M.M. Gianinni, J. Phys. **G15** (1989) 1605

62. J. Ryckebusch *et al.*, Nucl. Phys. **A568** (1994) 828

63. J. Ryckebusch *et al.*, Phys. Rev. **C49** (1994) 2704

64. R.C. Carrasco and E. Oset, Nucl. Phys. **A536** (1992) 445

65. R.C. Carraso, E. Oset and L.L. Salcedo, Nucl. Phys. A**541** (1992) 585

66. R.C. Carraso, E. Oset and L.L. Salcedo, Nucl. Phys. A**540** (1994) 701

67. P. Grabmayr, Proc. Workshop on Future Detectors for Photonuclear Experiments, Edinburgh, (1991), ed. D. Branford, (1992) p. 225

68. I.J.D. MacGregor, Proc. Workshop on Future Detectors for Photonuclear Experiments, Edinburgh, (1991), ed. D. Branford, (1992) p. 232

69. I. Anthony *et al.*, Nucl. Instr. Meth. **A301** (1991) 230

70. T. Hehl, Prog. Part. Nucl. Phys. **34** (1995) 385

71. S. Cierjacks *et al.*, Nucl. Instr. Meth. **192** (1992) 407

72. R. Brun, M. Hansroul and J.C. Lassalle, CERN (1982) unpublished

73. D.F. Measdey and C. Richard-Serre, CERN Report 69, (1969)

74. V.I. Kukulin *et al.*, Nucl. Phys. **A513** (1990) 221

75. R. Schneider, Proc. 2nd Workshop on Electromagnetically Induced Two-nucleon Knockout (Gent) 1995, p. 335

76. D.A. Jenkins, P.T. Debevec and P.D. Harty, Phys. Rev. **C50** (1994) 74

77. S. Klein, code MORGAINE, A2 internal report SFB201-A2-IR02/89 and PhD thesis, Univeristy Tübingen (1989), unpublished

78. P. Grabmayr, Proc. PANIC XIII, Perugia (1993), ed. A.Pascolini, (World Scientific, Singapore, 1994) p. 479,

79. N. Kolb, Proc. 2nd Workshop on Electromagnetically Induced Two-nucleon Knockout, Gent, 1995

80. J.C. McGeorge *et al.*, Phys. Rev. **C51** (1995) 1967

81. M Vanderhaegen *et al.*, Nucl. Phys **A580** (1994) 551

82. T.T.H. Yau, PhD Thesis in preparation, University Glasgow

83. P. Grabmayr *et al.*, Phys. Lett. **B370** (1996) 17

84. P.D. Harty *et al.*, Phys. Lett. Phys. Lett. **B380** (1996) 247

85. G.E. Cross *et al.*, Nucl. Phys. **A 593** (1995) 463

86. T. Lamparter *et al.*, Z. Phys. **A355** (1996) 1

87. P. Grabmayr *et al.*, Proc. of the XIV PANIC, Williamsburg (1996) (World Scientific, Singapore, to be published)

88. I.J.D. MacGregor *et al.*, Nucl. Phys. **A533** (1991) 269

NUCLEON-NUCLEON CORRELATIONS IN THE PIONIC DOUBLE CHARGE EXCHANGE

Heinz Clement

Physikalisches Institut
der Universität Tübingen
Auf der Morgenstelle 14
D-72076 Tübingen, Germany

INTRODUCTION

Due to its isospin-one nature the pion exists in three charge states π^+, π^0 and π^-. Hence, pion-induced nuclear reactions may lead to processes, where charges of up to two units are exchanged between pion and nucleus without changing the number of nucleons (N) in the nucleus. Such charge-exchange reactions are known also from light- and heavy-ion reactions, where in particular single-charge-exchange reactions have been studied since a long time. In nucleonic-induced reactions the charge exchange usually takes place by the exchange of nucleons between projectile and target and this process may involve either single nucleons or even clusters of them. Hence, a heavy-ion double-charge exchange reaction, where two units of charge are exchanged between target and projectile, may proceed both via cluster exchange in a single step and in a sequential process via successive exchange of single nucleons. Usually in such reactions at intermediate energies the cross section is found to be dominated by the one-step process.

In pionic charge exchange the situation is much different, since there no nucleons can be exchanged between projectile and target, and the fundamental mechanism is the pion-nucleon charge-exchange process due to the strong isovector part in the pion-nucleon interaction. The pionic double-charge-exchange reaction therefore has to proceed via successive charge exchanges on two like nucleons in the target mediated by the exchange of a pion or another meson between the two participating nucleons. This means that the pionic charge exchange is not spoiled by one-step cluster exchange as in the nucleonic case; it rather should depend sensitively on the separations between the two nucleons involved and thus provide an appropriate tool to study two-nucleon correlations in nuclei.

Though the unique features of the pionic double charge exchange had been known

already decades ago, it has not been until the end of the seventies, that first measurements of the pionic charge exchange to individual final nuclear states have become feasible, soon after meson factories had been commissioned at Los Alamos, USA (LAMPF),
Vancouver, Canada (TRIUMF) and Villigen, Switzerland (PSI). Meanwhile there exists
a broad base of data both for single and double charge exchange on many nuclei, ranging from the lightest to the heaviest ones in the periodic table, and covering incident
pion energies from $T_\pi = 19$ MeV up to 550 MeV. Some hopes originally put onto the
pionic charge exchange appear indeed to have been fulfilled, but there have been also
many unforeseen surprises uncovered in these measurements. In the following section
some basic features of the pion-nucleon interaction both on the free and on the bound
nucleon will be discussed. After that, the current status and perspectives of the pionic
single-(SCX) and double-charge-exchange (DCX) reactions on nuclei are presented. In
the last section we finally discuss a very special kind of short-range correlation, the possibility of a resonance in the πNN-system, for which evidence has been found recently
in the DCX-reaction at energies below the delta-resonance and very recently in other
reactions.

An excellent survey about the experimental and theoretical work on pionic charge
exchange may be found in the proceedings of the 1985 and 1989 LAMPF workshops[1, 2]
as well as in the proceedings of the conferences on pion-nuclear physics in Penyscola[3]
1991 and Dubna[4] 1994. Recent reviews of the pionic charge exchange are given in refs.[5]
and[6]. With the exception of section 5 this writeup of my lecture is basically an excerpt
of my review article[5] updated to the present status and with special emphasis on NN
correlations. For more details on earlier work see refs.[5, 6] and references therein.

PION-NUCLEON INTERACTION

The interaction of pions with nucleons is governed by the Δ- or (3,3)-resonance at
$T_\pi = 190$ MeV ($p_{Lab} = 300$ MeV/c, $\sqrt{s} = 1232$ MeV) in the pion-nucleon system. Due
to its large width of $\Gamma \approx 115$ MeV this resonance dominates the πN-cross section (see
fig. 1) over the whole range of pion energies currently available ($T_\pi \leq 550$ MeV) at the
meson factories LAMPF, PSI and TRIUMF. At resonance the peak $\pi^+ p$-cross section
reaches its unitarity limit of $\sigma_{tot} \approx 200$ mb. Elastic scattering is the only open channel
in the $\pi^+ p$-system up to the pion production threshold. In consequence the integral
elastic scattering cross section is identical to the total cross section over the resonance.
This is in contrast to the situation in the $\pi^- p$-system, where the single charge exchange
(SCX) reaction exceeds the strength of the elastic scattering already at lowest energies.
At resonance the observed relative cross sections of $\pi^+ p$, $\pi^- p$ elastic scattering and of
SCX in the ratio of 9 : 1 : 2 are in agreement with the $I = 3/2$ isospin nature of the
Δ resonance. The $J = 3/2$ spin nature of the resonance shows also up in the angular
distributions for cross section and vector-analyzing power, which clearly exhibit the
dominance of the $p_{3/2}$ partial wave.

In the following we first consider the free pion-nucleon interaction as it manifests
itself in the pion-nucleon system, before we discuss the medium corrections to the free
interaction as observed in pionic atoms and in the pion scattering off nuclei.

Free Interaction

The free pion-nucleon interaction can be deduced from phase-shift analyses of elastic $\pi^\pm p$-scattering and SCX. Therefore, the fundamental πN-system has been studied

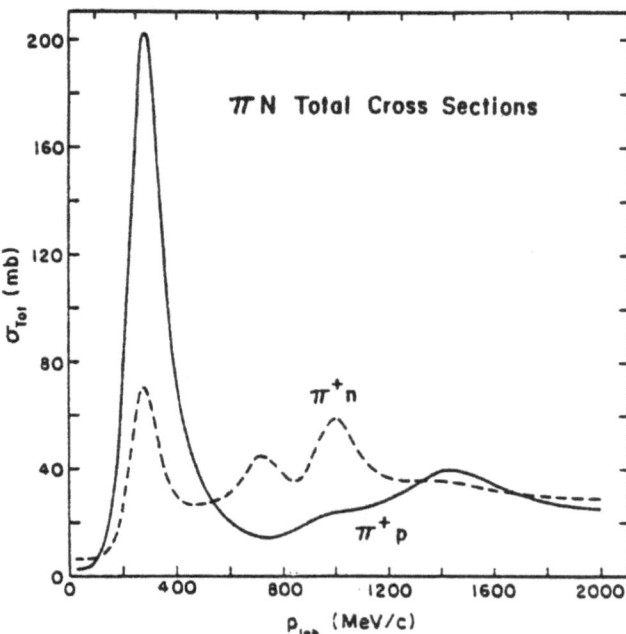

Figure 1. Total π^+p and $\pi^+n(\pi^-p)$ cross sections as a function of the incident pion momentum p_{Lab}.

very carefully in the past decades over a wide range of energies. For a recent review on both theoretical and experimental work, see Kluge [7, 8] as well as the proceedings of the most recent meson-nucleon conference at Blaubeuren[9].

A number of new high precision measurements of π^+p and π^-p scattering as well as $\pi^-p \to \pi^0n$ have been carried out in recent years below, at and above the Δ resonance, in order to clarify discrepancies between hitherto existing data sets. These new data sets are now in much better agreement with each other. In particular at low energies the situation has been clarified considerably[9, 10, 11].

Phase-shift analyses including also the more recent πN data have been carried out by Arndt[12]. The very elaborate phase-shift analyses of the Karlsruhe-Helsinki (KH) group [13, 14, 15] do not include the more recent data; however, they are the only ones which satisfy both, analyticity and unitarity. Thus they allow a unique extrapolation to the Cheng-Dashen point, where the πN-scattering amplitude Σ is closely related to the pion-nucleon sigma term $\sigma_{\pi N}$, which is a measure of the explicit chiral-symmetry breaking in the quark mass term of the QCD Lagrangian for the nucleon. Within chiral perturbation theory $\sigma_{\pi N}$ can be calculated from the experimental baryon mass spectrum. At present the values of $\sigma_{\pi N}$ and Σ are in reasonable agreement with each other and imply for the strange quark content in the nucleon a value of $y = 0.2 \pm 0.2$. For a more detailed discussion of this matter see refs. [7, 8].

Whereas πN-scattering in the region of the Δ resonance is dominated nearly exclusively by the resonant P_{33} partial wave, the region below the Δ resonance is affected by all s and p waves S_{11}, S_{31}, P_{11}, P_{13}, P_{31} and P_{33} — where the indices denote twice the isospin and twice the spin of the particular partial wave.

Following the notation of Koltun[16] the isospin decomposition of the scattering amplitude $f(\Theta)$ into isoscalar and isovector parts is then given by

$$f(\mathbf{k}, \mathbf{k}') = b_0 + \mathbf{t} \cdot \boldsymbol{\tau}\, b_1 + (c_0 + \mathbf{t} \cdot \boldsymbol{\tau}\, c_1)\mathbf{k} \cdot \mathbf{k}' + (d_0 + \mathbf{t} \cdot \boldsymbol{\tau}\, d_1)i\boldsymbol{\sigma} \cdot (\mathbf{k} \times \mathbf{k}') \qquad (1)$$

where $\mathbf{t}$ and $\boldsymbol{\tau}$ denote the isospins of pion and nucleon, respectively. The amplitudes b_0 and b_1 represent the isoscalar and isovector s-wave contributions, c_0 and c_1 give the corresponding p-wave non-spinflip amplitudes, whereas d_0 and d_1 refer to the p-wave spinflip contributions. Their numerical values may be derived from phase shifts. The s- and p-wave amplitudes most remarkably have opposite signs at low energies, which lead to a destructive interference of these amplitudes at forward angles (or backward angles depending on $\mathbf{t} \cdot \boldsymbol{\tau}$) both in isoscalar and isovector channels, leading to pronounced effects in the angular distributions of elastic and inelastic scattering as well as of charge exchange at low energies.

Effective Interaction

In case of pions interacting with nuclei the free πN interaction needs to be modified by medium effects caused by the surrounding nuclear medium. This effective interaction is discussed in the following in terms of a pion-nucleus optical potential, U_{opt}, which is meant to represent the pion-nucleus interaction in the elastic channel, i.e., for the target nucleus in its ground state. For a survey on this topic see e.g. Koltun[16] or Ericson and Weise[17]. The imaginary part of the potential accounts for the loss of pion flux due to absorption caused by target excitation and nucleon knock-out and in particular due to true absorption of pions (see below). The scattering process may then be described by the relativistic Klein-Gordon equation for spinless particles

$$(\omega - U_{opt})^2 p^2 c^2 = m_0^2 c^4 \quad , \qquad (2)$$

where ω is the total pion energy in the pion-nucleus system, m_0 the pion rest mass and p the particle momentum.

The connection of the optical potential to the pion-nucleon scattering amplitude discussed in the previous section is most simply achieved in the single-scattering and impulse approximations assuming the free pion-nucleon interaction to be valid also inside the nucleus:

$$U_N(q, \mathbf{k}, \mathbf{k}') \sim f(\mathbf{k}, \mathbf{k}')\tilde{\rho}(q) \quad . \qquad (3)$$

Here $\tilde{\rho}(q)$ represents the Fourier transform of the point nucleon density in the nucleus.

At low energies the pion-nucleon interaction on nucleons embedded in a nucleus is modified appreciably by the influence of the surrounding nucleons. These medium corrections due to correlations between nucleons have first been evaluated by Ericson and Ericson [18], and lead in zero-range approximation to the following major changes in the s- and p-wave amplitudes:

(i) Pauli correlations give rise to a repulsion in the isoscalar s-wave with amplitude b_0.

(ii) The polarization of the nucleonic medium by the pionic p-waves in analogy to the dipole polarization of an optical medium by light (Lorentz-Lorenz effect), gives rise to the so-called Lorentz-Lorenz-Ericson-Ericson (LLEE) effect, which leads to a substantial suppression of the pion-nucleon p-wave in the nucleus.

(iii) Target excitation and nucleon knock-out in the scattering process lead to a loss of pion flux in the elastic channel and thus give rise to increased imaginary parts in the single-nucleon parameters.

(iv) True pion absorption can take place only if at least two nucleons are present. Pion absorption on a free nucleon is kinematically forbidden, since energy and momentum cannot be conserved simultaneously for such a two-body process. By far the most dominant process appears to be the two-nucleon absorption on an isoscalar proton-neutron pair (quasi-deuteron absorption) with the absorption on $I = 1$ nucleon pairs being smaller by an order of magnitude[19]. Therefore, pion absorption can approximately be taken into account in the pion-nucleus potential by adding isoscalar s- and p-wave terms which are proportional to $\rho_n \rho_p$.

The single-nucleon and two-nucleon parameters in such a second-order optical potential have been deduced in global analyses both of pionic atom data and of π^+ and π^- scattering data at low energies. A recent analysis has been published by Meirav et al.[20]. Their so-called J4 potential provides good descriptions of the bulk of available π^+ and π^- data on differential as well as on total reaction cross sections for $T_\pi \leq 80$ MeV.

The optical model discussed so far contains both isoscalar and isovector terms. It thus is capable to describe both elastic scattering and SCX-reactions. In order to facilitate also a phenomenological description of DCX-data Johnson[21] extended the optical model to include also isotensor terms

$$U = U_0 + (\mathbf{t} \cdot \mathbf{I})U_1 + (\mathbf{t} \cdot \mathbf{I})^2 U_2 \ , \tag{4}$$

where $\mathbf{t}$ and $\mathbf{I}$ denote the isospin operators for pion and target nucleus, respectively, and U_0, U_1, U_2 stand for the isoscalar, isovector and isotensor parts in the potential. In the description of SCX data this ansatz has been used quite successfully. However, for DCX, in particular at low energies, it fails severely in giving a systematic description.

SINGLE CHARGE EXCHANGE

The isospin-one nature of the pion permits two types of charge-exchange reactions on nuclei:

(i) single-charge-exchange (SCX) reactions (π^+, π^0) and (π^-, π^0), which in the target nucleus cause the conversion of a neutron (proton) into a proton (neutron) and thus change the z-component of the target isospin by $\Delta I_z = \pm 1$.

(ii) double-charge-exchange (DCX) reactions (π^+, π^-) and (π^-, π^+), which result in a conversion of two neutrons (protons) into two protons (neutrons) within the target nucleus and consequently change the z-component of the target isospin by $\Delta I_z = \pm 2$.

In the (π^+, π^0)-reaction on a target with $I = I_z \geq 1$ isovector transitions are possible to states with isospin $I - 1, I, I + 1$ in the final nucleus, since the final nucleus has $I_z^f = I - 1$. In the (π^-, π^0)-reaction on the other hand the final nucleus has $I_z^f = I + 1$, hence $I^f \geq I + 1$ for all of its eigenstates. Therefore the (π^-, π^0)-reaction is extremely selective allowing only for isovector transitions to $I + 1$-states.

Within the isospin description the SCX process to the so-called Isobaric Analog State (IAS) can be viewed as a special case of elastic scattering with the primary process being $\pi^+ n \to \pi^0 p$ and $\pi^- p \to \pi^0 n$, respectively. Therefore, we start the discussion of charge exchange on nuclei by examining first the charge exchange on the free nucleon.

From eq. (1) we find for this process

$$
\begin{aligned}
f_{SCX}(\mathbf{k}, \mathbf{k}') &= b_1 + c_1 \mathbf{k} \cdot \mathbf{k}' + id_1 \boldsymbol{\sigma} \cdot \mathbf{k} \times \mathbf{k}' \\
&= b_1 + c_1 k^2 \cos \Theta + id_1 \boldsymbol{\sigma} \cdot \mathbf{n} k^2 \sin \Theta
\end{aligned}
\tag{5}
$$

Due to the $\sin \Theta$ dependence spinflip contributes only at angles around $90°$ significantly. In the Δ-resonance region the cross section is dominated by the P_{33}-partial wave causing charge exchange to have the same angular dependence as elastic πN-scattering. Above the Δ-resonance d-waves start contributing significantly and the cross section gets more and more forward peaked. At low energies the repulsive isovector s-wave plays an important role and leads to a strong destructive sp-interference in the forward angle region of the (π^+, π^0) cross section, in particular near $T_\pi \approx 50$ MeV. Near this energy the $0°$ cross section declines by three orders of magnitude. The experimental SCX cross section at $T_\pi \approx 48$ MeV is found to be as small as 4μb/sr at $0°$ and exhibits a strongly backward peaked angular distribution.

Turning now to the charge exchange on nuclei we expect for the (π^+, π^0) reaction leading to the IAS in the final nucleus

$$
\sigma_{PWIA}(q) = (N - Z)\sigma_{\pi N}(q)|F(q)|^2 \; ,
\tag{6}
$$

if the impulse approximation is valid. Here, $F(q)$ denotes the nuclear form factor for the SCX transition at momentum transfer q and $(N - Z)$ is the number of valence nucleons available for this process. For the IAS transition on $I = 1$ nuclei the zero-degree cross section thus becomes:

$$
\sigma_{PWIA}(0°) = 2\sigma_{\pi N}(0°) \; .
\tag{7}
$$

Indeed, the measured energy excitation function for the forward-angle cross section of (π^+, π^0) on ^{14}C does exhibit a deep minimum at $T_\pi = 48$ MeV following in magnitude and in shape very closely the free πN-cross section (multiplied by two) up to $T_\pi \approx 80$ MeV. This observation impressively demonstrates that at low energies both s- and p-waves contribute to the charge exchange with values nearly unchanged by the nuclear medium. In the region of the Δ-resonance, however, medium effects get very substantial, lowering the nuclear cross section by an order of magnitude, compared to the one on the free nucleon. Distorted wave impulse approximation (DWIA) calculations[22, 23] which properly include medium corrections account very well for the ^{14}C data also in this region. The dramatic energy dependence of $\sigma(0°)$ near 50 MeV persists qualitatively even in SCX on heavy nuclei.

Naturally the transition to that state in the final nucleus, which is the isobaric analog to the target ground state ("analog transition"), is strongly favoured in the SCX reaction. Since this IAS in the final nucleus has a wave function, which is identical to the target ground state regarding space- and spin-parts, the overlap of their wave functions in the form factor is optimal if Q-value effects can be neglected. Therefore, the analog transition is expected to dominate the spectrum in (π^+, π^0). For (π^-, π^0) there is no analog transition, since there the final nucleus has only states of higher isospin.

The neutral pion produced in SCX reactions decays with a lifetime of 10^{-16}s and with a probability of 99% into two gammas. With regard to the time resolution of detectors this decay is instantaneous with the π^0 production. The detection of a π^0 therefore relies on a coincident detection of the two emitted photons.

The bulk of presently available high-quality SCX data have been obtained with the π^0-spectrometer which had been commissioned at LAMPF in 1979 [24]. Its novel design provides an energy resolution of 2 - 5 MeV with high efficiency for neutral pions with kinetic energies of 40 - 500 MeV. The spectrometer consists of two arms and measures energy and direction of the neutral pions by measuring the directions and energies of the two gamma rays.

The IAS peak has been the only transition with the exception of giant resonances that could be identified unambiguously so far in measurements with the π^0-spectrometer. There have been only weak indications for transitions to other individual states. By far the best investigated analog transition in SCX is the one in ^{14}C which has been measured in the forward angular region at each of 12 energies between 20 and 500 MeV. The bulk of investigations deals with the mass and energy dependence of forward-angle cross sections for nuclei reaching from ^{7}Li up to ^{208}Pb.

The mass dependence of the forward-angle cross section follows a very simple scaling law

$$\sigma(0°, T_\pi) \sim (N - Z)A^{-\alpha(T_\pi)} \tag{8}$$

at all energies. Near $T_\pi \approx 50$ MeV the exact location of the minimum in the cross section excitation function depends crucially on the individual reaction Q-value. To account for this it is better in this energy region to consider the minimum cross section rather than plotting the cross section at a fixed energy. This way one obtains a very smooth and flat A-dependence with $\alpha \approx 0$, which is consistent with the PWIA result, eq. (6), indicating that distortions are small at low energies. On the contrary, in the region of the Δ-resonance the observed mass dependence is well described with $\alpha = 4/3$. Actually such a mass dependence of the SCX had been predicted by Johnson[21] prior to the availability of data on the basis of a geometrical model, assuming that in the region of the Δ-resonance the nucleus acts like a black disc allowing the SCX reaction to take place only at the circumference. In this simple picture the charge of a pion is exchanged on a specific nucleon out of A nucleons in the target, thus in this respect the amplitude for SCX should go inversely proportional to A. On the other hand for black disc scattering due to strong absorption the actual reaction volume reduces to the circumference of a disc, i.e., is proportional to some radius $R \sim A^{1/3}$ of the nucleus. Hence, the resulting amplitude for SCX should go like $f_{SCX} \sim A^{-1}R \sim A^{-2/3}$, thus in the black disc limit $\alpha = 4/3$.

In the Δ-resonance region the measured angular distributions show a typical diffraction pattern of a black disc scattering, also in accordance with the predictions of ref.[21]. At low energies the angular distributions are characterized by the $s - p$ interference in the πN system. Here, PWIA-calculations give already a quite good qualitative account of the measured angular distributions. For a more detailed description the approach of Johnson and Siciliano [25] based on the isospin-invariant optical model has been used with much success. For more details see, e.g., ref.[5] and references therein.

DOUBLE CHARGE EXCHANGE

The pionic double charge exchange changes the charge of the target nucleus by two units while leaving the number of nucleons unchanged. Charge conservation ensures that at least two nucleons must be involved in a DCX reaction on a nucleus. In pionic DCX also the process cannot proceed via exchange of nucleons in projectile and target as is the case usually in DCX reactions with heavy ions, but must be mediated by meson exchange between participating nucleons. Therefore the pionic DCX process should crucially depend on spatial correlations between respective nucleons.

Discussions of the appealing features of pionic DCX date back to the early 1960s. Since then a large basis of data has been obtained though the DCX cross sections turned out to be very small, in the region of nb/sr up to μb/sr. These data taken on many nuclei and many energies supply rich information on mass, energy and angular dependences of the pionic DCX. From the discussions in the previous chapter we expect that transitions connecting analog states should be prefered, and that the DCX process should proceed via two sequential steps of single charge exchange. Indeed, transitions to the so-called Double Isobaric Analog State (DIAS) in the final nucleus are found to dominate the DCX spectrum — at least at energies above the Δ-resonance and to some extent also at low energies. From this point of view it was quite a surprise, when first measurements of non-analog transitions performed near the Δ-resonance revealed that these cross sections are comparable in magnitude to the ones obtained for the DIAS transitions. Later measurements uncovered completely different energy dependences for DIAS and non-analog transitions. For the DIAS transitions the cross section decreases with decreasing energy until some minimum is reached at or somewhat below the Δ-resonance. To the contrary the non-analog transitions peak at $T_\pi \approx 140-160$ MeV and exhibit a resonance-like energy dependence in resemblance of the Δ-resonance in the free πN-system. Clearly the reaction mechanisms that dominate DIAS and non-analog transitions must be very different there. To account for the behavior of the non-analog transitions a double Δ-excitation on a single nucleon has been proposed[21] in contrast to the sequential charge exchange for the DIAS transitions.

It came as another big surprise, when in 1984 the first DCX measurements below the Δ-resonance were conducted. Navon et al.[26] and later Leitch et al.[27] measured the DIAS transition on ^{14}C at an incident pion energy of $T_\pi = 50$ MeV and found that there the forward-angle cross section for this transition is as large as at $T_\pi \approx 300$ MeV and even close to the forward-angle cross section for SCX on ^{14}C at $T_\pi = 50$ MeV. This observation was in clear contradiction to expectations. Since the destructive interference of s- and p-waves in the πN-system causes the SCX cross section to undergo a deep minimum near $T_\pi = 50$ MeV, one could have speculated that also something similar would happen for the DCX process.

Due to the very different behavior of the DCX process in different energy regions I shall discuss this reaction in the following, first in the region of the Δ-resonance and then at lower energies. Before doing so I would like to comment briefly on the instrumentation used for the DCX measurements.

DCX cross sections are in the range of nb/sr to μb/sr, while competing processes, in particular elastic scattering, are larger by many orders of magnitude in the range of mb/sr to many b/sr. Thus a DCX experiment has the task to find a single negative pion from the (π^+, π^-) reaction amongst $10^6 - 10^9$ elastically scattered positive pions. This can only be achieved reliably by using magnetic spectrometers, where particles can easily be separated according to their charge and their momentum. However, due to the tiny DCX cross sections, these spectrometers must have a very efficient background suppression in addition to a good energy resolution and a large solid angle. Due to the pion decay the path length in magnetic spectrometers must also be limited to a practical size. Hence, except for the very first DCX measurements, all other DCX data have been obtained at LAMPF, PSI and TRIUMF by use of dedicated magnetic spectrometers.

By far the largest part of presently available DCX data has been collected at incident pion energies between 100 MeV and 300 MeV. The (π^+, π^-) reaction to individual final states has been measured very systematically on nuclei from ^{9}Be up to ^{209}Bi. In addition to these studies there have been measurements of the (π^-, π^+) reaction on several nuclei mainly devoted to the study of exotic nuclei and of giant resonances. Here we concentrate on transitions to the DIAS (DIAT) as well as to non-analog ground

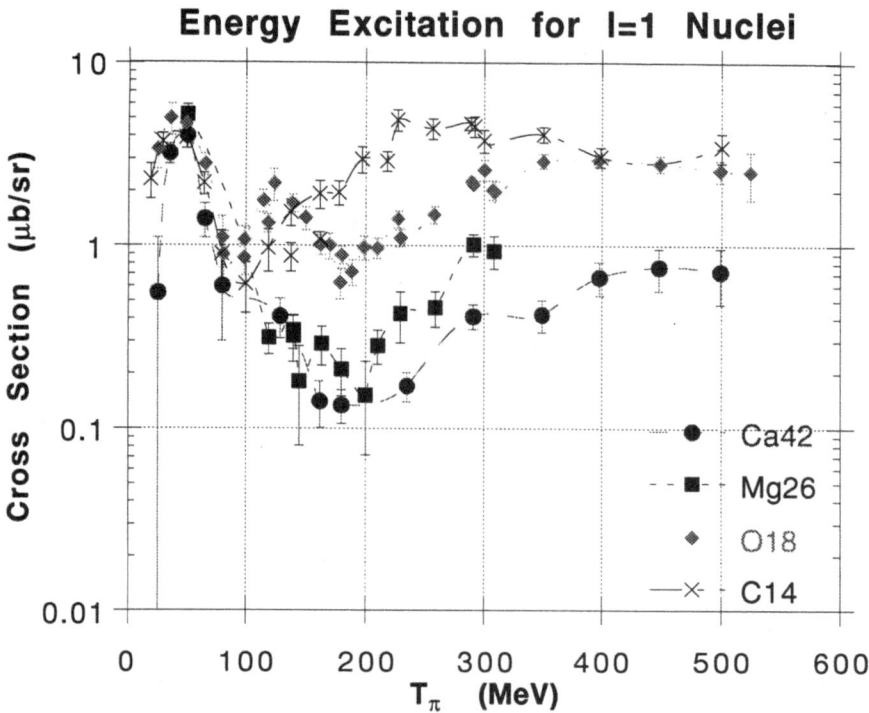

Figure 2. Energy dependence of the forward-angle cross sections for the DIATs on $I = 1$ nuclei.

state transitions (GST). The bulk of data on these transitions consists of forward-angle measurements at $\Theta_{Lab} = 5°$. For a representative number of nuclei there exist also angular distributions, predominantly at $T_\pi = 164$ MeV and $T_\pi = 292$ MeV.

The data collected for the DIATs exhibit a pronounced A-dependence. As for the SCX data, also the bulk of DIATs follows closely the predictions of Johnson[21] on the basis of a strong absorption model, which predicts for the mass dependence

$$\sigma(0°) \sim \binom{N - Z}{2} A^{-10/3} \tag{9}$$

where $\binom{N-Z}{2}$ gives the number of possible combinations of valence neutron pairs available in the target nucleus (N, Z) to undergo the DIAT. The model assumes the DIAT to proceed via two successive SCX steps, each of which go like A^{-1} as discussed in the previous section. Since the reaction volume is restricted again to the circumference of a disc due to strong absorption, we have $f_{DCX} \sim A^{-2}R \sim A^{-5/3}$, which then gives the A-dependence of eq. (9).

The forward-angle cross sections measured for the DIATs in $I = 1$ nuclei, where the DIAS in the final nucleus is identical to the ground state there, is shown in fig. 2. For reference of the data see ref.[5]. The data extend from $T_\pi = 19$ - 550 MeV. For these as well as for the other cases the cross sections are lowest near the Δ-resonance energy and increase then with increasing beam energy. The latter behavior is qualitatively in agreement with the simple strong absorption model which predicts a k^2-dependence.

The surprising observation of Holt et al.[28], that the "forbidden" ground-state transition on the $I = 0$ nucleus ^{16}O is not small compared to the favoured DIAT on

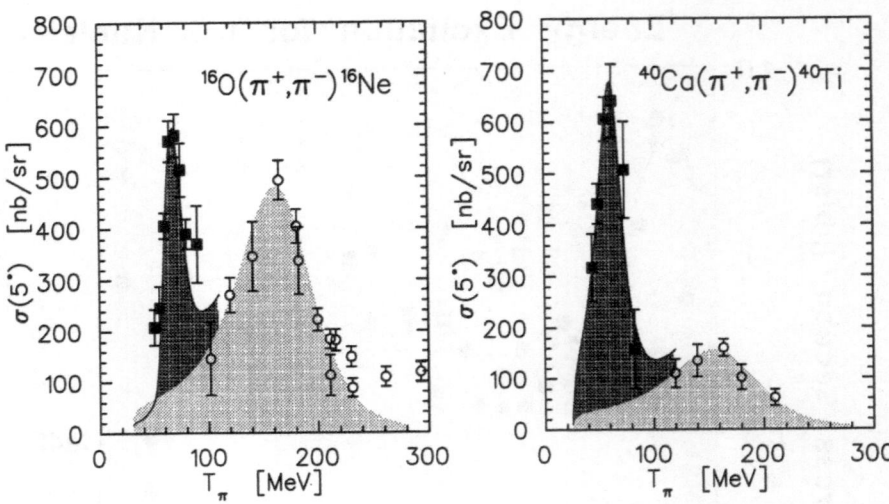

Figure 3. Energy dependence of the DCX ground-state transitions on ^{16}O and ^{40}Ca (from ref. [39]).

^{18}O, initiated a systematic study of non-analog transitions. All these data exhibit a pronounced resonance-like behavior in the energy dependence (see fig. 3) peaking near $T_\pi \approx 160$ MeV with a width of $\Gamma \approx 80$ MeV somewhat smaller than that of the Δ-resonance in the free πN-system. This behavior led to the conclusion that Δ-degrees of freedom must be dominantly involved in these processes.

The mass dependence of non-analog transitions is quite different from that observed for the DIATs. The non-analog ground-state transitions on even-even nuclei obey very closely a $A^{-4/3}$ dependence for the forward-angle cross sections. The agreement of the data with the $A^{-4/3}$ behavior gets even better, if the different Q-values of the ground-state transitions are accounted for and the relevant cross sections are taken at the energies, where the individual excitation functions peak. Both (π^+, π^-) and (π^-, π^+) ground-state transitions show these particular mass and energy dependences. The angular distributions for 0^+ ground-state transitions exhibit a clear diffractive behavior very close to a $J_0^2(qR)$-behavior, corresponding to that for a black disc.

These features for mass and angle dependence are identical to those observed for the SCX and indicate that the underlying reaction mechanism for non-analog DCX must be basically a single-step process. We recall that the $A^{-10/3}$ dependence in the strong absorption model arises because the DCX process for the DIAS transition is assumed to be two-step, each step of which has an amplitude proportional to A^{-1} leading in the cross section to an overall A^{-4} factor. This then multiplies the basic cross section of $R^2 \sim A^{2/3}$ for a reaction to contribute only at the circumference. Thus a possibility for obtaining a $A^{-4/3}$ behavior would be an effective single-step process with an amplitude being proportional to A^{-1}.

The $A^{-4/3}$-mass dependence of the non-analog transitions and the fact, that these cross sections are not small compared to the DIATs and exhibit also an energy behavior resembling that of the Δ-resonance, demand a resonant DCX process which is non-sequential, proceeds in a quasi-single step and involves Δ-degrees of freedom. Such a process has been found in the double or successive delta (DINT) mechanism [25, 29], which is able to account for mass, energy and angle dependences of non-analog transitions over the whole region of the Δ-resonance. Some authors argue that also single-step processes involving preexisting Δ-h-components in nuclei might be of importance.

Since the DCX data follow quite universal mass dependences as predicted by the strong absorption diffraction model, correlations between valence nucleons due to the nuclear structure of individual nuclei apparently do not play a dominant role for the DCX process at resonance energies. They rather seem to be buried underneath the dominating strong absorption effects, and only deviations from the simple A-dependences would be indicative for an influence of such correlations. Evidence for such deviations have been found at $T_\pi = 292$ MeV, which in the energy range discussed here is the energy farthest away from the Δ-resonance. There, the observed $A^{-7/3}$-dependence for $I = 1$ nuclei could be a clear signal for the influence of short-range correlations between valence nucleons. A much stronger evidence for the role of nucleon-nucleon correlations has been found recently in the DCX process at low energies as will be discussed in the following.

From the trend in the data collected in the energy range of the Δ-resonance as well as from theoretical calculations it had been expected that at low energies DCX cross sections should become very small. Therefore, it came quite as a surprise when in 1984 the first data taken on ^{14}C showed that these cross sections were quite sizeable at this energy. The forward-angle cross section of 4μb/sr at $T_\pi = 50$ Mev is substantially larger than at resonance and as large as at the highest measured energies (fig. 2). This large increase in the DCX cross section at 50 MeV contrasts sharply with the energy dependence of SCX cross sections both for the free nucleon and for ^{14}C. For the reaction $\pi^- p \to \pi^0 n$ the forward-angle cross section decreases by more than three orders of magnitude between 160 MeV and 50 MeV due to the s-p waves interference, and the integrated cross section decreases by a factor of about 10. For ^{14}C, the decrease in the forward-angle cross section over the same energy interval is even more than a factor of 400. This totally different behavior of SCX and DCX cross sections on ^{14}C is reflected in the ratio of the respective forward-angle cross sections which runs from a value of about 2000 at resonance to less than 1.5 at 50 MeV.

Besides the completely different energy dependences also the angular distributions for SCX and DCX are vastly different at $T_\pi = 50$ MeV. Whereas SCX angular distributions exhibit a deep minimum at 0° due to the destructive interference of isovector s- and p-waves, the DCX cross section is strongly forward peaked.

The highly surprising features of the first DCX data at low energy immediately prompted speculations about their origin, in particular about the role of short-range correlations between the nucleons involved in the DCX process:

If two nucleons are tightly correlated in space, then we might expect that a π^0 resulting from SCX on the one nucleon finds a good chance to initiate a second SCX process on the nearby partner nucleon. The angular distribution of the DCX process also gets strongly forward peaked in such a case. To see this we consider the DCX process in the impulse approximation, which basically is valid at low energies as we have discussed in the preceding sections. Then the angular dependence of the DCX process is determined essentially by the two-nucleon form factor which is the Fourier transform of the distribution of the centroid of the nucleon pair participating in the DCX transition. When the nucleons of the pair are closely spaced, the distribution of their centroid is similar to that of the nucleons themselves. Since their wave functions usually extend over the whole nuclear volume, the centroid of a closely spaced pair is not strongly localized and consequently has a sharply peaked form factor. On the other hand, if the nucleons are not closely correlated in space they are on average much further apart, and the centroid of such a pair tends to be concentrated at smaller radii giving rise to a form factor which is flat in momentum transfer and hence also in scattering angle.

If nucleons come very close together, say less than 1 fm, then the meson-nucleon picture may get inadequate and it may be more appropriate to treat closely spaced nucleons in terms of quark clusters. Miller[30] proposed the DCX process at low energy to take place directly on six-quark clusters, turning two down quarks into two up quarks. Assuming such clusters to exist in the valence nucleon wave functions with a probability of a few percent he, indeed, obtained a large and sharply forward-peaked DCX cross section in rough agreement with the data. However, this seemingly clear evidence for six-quark clusters in nuclei was put seriously in doubt when Karapiperis and Kobayashi[31] succeeded to explain the DIAT data on ^{14}C at $T_\pi = 50$ MeV also by conventional mechanisms in the framework of multiple-scattering theory by considering the contribution of non-analog routes.

At the same time Gibbs and coworkers[23] have demonstrated that the low-energy DCX crucially depends on spatial correlations between nucleons at distances of 1 fm and less. Small separations between nucleons occur in particular if two nucleons move in a relative s-state with $S = 0$ and $I = 1$, as also pointed out by Bleszynski and Glauber[32]. These authors also stress the equivalence between statements concerning the influence of nucleon-nucleon correlations, formulated from the view point of nuclear structure, and statements concerning the influence of non-analog routes to the DCX process, formulated in terms of the reaction mechanism.

A very essential step towards a better understanding of low-energy DCX was made when the first data on Ca-isotopes [33] clearly demonstrated that this type of reaction is dominated by nucleon-nucleon correlations. In abscence of such correlations we would expect that the cross section for the DIAT would be proportional to the number of all possible combinations of valence neutron pairs, i.e. to $\binom{N-2}{2}$. For ^{48}Ca and ^{42}Ca with eight and two valence neutron pairs, respectively, we thus would expect a ratio of 28 : 1. The measurements at $T_\pi = 35$ MeV, however, yielded a ratio of roughly 1 : 1. Auerbach et al.[34, 45] as well as Bleszynski and Glauber [35] could quantitatively demonstrate that the missing factor of 28 is largely due to the influence of spatial correlations, as incorporated in the shell model via Pauli principle and pairing. In ^{48}Ca the neutron shell is closed and thus the valence neutrons are on average much further apart from each other than in ^{42}Ca, where there is only a single valence nucleon pair. These medium-range shell-model correlations are opposed at very small NN distances ($\lesssim 0.5$ fm) by short-range correlations due to the repulsive core in the NN interactions. The latter also lead to a sizeable effect in the DCX cross section counteracting[6, 38] somewhat the effect of the pairing correlations.

ADCXnother piece of evidence for the sensitivity of the DCX to short-range correlations comes from systematics of DIATs. As noticed by several groups (see e.g. refs.[6, 34, 45] the DCX cross section can theoretically be split into two parts, belonging to a long-range and a short-range part of the DCX operator. In ref. [36] it has been pointed out that at high energies ($T_\pi = 295$ MeV) the systematics of the DIAT is in accord with a dominance of the long-range part. To the contrary the DIAT data of low pion energy show that the short-range part dominates[36, 37] — which is in full agreement with the findings about the sensitivity of the low-energy DCX to short- and medium-range NN-correlations discussed above. Thus, the hope of some twenty-five years ago that the pion DCX reaction would provide quantitative information about two-nucleon correlations in nuclei, appears to be fulfilled in DCX at low pion energies.

SPECIAL SHORT-RANGE CORRELATIONS — THE CASE OF DIBARYONS

The typical energy, where the effect of short- and medium-range correlations on the DCX cross section could be demonstrated to work on a quantitative level in comparison with data has been $T_\pi = 35$ MeV. At lower energies hardly any data exist, since most of those slow pions emerging from the target decay before they can be detected. At higher energies, on the other hand, the data show an unexpected energy dependence. All hitherto measured DCX transitions show in their forward-angle cross section a resonance-like excursion in the energy range $T_\pi = 40$ - 90 MeV — with the exception of the DIAT in ^{48}Ca. For the DIATs the typical behavior is shown in fig. 2, where the forward-angles cross sections are plotted for $I = 1$ nuclei. For non-analog ground-state transitions the situation is even more striking. As an example our recent measurements[39] on the doubly-closed-shell nuclei ^{16}O and ^{40}Ca are shown in fig. 3 together with LAMPF measurements at $T_\pi \geq 100$ MeV[40, 41].

The energy dependence is characterized by a pronounced and quite narrow resonance-like structure peaking near $T_\pi = 65$ MeV as well as by a broad bump near $T_\pi = 160$ MeV. The latter is due to the double-delta mechanism[25, 29] as already discussed above. The first structure has recently been associated[42, 43] with yet another, still hypothetical resonance. In the πN-system there is no resonance below the Δ-resonance, hence this resonance at low pion energies cannot happen in the πN-system. Since the angular distributions are smooth, showing no dependence proportional to $P_J^2(\cos\Theta)$, this structure cannot be due to a π-nucleus resonance with angular momentum J either. Hence, the only possibility left for a resonance is one which is in the πNN-system, the basic system of the DCX process. It has been shown[42, 43] that both energy and angular dependences of DIATs and GSTs at low energies can be well described if such a resonance, called d', has quantum numbers $I(J^P) = $ even (0^-), a total πNN decay width $\Gamma_{\pi NN} \approx 0.5$ MeV, a mass $m = 2.06$ GeV, and a spreading width in the nuclear medium of $\Gamma_{\text{spread}} \approx 5$ MeV. The heavily shaded areas in fig. 3 show the calculation for the d' resonance which interferes with the tail of the double-delta process (light-shaded areas). Due to its quantum number this resonance cannot decay into the NN-channel. From unsuccessful dibaryon searches in other reactions, in particular in $\pi^- d \to nn\pi^+\pi^-$ and $pp \to \pi^- X$ the isospin $I = 0$ appears to be most likely for d'[44], otherwise it should have been seen there.

As mentioned above, the only known case where no peak structure has been observed in the energy dependence of the forward-angle DCX cross sections is the DIAT in ^{48}Ca — in agreement also with the d' hypothesis[43]. The reason for this is that both initial and final states have a doubly-closed-shell structure, which minimizes the correlation effects at small NN-distances[34, 35, 45]

Another hint on the nature of this peculiar resonance-like structure at low energies may be obtained from the dependence of its peak energy on the reaction Q-value. If there were no Q-dependence, then the effect could be associated with an initial-state interaction. If the peak energy changes just in accordance with the Q-value, then the structure could be due to a final-state-interaction effect. The low-energy data on GSTs and DIATs, however, exhibit roughly a $Q/2$-dependence of the peak energy, which again is in favor of a resonance in the intermediate πNN system.

The assumption of a NN-decoupled d' resonance is supported by all hitherto measured low-energy DCX-data, which range from ^{7}Li to ^{56}Fe. However, since this phenomenon is observed in nuclei, subtle nuclear medium effects as a possible origin for the peculiar energy dependence cannot be excluded definitely — though ^{7}Li with the cluster partition $t + \alpha$ constitutes already the DCX on a three-nucleon system, if the α core can be considered as a pure spectator. ^{7}Li is the lightest nucleus, where the DCX to a discrete final state can be observed. The DCX on the lightest nuclei possible at all, i.e. on ^{3}He and ^{4}He, leads no longer to any discrete final state but only to a continuum of 3 and 4 identical nucleons. In this case the d' resonance can no longer be formed in the s-channel, however, it may be produced in the course of this reaction. This production starting at the d' production threshold could be observed either by a steep rise in the energy dependence of the integral cross section[46] or by an exclusive measurement of the invariant πNN mass in this reaction. Available data on integral cross sections for the DCX on ^{4}He are again in support of the d' hypothesis[46], but the quality and significance of these data in the energy region of interest is still too low for any definite conclusions. New and more precise DCX-measurements on ^{4}He, both inclusive and exclusive, are presently carried out at TRIUMF. First preliminary results[47] again are in favor of the d' resonance.

These findings in the DCX data show evidence for the existence of a πNN resonance link directly, of course, to the longstanding problem of possible resonances in the system of two baryons. Hence, in the following a short review of the so far not particularly glorious history of dibaryon searches is given before we return again to the discussion of other evidences for the existence of d'.

History of Dibaryons

With the realization of quarks (q) being the basic building blocks of hadrons also the idea emerged in the late seventies that this substructure of baryons should cause nontrivial resonances in the $B = 2$ system — apart from the only and very loosely bound nucleon-nucleon (NN) state, the deuteron, the features of which can be understood very well by meson-exchange without requiring directly any hadron substructure. With the establishment of the quantum chromodynamics (QCD) as the appropriate theory of the strong interaction first QCD-based model calculations for $6q$-systems appeared[48, 49, 50] in the late seventies and early eighties, triggered by the paper of Jaffe[48] on a possible H dibaryon or dihyperon with strangeness $S = -2$. These predictions of a large number of $6q$-states, part of which is shown in fig. 4 as an example, caused a rush for experimental dibaryon searches in the years to follow. Unfortunately, despite a vast number of dedicated experiments (for a review see, e.g., [51, 52, 53, 54, 55, 56]) the hunt for dibaryons has been without much success up to now, and as of yet no unambiguous evidence for their existence has been found.

The bulk of such experiments has been dedicated to searches of dibaryons coupled to NN or to $N\Delta$ channels, which also constitute by far the most dominant species of predicted dibaryon states. Experimentally the search for these states can be undertaken conveniently by studying NN or πd collisions, or other similar processes. However, in these NN- or $N\Delta$-coupled cases it can be expected that probabilities of such fall-apart decays of dibaryon states are not small. Their widths should rather be even large compared to those of usual baryon resonances, which already are in the order of 100-200 MeV. The existence of such very broad dibaryon resonances is extremely difficult to establish in the experiment and the failure of the hitherto dibaryon searches in this sector could be ascribed to their fall-apart nature.

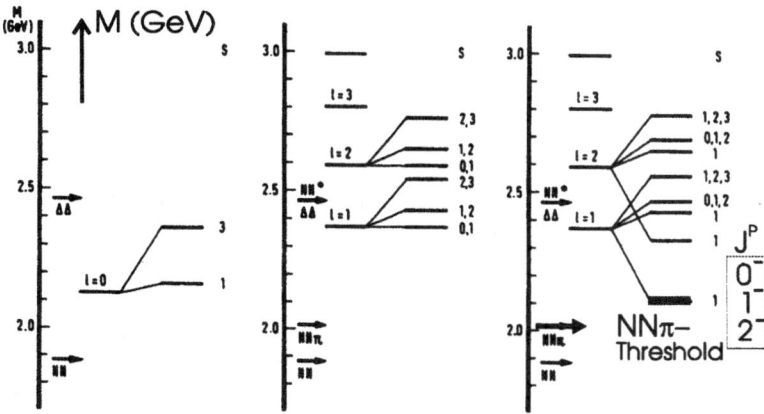

Figure 4. Example of a predicted mass spectrum for nonstrange and isoscalar ($I = 0$) dibaryon resonances (from Mulders et al., 1980). To find the J^P assignment for a particular state, spin S and orbital angular momentum l have to be combined. E.g., the lowest lying combination of $l = 1$ and $S = 1$ gives rise to a (in this calculation degenerate) triplet of states with $J^P = 0^-, 1^-, 2^-$, two of which are NN-decoupled.

NN- and NΔ-Decoupled Resonances

The situation is very different for NN- and NΔ-*decoupled* dibaryon states, i.e. for resonances with quantum numbers $I = 0$ and $J^P = 0^+, 0^-, 2^-, 4^-, \ldots$. In this case the only decay channels are γNN and πNN. If in addition the resonance energy is not far above the πNN-threshold, then we expect a very narrow width in the order of MeV only. As seen in fig. 4, indeed such low-lying states with $I = 0$ and $J^P = 0^-, 2^-$ have been predicted[49, 50, 57], to actually be the lowest-lying dibaryon states. On the other hand the small width, which is very desirable from the experimental point of view, implies necessarily small formation and production cross sections in the suitable reactions which will be discussed below. Another prerequisite for the selection of a suitable reaction is that the cross section due to conventional processes is as small as possible, so that an exotic process like the formation or production of a dibaryon resonance has a chance to be observed. For the interpretation of a possible dibaryon signature it is also desirable that the reaction, if possible, only involves the minimal number of particles necessary to form or produce such a resonance.

If the resonance energy is below the πNN-threshold, then γNN is the only decay channel. The only basic reaction in this case would be $\gamma d \to R \to \gamma np$ or γd, with R denoting the resonance. Another possibility would be to look for such a resonance in more complicated reactions like the two-photon emission in NN-Bremsstrahlung, e.g., $NN \to R\gamma \to NN\gamma\gamma$, where the signal-to-background ratio could be more favorable.

If the resonance energy is above the πNN-threshold, then the most dominant decay channel will by far be πNN. In this case there are basically three types of reactions suitable for the dibaryon search under the conditions discussed above:

(i) Pionic Double Charge Exchange:

   Since charge conservation ensures that this reaction is a genuine 2N-process, the cross section is correspondingly very small in the range of nb/sr to μb/sr and has a high sensitivity to NN-correlations of short range — as pointed out in the section on DCX. This way it is an ideal reaction to look for exotic mechanisms. Its only disadvantage is that this reaction cannot be performed on the basic NN system as target, since both pp and nn are unbound and on the deuteron DCX is not possible. Hence this reaction can only be measured on nuclei with $A \geq 3$, i.e. on NN-pairs embedded in a nuclear medium, where subtle many-body effects cannot be excluded definitely for alternative explanations.

(ii) Photo-pion production:

This process has the great advantage that it can be carried out on the basic system, the deuteron. The drawback here is, however, that the photo-pion production basically is a single-nucleon process proceeding via the Δ-resonance. Hence, we face large cross sections of up to 300 μb depending strongly on the γ energy of interest. For energies well below the Δ-resonance, i.e., 140 MeV $< E_\gamma < 200$ MeV the π^0 production has a cross section substantially lower than the $\pi^\pm$ production and offers thus a window to look for exotic processes with an improved sensitivity. Conventional pion production on the deuteron may proceed either incoherently, i.e. in a quasifree process on the constituent nucleons, or in a coherent process, where the deuteron stays in its bound state. For the quasifree process, for which there are as of yet no data available in the energy range of interest, we may get some rough estimate from the underlying free photonucleon process. This leads to about 120, 150 and 40 μb for π^+, π^- and π^0 production, respectively, on the deuteron near $E_\gamma = 200$ MeV, where the d' resonance is expected to show a signal. Hence the relative smallness of the π^0 production cross section, which is due to the suppression of the E_{0+} amplitude, makes this channel by far the most suited one to look for the d' signal. For the latter the cross section has been estimated[58] to be at most 1 μb or below. First measurements of the reaction $\gamma d \rightarrow np\pi^0$ at MAMI using tagged photons and the TAPS setup, which is capable of detecting both π^0 and n, have been carried out. The data analysis is in progress.

(iii) 2π production in NN-collisions:

This reaction constitutes the most basic hadronic reaction for reaching dibaryon states decoupled from the NN channel. The second pion produced associatedly to the πNN system in the exit channel allows, in principle, any spin-parity combination of the latter. Also the simultaneous investigation of the invariant masses of $pp\pi^-$ and $pp\pi^+$ in the reaction $pp \rightarrow pp\pi^-\pi^+$ permits a distinction between $I = 0$ and $I = 2$ resonances, depending on whether such a resonance is seen in $pp\pi^-$ only or in both $pp\pi^-$ and in $pp\pi^+$. The drawback here is that the cross section of the conventional process is with $10 - 100\mu$b in the region of interest already quite substantial. Another problem is that this reaction with 4 particles in the exit channel is already rather complex and a careful exclusive measurement requires a very sophisticated experimental setup as well as data analysis.

For the d' resonance the production threshold is at $T_p \approx 710$ MeV, i.e., roughly 100 MeV above the 2π threshold. From the width $\Gamma_{\pi NN} \approx 0.5$ MeV as obtained from the analysis of the DCX data, we deduce[59] a d' production cross section of about 1 μb at $T_p \approx 730$ MeV in $pp \rightarrow pp\pi^-\pi^+$. The prediction is expected to be accurate to within about 30%, due to the approximations used. However, the absolute value depends substantially on the coupling constant $g^2_{\omega NN}$, which is crucial for the description of the repulsive initial-state interaction. The above-quoted cross-section value is obtained for $g^2_{\omega NN} = 6$, which is the value in the SU(3)-limit, if $g^2_{\phi NN} = 0$. However, if we use $g^2_{\omega NN} = 12$, which is the more realistic value obtained in the Bonn potential[60], then the prediction drops roughly by a factor of three.

There are only a few measurements of the integral cross section of the reaction $pp \rightarrow pp\pi^-\pi^+$ in the energy range of interest. They show that the cross section of

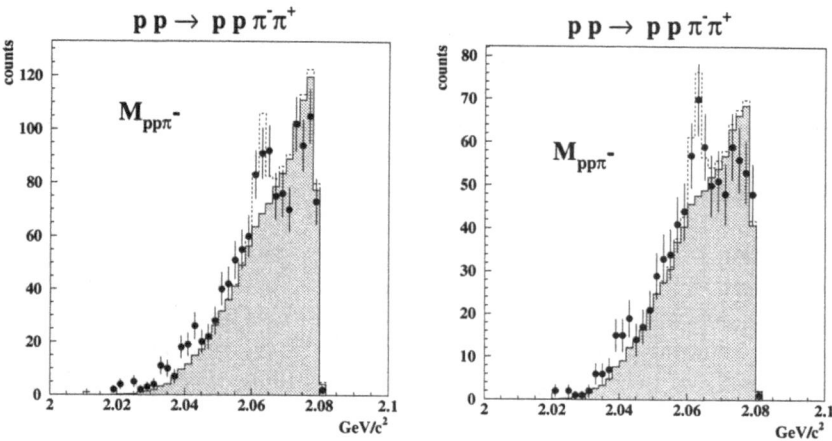

Figure 5. Invariant-mass spectrum $M_{pp\pi^-}$ of the reaction $pp \to pp\pi^-\pi^+$ at $T_p = 750$ MeV with (right) and without (left) constraints on the missing-mass spectrum $MM(pp\pi^+)$. The solid histograms (shaded) show the MC simulations of the conventional 2π production process, the dashed ones show the result with inclusion of the d' production process (see ref. [62]).

the conventional process is larger than the predicted d' cross section by an order of magnitude. To overcome somewhat this problem we had proposed[59] to utilize the final-state effect in the d' decay, which leads to an enhancement in the invariant mass M_{pp} of the two protons belonging to d'. Since in the conventional process such a strong final-state effect is absent, a cut on low M_{pp} masses could lead to a substantial suppression of the conventional mechanism compared to the d' production.

In a first exclusive measurement carried out at ITEP at $T_p = 920$ MeV[61] no sign of a narrow resonance could be found in the unconstrained $M_{pp\pi^-}$ and $M_{pp\pi^+}$ spectra. However, with a cut on low M_{pp} masses a bump near 2.06 GeV is observed in $M_{pp\pi^-}$, while none is seen in $M_{pp\pi^+}$. Its width of approximately 10 MeV is attributed to the experimental resolution. If this observation is correct, then it would mean that d' has isospin $I = 0$ — as already anticipated from the discussion above. Since the effect seen in the ITEP data, which result from a difference measurement on CH_2 and C targets, is not yet fully convincing and beyond any doubts, we started a series of exclusive measurements at the storage ring CELSIUS, using the WASA/PROMICE setup with a H_2 cluster jet target.

The results of a first analysis have already been published [62] and part of them is shown in fig. 5. For the data presented there only events registered in the forward detector, a segmented scintillation calorimeter covering the angular range $4° \le \theta \le 21°$ have been used. Since the present setup contains no magnetic field, the delayed pulse technique has been installed for π^+ identification. The trigger has been set to at least 3-prong events of charged particles in the forward detector, thus suppressing events from single pion production. For the selected 2π-production events we demand two identified protons and one identified π^+ particle. The missing mass spectrum of these identified $pp\pi^+$ events peaks as expected at the mass of the unidentified π^- particle.

Using the momentum and energy information of the identified π^+ from these selected $pp\pi^+$ events together with the corresponding information of the beam, we obtain the $M_{pp\pi^-}$-spectrum as shown on the left of fig. 5.

If in addition we require the missing mass of the registered $pp\pi^+$ event, i.e. the mass of the unobserved π^-, to be in the range of 130 MeV - 150 MeV, then we obtain the $M_{pp\pi^-}$-spectrum as shown on the right of fig. 5. On top of a smooth distribution we observe in both cases a narrow anomaly around 2.063 GeV with a statistical significance of $(4-5)\sigma$ relative to the Monte-Carlo (MC) simulations for the conventional 2π-production. For these simulations a pure phase-space distribution of the conventional 2π production process has been assumed. The width of the anomaly is in accordance with the experimental resolution. For details see ref.[62]. Further measurements are in progress.

We note that the $M_{pp\pi^+}$ spectrum[62] does not show any irregularity beyond statistical fluctuations. Since the resolution there is not as good as in $M_{pp\pi^-}$ the negative result in $M_{pp\pi^+}$ does not yet exclude $I=2$ for d' definitely. However, in case of isospin being a good quantum number, we would expect[44, 63] the resonance effect to be 36 times larger in $M_{pp\pi^+}$ than in $M_{pp\pi^-}$ in case of $I=2$. Hence, an $I=2$ assignment for d' appears to be extremely unlikely.

Outlook — renaissance of 6q searches?

To summarize, for the existence of the πNN resonance d' evidence has been found so far in three different channels

(i) in the DCX on nuclei to discrete final states:
   The signature found there is in agreement with a formation of d'.

(ii) in the DCX on ^{4}He leading to a 4N continuum:
   The sharp rise in the energy dependence of the integral cross section is in accordance with the opening of the production threshold of d' in this reaction

(iii) in the 2π production im pp-collisions:
   The peak observed in $M_{pp\pi^-}$ (fig. 5) constitutes the first evidence for the free d' production on the basic NN system.

Hence, we have reason to believe that d' formation and production has been observed in these reactions in full agreement with predictions based on the d' hypothesis. Consequently, that would mean that there are, indeed, more states in the dibaryon system than just the trivial meson-dominated deuteron ground state and the virtual $^1S_0$ level.

A final proof of the existence of d' will have a number of implications. Since it is unlikely that d' would be singular in the dibaryonic world, the search for further resonances could lead to a revival of the hunt for 6q states. The existence of such states will provide a unique opportunity to study for the first time the QCD of baryonic systems with more than three quarks. Among others this could give new insights into the problems of confinement[64]. Another point of interest would be the impact of dibaryonic resonances, in particular of their bosonic nature, on the behavior of nuclear matter under extreme conditions.

REFERENCES

1. Proc. LAMPF Workshop on Pion Double Charge Exchange H.W. Baer and M.J. Leitch, eds., Los Alamos, New Mexico, USA, (1985) Los Alamos National Laboratory Report No. LA-10550
2. Proc. Second LAMPF Workshop on Pion-Nucleus Double Charge Exchange, W.R. Gibbs and M.J. Leitch, eds., Los Alamos, New Mexico, USA, (1989) (World Scientific, Singapore, 1990)

3. Proc. Int. Workshop on Pions in Nuclei, E. Oset, M.J. Vicente- Vacas, C. Garcia-Recio, eds., Penyscola, Spain (1991) (World Scientific, Singapore, 1992)

4. Proc. Int. Conf. on Mesons and Nuclei at Intermediate Energies, M.Kh. Khankhasayev, Zh. B. Kurmanov, eds., Dubna, Russia, (1994) (World Scientific, Singapore, 1995)

5. H. Clement, Prog. Part. Nucl. Phys. **29** (1992) 175

6. M.B. Johnson and C.L. Morris, Ann. Rev. Nucl. Part. Sci. **43** (1993) 165

7. W. Kluge, Rep. Prog. Phys. **54** (1991) 1251

8. W. Kluge, Prog. Part. Nucl. Phys. **36** (1996) 239

9. Proc. VI. Int. Symp. on Meson-Nucleon Physics and the Structure of the Nucleon, G.J. Wagner, R. Bilger and T. Hehl, eds., Blaubeuren, Germany (1995), πN Newsletter **10** (1995) and **11** (1996)

10. Ch. Joram et al., Phys. Rev. **C51** (1995) 2144 and 2159

11. R. Wieser et al., Phys. Rev. **C**, accepted for publication

12. R.A. Arndt, I.I. Strakovski, R.L. Workman and M.M. Pavan, Phys. Rev. **C52** (1995) 2120

13. R. Koch and E. Pietarinen, Nucl. Phys. **A336** (1980) 331

14. R. Koch, Z. Phys. **C15** (1982) 161, and Nucl. Phys. **A448** (1986) 707

15. G. Höhler (1983), Landolt-Börnstein (H. Schopper, ed.) Vol. I9b2. Springer, Heidelberg, Berlin

16. D.S. Koltun, Adv. Nucl. Phys. **3** (1969) 71, M. Baranger, E. Vogt, eds. (Plenum Press, New York, 1969)

17. T.E.O. Ericson and W. Weise , "Pions and Nuclei" (Clarendon Press, Oxford, 1988)

18. M. Ericson and T.E.O. Ericson, Ann Phys. (N.Y.) **36** (1966) 323

19. H.J. Weyer, Phys. Rep. **195** (1990) 295

20. O. Meirav, E. Friedman, R.R. Johnson, R. Olzewski and P. Weber, Phys. Rev. **C40** (1989) 843

21. M.B. Johnson, Phys. Rev. **C22** (1980) 192

22. T. Karapiperis and T. Takaki, Nucl. Phys. **A518** (1990) 752

23. W.R. Gibbs, W.B. Kaufmann and P.B. Siegel, in: "Proc. LAMPF Workshop on Pion Double Charge Exchange", H.W. Baer and M.J. Leitch, eds., Los Alamos, USA, LANL report No. LA-10550 C90

24. H.W. Baer, R.D. Bolton, J.D. Bowman, M.D. Cooper, F.H. Cverna, R.H. Heffner, C.M. Hofman, N.S.P. King, J. Piffaretti, J. Alster, A. Doron, S. Gilad, M.A. Moinester, P.R. Bevington and E. Winkelmann, Nucl. Instr. Meth. **180** (1981) 445

25. M.B. Johnson and E.R. Siciliano, Phys. Rev. **C27** (1983) 730, ibid. 1647

26. I. Navon, M.J. Leitch, D.A. Bryman, T. Numao, P. Schlatter, G. Azyelos, R. Poutisson, R.A. Burnham, M. Hasinoff, J.M. Poutisson, J.A. Macdonald, J.A. Spuller, C.K. Hargrove, H. Mes, M. Blecher, K. Gotow, M. Moinester and H. Baer, Phys. Rev. Lett. **52** (1984) 105

27. M.J. Leitch, E. Piasetzky, H.W. Baer, J.D. Bowman, R.L. Burman, B.J. Dopesky, P.A.M. Gram, F. Irom, D. Roberts, G.A. Repka jr., J.N. Knudson, J.R. Comfort, V.A. Pinnick, D.H. Wright and S.A. Wood, Phys. Rev. Lett. **54** (1985) 1482

28. R.J. Holt, B. Zeidman, D.J. Malbrough, T. Marks, B.M. Preedom, M.P. Baker, R.L. Burman, M.D. Cooper, R.H. Heffner, D.M. Lee, R.P. Redwine and J.E. Spencer, Phys. Lett. **69B** (1977) 55

29. M.B. Johnson, E.R. Siciliano, H. Toki and A. Wirzba, Phys. Rev. Lett. **52** (1984) 593

30. G.A. Miller, Phys. Rev. Lett. **53** (1984) 2008

31. T. Karapiperis and M. Kobayashi, Phys. Rev. Lett. **54** (1985) 1230

32. M. Bleszynski and R.J. Glauber, Phys. Rev. **C36** (1987) 681

33. H.W. Baer, M.J. Leitch, R.L. Burman, M.D. Cooper, A.Z. Cui, B.J. Dropesky, J.N. Knudson, J.R. Comfort, D.H. Wright and R. Gilman, Phys. Rev. **C35** (1987) 1425

34. N. Auerbach, W.R. Gibbs and E. Piasetzky, Phys. Rev. Lett. **59** (1987) 1076

35. E. Bleszynski, M. Bleszynski and R.J. Glauber, Phys. Rev. Lett. **46** (1988) 1483

36. H. Ward et al., Phys. Rev. **C47** (1993) 687

37. K. Föhl et al., Phys. Rev. **C53** (1996) 2033

38. M.B. Johnson, E.R. Siciliano and H. Sarafian, Phys. Lett. **B243** (1990) 18

39. K. Föhl et al., Proc. Int. Conf. on Particles and Nuclei, Williamsburg, VA, USA (1996) and K. Föhl, PhD thesis, Univ. Tübingen 1996

40. L.C. Bland, R. Gilman, M. Carchidi, K. Dhuga, C.L. Morris, H.T. Fortune, S.J. Greene, P.A. Seidl and C.F. Moore, Phys. Lett. **128B** (1983) 157

41. D.P. Beatty et al., Phys. Rev. **C48** (1993) 1428, and references therein

42. R. Bilger, H. Clement, K. Föhl, K. Heitlinger, C. Joram, W. Kluge, M. Schepkin, G.J. Wagner, R. Wieser, R. Abela, F. Foroughi and D. Renker, Z. Phys. **A343** (1992) 491

43. R. Bilger, H.A. Clement and M.G. Schepkin, Phys. Rev. Lett. **71** (1993) 42

44. R. Bilger, H.A. Clement and M.G. Schepkin, Phys. Rev. Lett. **72** (1994) 2972

45. N. Auerbach, W.R. Gibbs, J.N. Ginocchio and W.B. Kaufmann, Phys. Rev. **C38** (1988) 1277

46. H. Clement, M. Schepkin, G.J. Wagner and O. Zaboronsky, Phys. Lett. **B337** (1994) 43

47. R. Meier et al., Proc. Int. Conf. on Particles and Nuclei, Williamsburg, VA, USA (1996)

48. R.L. Jaffe, Phys. Rev. Lett. **38** (1977) 195

49. P.J. Mulders, A.T. Aerts and J.J. de Swart, Phys. Rev. Lett. **40** (1978) 1543

50. P.J. Mulders, A.T. Aerts and J.J. de Swart, Phys. Rev. **D21** (1980) 2653

51. K.K. Seth in: "Proc. Conf. Medium and High Energy Nuclear Physics", W.Y.P. Hwang et al., eds., Taiwan (1988) (World Scientific, Singapore, 1989)

52. K.K. Seth in: "Proc. Int. Workshop on Pions in Nuclei", E. Oset, ed., Penyscola, Spain (1995) (World Scientific, Singapore, 1992) p. 205

53. E.N. Komarov, St. Petersburg preprint # 1853 (1993)

54. I.I. Strakovskii, Sov. J. Part. Nucl. **22** (1991) 296

55. B. Tatischeff, M.P. Comets, Y. Le Bornec and N. Willis (1991), Orsay preprint IPNO-DRE 91-18

56. H. Clement et al., Prog. Part. Nucl. Phys. **36** (1996) 369

57. L.A. Kontratyuk, B.V. Martemyanov and M.G. Schepkin, Sov. J. Nucl. Phys. **45** (1987) 776

58. R. Bilger et al., Nucl. Phys. **A596** (1996) 586

59. M. Schepkin, O. Zaboronsky and H. Clement, Z. Phys. **A345** (1993) 407

60. K. Holinde, Prog. Part. Nucl. Phys. **36** (1996) 311

61. L. Vorobyev et al., JETP Lett **52** (1994) 75 (77)

62. W. Brodowski et al., Z. Phys. **A355** (1996) 5

63. M. Schepkin et al., priv. comm.

64. A.J. Buchmann et al., Prog. Part. Nucl. Phys. **36** (1996) 383

ELEMENTS OF QUANTUM TRANSPORT THEORY

Rudi Malfliet

Institute for Nuclear Theory, University of Washington,
Box 351550, Seattle, Wa 98195, USA
and
Kernfysisch Versneller Instituut, University of Groningen,
9747 AA Groningen, The Netherlands

INTRODUCTION

Quantum effects occur in all physical systems, when typical wavelengths become of the order of macroscopic length scales. A simple system to illustrate this, is the scattering of light through a slit, where for wavelengths smaller than the slit diameter we have the geometrical optics limit. For wavelengths of the order of the slit diameter or larger we encounter wave phenomena like diffraction which are due to interference. In these lectures we will focus on non-stationnary, non-equilibrium phenomena in quantum systems which can be characterized by their transport properties. These are expressed by transport coefficients like viscosity, heat conductivity and the diffusion constant. It is well-known that most classical systems show diffusive behaviour, i.e. the time evolution of the number density is governed by the diffusion equation. A prime example is Brownian motion, a heavy particle moving through a system of random scatterers.

The first and very intriguing result for quantum systems was obtained by Anderson[1], who demonstrated 'Absence of diffusion in certain random lattices'. While diffusion implies that the initial state spreads through the whole system leading to an extended state, the absence of diffusion implies localization. If we denote the probability amplitude for occupying a site k as $f_k(t)$ with initial condition $f_k(t) = 1$ at $t = 0$, then the time-dependent Schrödinger equation reads :

$$i\frac{\partial}{\partial t}f_k(t) = E_k f_k + \sum_{k \neq j} V_{kj} f_j .$$

(1)

Diffusion corresponds to the situation where $\lim_{t \to +\infty} f_k(t) = 0$ and the occupation amplitude diffuses away through energy-conserving interactions. Localization takes place when $\lim_{t \to +\infty} f_k(t) = \frac{1}{1+K}$, with K measuring how much the amplitude of the state has spread to neighbouring states through virtual collisions. If we denote $|V|$

as the strength of the interaction and ΔE as the amount of randomness or energy bandwidth, we have the following situations: if $\frac{\Delta E}{|V|}$ is small we have many possible energy-conserving transitions and thus diffusion; if $\frac{\Delta E}{|V|}$ is large, the nearby states are very different in energy and only virtual transitions can occur, leading to localization; if we are in an intermediate situation we may have diffusion through quantum jumps between localized states[1].

Since Anderson's observation there is considerable interest in quantum aspects of propagation, multiple scattering and localization of particles and waves in disordered media. Especially, the importance of quantum-interference effects for the transport properties has been highlighted. In their pioneering work, Vollhardt and Wölfle[2] have proposed a self-consistent diagrammatic approach (SCDA) for the diffusion constant and demonstrated the occurrence of localization based on the inclusion of maximally-crossed diagrams. This approach was first developed for electron transport, but has also been applied for classical waves[3] and, particularly interesting, for the study of localization of light[4].

THE BOLTZMANN EQUATION

In classical physics transport properties are commonly described[5] by the Boltzmann equation for the one-particle distribution function $f(\vec{r}, \vec{p}, t)$. This distribution function represents the occupation probability of the phase-space element at $(\vec{r}, \vec{p})$ at time t:

$$\frac{\partial f}{\partial t} + \frac{\partial \vec{r}}{\partial t}\frac{\partial f}{\partial \vec{r}} + \frac{\partial \vec{p}}{\partial t}\frac{\partial f}{\partial \vec{p}} =$$
$$\int \int \int d^3 p_1' d^3 p_2' d^3 p_2 W(\vec{p}\vec{p_2}|\vec{p_1}'\vec{p_2}')[f(\vec{r},\vec{p_1}',t)f(\vec{r},\vec{p_2}',t) - f(\vec{r},\vec{p},t)f(\vec{r},\vec{p_2},t)] (2)$$

The left-hand side of the equation corresponds to mean-field propagation by virtue of Hamilton's equations $\frac{\partial \vec{r}}{\partial t} = \frac{\partial h}{\partial \vec{p}}$ and $\frac{\partial \vec{p}}{\partial t} = -\frac{\partial h}{\partial \vec{r}}$, with h the single-particle hamiltonian energy which may itself be a functional of f as in the Vlasov equation[5]. In the conventional Boltzmann equation h is equal to the kinetic energy of the particle since it is assumed that the particle moves unperturbed in between collisions. On the right-hand side of the Boltzmann equation we have the collision term (2-body collisions only) with collision probability W, which is related to the differential cross section for the process $\vec{p} + \vec{p_2} \to \vec{p_1}' + \vec{p_2}'$ and includes energy-momentum conservation.

What are the criteria of validity? Consider the following typical length scales. First we have r_0, the average distance between the scatterers determined by the density n; then there is the interaction range a; the mean free path $l = \frac{1}{n\sigma}$ with σ the total cross section, describes the mean distance a particle can travel without collisions; finally we have the wavelength λ which is related to the particle's momentum p. The Boltzmann equation is valid under the condition:

$$\lambda < a < r_0 < l, \tag{3}$$

which implies a classical dilute system with short-range interactions, also called the weak-coupling limit. The collisions are local in time and space (i.e. Markovian) and the distribution function is ensemble-averaged (or coarse-grained). The ensemble average can be taken over the different samples of the same initial condition (being a random distribution of scatterers). In order to obtain the Boltzmann equation, the molecular chaos assumption has to be applied which involves the following approximations in the ensemble-averaging procedure:

$$\langle h(f) \rangle \rightarrow h(\langle f \rangle) , \tag{4}$$

$$\langle W f_1 f_2 \rangle \rightarrow W \langle f_1 \rangle \langle f_2 \rangle . \tag{5}$$

The Boltzmann transport equation has the well-known Maxwell-Boltzmann distribution function as its equilibrium solution, i.e. when the collision term is zero. There is then a balance between the collisions 'out' and the ones 'in'. We denote the equilibrium solution as $f^{(0)}$.

The Diffusion Constant

From the Boltzmann equation the macroscopic conservation laws for the particle density and current, the momentum density and current, and the energy density and current[5] can be deduced. These are simply obtained by multiplying Eq. (2) with $1, \vec{p}, h(p)$ respectively and integrate over $\vec{p}$. For the particle density we find:

$$\frac{\partial n(\vec{r}, t)}{\partial t} + \vec{\nabla}_r \vec{J}(\vec{r}, t) = 0 , \tag{6}$$

with the particle density $n(\vec{r}, t)$ and the current $\vec{J}(\vec{r}, t)$:

$$n(\vec{r}, t) \equiv \int d^3p f(\vec{r}, \vec{p}, t) , \tag{7}$$

$$\vec{J}(\vec{r}, t) \equiv \int d^3p (\frac{\vec{p}}{m}) f(\vec{r}, \vec{p}, t) . \tag{8}$$

The diffusion constant D is defined by Fick's law:

$$\vec{J}(\vec{r}, t) = -D \vec{\nabla}_r n(\vec{r}, t) , \tag{9}$$

which relates a 'disturbance' in the particle density to the resulting current. Combining Eq. (9) with the conservation law Eq. (6) results in the diffusion equation:

$$\frac{\partial n(\vec{r}, t)}{\partial t} - D \nabla_r{}^2 n(\vec{r}, t) = 0 . \tag{10}$$

The diffusion constant will in general be a function of $\vec{r}$ and t. Its macroscopic limit is obtained for $\vec{r}, t \rightarrow +\infty$.

We will now derive an expression for the diffusion constant using the Boltzmann equation. First we rewrite the equation using the 'relaxation-time' approximation for the collision term in Eq. (2):

$$\frac{\partial f}{\partial t} + \frac{\partial h}{\partial \vec{p}} \frac{\partial f}{\partial \vec{r}} - \frac{\partial h}{\partial \vec{r}} \frac{\partial f}{\partial \vec{p}} = -\frac{(f - f^{(0)})}{\tau} . \tag{11}$$

In this equation the expression $\frac{\partial h}{\partial \vec{p}}$ can be recognized as the 'group velocity' $\vec{v}_{gr}$ and τ corresponds to the relaxation time which we will specify later. We can now eliminate the gradient of the density in Eq. (9) by virtue of Eq. (11) and obtain for D:

$$D = -\frac{\vec{J}(\vec{r}, t)}{\vec{\nabla}_r n(\vec{r}, t)} = -\frac{\int d^3p (\frac{\vec{i} \cdot \vec{p}}{m}) \bar{f}}{\int d^3p (\vec{i} \cdot \vec{\nabla}_r) \bar{f}} = \frac{\int d^3p (\frac{\vec{i} \cdot \vec{p}}{m}) \bar{f}}{\int d^3p \frac{1}{(\vec{i} \cdot \vec{v}_{gr}) \tau} \bar{f}} , \tag{12}$$

with

$$\bar{f} = f - f^{(0)} . \tag{13}$$

101

In the macroscopic limit only the radial component of D remains ($\vec{i}$ is the unit vector in the radial direction). In deriving the result Eq. (12) it is assumed that $\vec{v}_{gr}$ is independent of $\vec{r}$ and t.

We now proceed to derive explicitly the expression for the relaxation time τ. The collision term which we denote as $(\frac{\partial f}{\partial t})_{coll}$ has the familiar form[5]:

$$(\frac{\partial f}{\partial t})_{coll} = n_s \int d^3 p' \delta(\epsilon_{p'} - \epsilon_p)[\mid T_{\vec{p}\vec{p}'} \mid^2 f(\vec{p}) - \mid T_{\vec{p}\vec{p}'} \mid^2 f(\vec{p}')] . \qquad (14)$$

Here n_s represents the concentration of fixed scatterers and the test particle under consideration scatters elastically from momentum state $\vec{p}$ to $\vec{p}'$ (and vice-versa) with probability $\mid T_{\vec{p}\vec{p}'} \mid^2$ where $T_{\vec{p}\vec{p}'}$ is the corresponding T-matrix. ϵ_p is the kinetic energy of the test particle. Writing $f(\vec{p})$ as $f^{(0)}(p) + \vec{p}.\vec{i}_z C(p)$ the expression for the relaxation time becomes:

$$\frac{1}{\tau} = -n_s \int d^3 p' \delta(\epsilon_{p'} - \epsilon_p) \mid T_{\vec{p}\vec{p}'} \mid^2 (1 - cos\theta_{pp'}) . \qquad (15)$$

The result for the diffusion constant D as given by Eq. (12), with the relaxation-time expression Eq. (15), has the familiar form $\frac{vl_{tr}}{3}$ with v the average velocity and l_{tr} the transport mean free path. The latter is not the same as the mean free path l discussed before. It is the average length over which the memory of the direction of the particle velocity remains preserved.

QUANTUM FORMALISM

What are the consequences of quantum effects for the diffusion process? Let us first explain which effects we are considering here. They are two-fold: first there is the Pauli principle for either bosons or fermions; secondly interference effects between waves comes into play. While the first constraint can be incorporated heuristically in the Boltzmann equation, leading to the so-called BUU-equation or Boltzmann-Uehling-Uhlenbeck equation[5], this cannot be achieved at all for the second aspect. Therefore, a completely different formalism is called for. Before introducing this formalism let us first reconsider the criteria mentionned before in relation to the validity of the Boltzmann equation. The major difference now is the fact that the wavelength λ becomes larger than some or all of the other relevant length scales. Thus a scattering wave may interact coherently with many scatterers and a scattering wave leaving one scatterer will not have reached its asymptotic limit when undergoing subsequent collisions. The general problem to analyse is, how phase-relations are maintained in the scattering process. If we associate a typical length scale to the phase-braking phenomena, then we are in the so-called mesoscopic regime if this phase-breaking length scale is of the order or larger than the macroscopic size of the system.

Where do the essential differences between the classical and the quantum situation arise? Consider a physical system with eigenstates $|\vec{k}>$. Classically, the probability that a particle occupies a state $|\vec{k}>$ is given by the distribution function $f(\vec{k})$ which we discussed before. Quantummechanically, the state of a particle is given by a super-position $\sum_{\vec{k}} \Psi_{\vec{k}} |\vec{k}>$ and the probability for occupying a state $|\vec{k}>$ is simply $\mid \Psi_{\vec{k}} \mid^2$. However the wave function $\Psi_{\vec{k}}$ contains more information like its phase which is relevant in case of interference. The appropriate quantity now is the so-called density matrix $\rho(\vec{k} \mid \vec{k}') = \Psi_{\vec{k}} \Psi_{\vec{k}'}^*$ where the diagonal matrix elements are related to the classical distribution function and classical probabilities. While the Boltzmann equation describes the dynamical evolution of the diagonal elements only, a genuine quantum formalism should take care of the full matrix structure including the phase information. Then the evolution and interaction of both the amplitudes and phases together have to be followed in time and space.

Green's Functions

In the following sections, describing the basic formalism we will follow closely references 9. and 15. We describe our system (considered here to be fermions) by second-quantized field operators in Heisenberg representation, $\psi^+(\vec{r}, t)$, and their hermitian conjugates $\psi(\vec{r}, t)$. The Heisenberg representation ensures that the full time dependence is incorporated in the field operators and not in their expectation values which are needed to study observables. These field operators of which $\psi^+(\vec{r}, t)$ corresponds to particle creation and $\psi(\vec{r}, t)$ to particle annihilation, satisfy equal-time anti-commutation relations $[\psi(\vec{r}, t), \psi^+(\vec{r}', t)] = \delta(\vec{r} - \vec{r}')$. If we restrict ourselves to two-particle interactions only, then the hamiltonian H can be written in terms of the field operators as:

$$
\begin{aligned}
H = {} & \int d^3 r \, \psi^+(\vec{r}, t)(-\frac{\hbar^2 \nabla^2}{2m})\psi(\vec{r}, t) \\
& + \frac{1}{2} \int d^3 r d^3 r' \, \psi^+(\vec{r}, t)\psi^+(\vec{r}', t)V(\vec{r} - \vec{r}')\psi(\vec{r}', t)\psi(\vec{r}, t) .
\end{aligned}
\tag{16}
$$

The equation of motion of an operator $A(t)$ in Heisenberg representation is given by $i\hbar \frac{\partial}{\partial t} A(t) = [A(t), H]$ from which the equations of motion for the field operators can be obtained:

$$
(i\hbar \frac{\partial}{\partial t} + \frac{\hbar^2 \nabla^2}{2m})\psi(\vec{r}, t) = \int d^3 r' V(\vec{r} - \vec{r}')\psi(\vec{r}', t)\psi(\vec{r}, t)\psi^+(\vec{r}', t) .
\tag{17}
$$

and a similar equation for $\psi^+(\vec{r}, t)$.

The fundamental quantities for the theory to be set up are Green's functions which are directly related to observables. These functions are expectation values of bilinear products of field operators, plus some time-ordering convention because the field operators do not commute. Let us first consider the conventional causal Green's function:

$$
g^c(1, 1') = \frac{1}{i\hbar} < T^c[\psi(1)\psi^+(1')] > .
\tag{18}
$$

Here T^c represents the usual chronological time-ordering operator, and the expectation value is related to the ensemble of initial conditions. The label 1 indicates $\vec{r}_1, t_1$. In a general non-equilibrium situation as considered here, the time-evolution of the fields in forward or backward time-direction is not the same. In order to generalize Eq. (18) to non-equilibrium it will be convenient to introduce four different Green's functions taking into account the different directions of time evolution of the field operators, following a technique originally proposed by Schwinger[6] and developed further by Keldysh[7] and Craig[8]. Since we have to distinguish fields given their direction of temporal evolution, we define the fields to be situated on a directed contour $C(t)$ in time. This contour has two branches, one called the upper one (denoted by $(+)$) and where time is directed from $-\infty$ to $+\infty$, the other branch called the lower one (denoted by $(-)$) and where time is directed from $+\infty$ to $-\infty$. The directed time contour $C(t)$ connects both and starts on the upper branch at $-\infty$ and continues on to $+\infty$, to return by the lower branch from $+\infty$ to $-\infty$. When considering a bilinear product of these fields it is clear that we can distinguish four different combinations. These are grouped together in one general Green's-function *matrix* $\underline{G}$ defined on the contour $C(t)$ as:

$$\underline{G}(1,1') = \frac{1}{i\hbar} < P[\psi(1)\psi^+(1')] > .\tag{19}$$

The path-ordering operator P defined on $C(t)$ orders the field operators in such a way that field operators further on the contour $C(t)$ are put to the left of those less further, and there is a sign change involved for every permutation of the field operators necessary to achieve the proper ordering. The Green's-function matrix $\underline{G}$ for fermions has the following structure:

$$\underline{G}(1,1') = \begin{pmatrix} G_{++}(1,1') & G_{+-}(1,1') \\ G_{-+}(1,1') & G_{--}(1,1') \end{pmatrix} ,\tag{20}$$

with

$$G_{++}(1,1') \equiv g^c(1,1') = \frac{1}{i\hbar} < T^c[\psi(1)\psi^+(1')] >\tag{21}$$

$$G_{--}(1,1') \equiv g^a(1,1') = \frac{1}{i\hbar} < T^a[\psi(1)\psi^+(1')] >\tag{22}$$

$$G_{-+}(1,1') \equiv g^>(1,1') = \frac{1}{i\hbar} < [\psi(1)\psi^+(1')] >\tag{23}$$

$$G_{+-}(1,1') \equiv g^<(1,1') = \frac{-1}{i\hbar} < [\psi^+(1')\psi(1)] > .\tag{24}$$

where we recognize again the causal Green's function g^c and the acausal one g^a, involving the anti-chronological time-ordering operator T^a. The $(++), (--), (-+)$ and $(+-)$ subscripts are associated with the location of the field operators on the time-directed branches and specify the time direction. The two off-diagonal Green's-function matrix elements, denoted as $g^>$ and $g^<$ are called the correlation Green's functions. They do not have time-ordering operators since the fields are either on the upper $(+)$ or lower branch $(-)$ and thus ordered on $C(t)$ by definition. The correlation Green's functions represent the central quantities in the formalism as we can see for instance through the relation:

$$\lim_{t'_1 \to t_1} g^<(1,1') = \frac{-1}{i\hbar} \rho(\vec{x}_1, \vec{x}'_1, t_1) ,\tag{25}$$

with ρ the density matrix.

In order to understand better the physical meaning of the different Green's functions introduced so far, let us consider the case of non-interacting particles in equilibrium. In an infinite homogeneous system the Green's functions depend only upon the difference of the arguments $\vec{r}_1 - \vec{r}'_1, t_1 - t'_1$ which are then Fourier-transformed (We denote the Fourier transformed variables as $p = \vec{p}, \omega$ respectively). By substituting the appropriate expressions for the field operators , which can be found in any textbook on field theory[10] the following result can be derived:

$$i\hbar g^<(p) = -2\pi\hbar n(p)\delta(\omega - \epsilon_p)\tag{26}$$

$$i\hbar g^>(p) = 2\pi\hbar[1 - n(p)]\delta(\omega - \epsilon_p)\tag{27}$$

$$g^c(p) = \frac{1 - n(p)}{\omega - \epsilon_p + i\eta} + \frac{n(p)}{\omega - \epsilon_p - i\eta}\tag{28}$$

$$g^a(p) = \frac{1 - n(p)}{\omega - \epsilon_p - i\eta} + \frac{n(p)}{\omega - \epsilon_p + i\eta} .\tag{29}$$

Here $\epsilon_p = \frac{p^2}{2m}$ is the single-particle energy, $n(p)$ is the momentum-occupation density which can be shown to be a Fermi-Dirac distribution function and the limit $\eta \to 0^+$ is assumed. We see that $g^<$ is proportional to the occupation number of particles times a delta-function. The latter tells us that the energy parameter ω equals the kinetic single-particle energy. This delta-function corresponds to the non-interacting limit of the so-called spectral function, which describes the weigth associated with a particular choice of the momentum $\vec{p}$ and the energy parameter ω. We use here the terminology 'energy parameter' since ω can be positive as well as negative. The Green's function $g^>$ is proportional to the occupation number of holes (the absence of particles) times the spectral function.

Dyson equations and Selfenergies

From the equations of motion for the field operators we can derive the corresponding equations for the different Green's functions. Inspecting Eq. (17) we can easily understand that any equation for a particular g will mix with all other g of the matrix $\underline{G}$, and furthermore will contain higher-order Green's functions i.e. two-particle Green's functions through the interaction terms. It can be shown[9] in a straigthforwardly manner that the equations for the Green's-function matrix $\underline{G}$ have the following general form:

$$s(1)\underline{G}(1,1') = \underline{\delta}(1-1') - i\hbar \int_C d2d1''d2'' <12|\underline{V}|1''2''> \underline{G}(1''2,1'2^+), \quad (30)$$

$$s^*(1')\underline{G}(1,1') = \underline{\delta}(1-1') - i\hbar \int_C d2d1''d2''\underline{G}(12^-,1''2'') <1''2''|\underline{V}|1'2>, \quad (31)$$

where $s(1) = (i\hbar\partial_{t_1} + \frac{\hbar^2\nabla_1^2}{2m})$ and $\underline{\delta}(\text{time}) = \delta(\text{time})$ if both arguments $(t_1,t_{1'})$ are on the upper branch of C, $\underline{\delta}(\text{time}) = -\delta(\text{time})$ if both arguments $(t_1,t_{1'})$ are on the lower branch of C, and $\underline{\delta}(\text{time}) = 0$ otherwise. The $+$ and $-$ superscripts indicate a time infinitesimally later or earlier on the contour than implied by the time variable involved.

The integrals in Eqs. (30,31) are over space and time and the time integrals are over the full directed contour C(t), combining upper and lower branch integrations. These are selected by the appropriate components of $\underline{\delta}$. Remark that an integration over the lower branch runs from $+\infty$ to $-\infty$. We can formally rewrite the equations for $\underline{G}$ as Dyson equations, introducing the selfenergy or mass operator $\underline{\Sigma}$ as follows:

$$s(1)\underline{G}(1,1') = \underline{\delta}(1-1') + \int_C d1''\underline{\Sigma}(1,1'')\underline{G}(1'',1'), \quad (32)$$

$$s^*(1')\underline{G}(1,1') = \underline{\delta}(1-1') + \int_C d1''\underline{G}(1,1'')\underline{\Sigma}(1'',1'). \quad (33)$$

By comparing Eqs. (32,33) with Eqs. (30,31) we obtain the definition of the selfenergy $\underline{\Sigma}$. This selfenergy $\underline{\Sigma}$ is also a two by two matrix with components Σ^c, Σ^a, $\Sigma^<$ and $\Sigma^>$. It is defined in terms of one-particle Green's functions and two-particle Green's functions. Equations for the latter can be derived[9] by combining again the equations of motion for the field operators. It results in a nested hierarchy of increasingly higher-order Green's functions. This set cannot be solved and one must instead try to find a suitable and physically acceptable approximation for the two-particle Green's function as a functional of the interaction and the one-particle Green's function itself. This leads to a particular approximation for the selfenergy $\underline{\Sigma}$ expressed in the interaction V and the one-particle Green's function $\underline{G}$. Finally it results in a self-consistent set of non-linear equations for the one-particle Green's function exclusively. We will not discuss any specific approximation for the selfenergy but refer to the litterature instead[9,11,12,13].

As an illustration of the Dyson equations let us write out one of them, i.e. the equation for $g^<(1, 1')$:

$$s(1)g^<(1, 1') = \int_{-\infty}^{+\infty} d1''[\Sigma^c(1, 1'')g^<(1'', 1') - \Sigma^<(1, 1'')g^a(1'', 1')] . \quad (34)$$

Here we see clearly that the different components are mixed together. One can however define new quantities for wich the Dyson equations take on a diagonal form, i.e. no mixing occurs. These quantities are the retarded and advanced Green's functions $g^{(+)}$, $g^{(-)}$ and selfenergies $\Sigma^{(+)}$, $\Sigma^{(-)}$ and are defined in terms of the other components as:

$$F^{(+)} \equiv F^c - F^> \qquad F^{(-)} \equiv F^c - F^< , \quad (35)$$

with F either a Green's function or selfenergy. The associated Dyson equations are:

$$s(1)g^{(+)}(1, 1') = \delta(1 - 1') + \int_{-\infty}^{+\infty} d1'' \Sigma^{(+)}(1, 1'')g^{(+)}(1'', 1') , \quad (36)$$

with a similar equation for $g^{(-)}$. The retarded and advanced quanties have the property that they are zero if $t_1 - t_{1'} < 0$ or $t_1 - t_{1'} > 0$ respectively.

The relation of the different Green's functions with observables can be found by first defining these observables in terms of the field operators which create or annihilate particles and holes. For instance for the total number of particles $N(t)$ and the total energy $E(t)$ we have[9]:

$$N(t) = -i\hbar Tr_{(1)}g^<(t, t) , \quad (37)$$

$$E(t) = -\frac{1}{4}i\hbar Tr_{(1)}[\; [i\hbar(\partial_{t_1} - \partial_{t_{1'}}) + 2K_1] \; g^<(t_1, t_{1'}) \;]_{t_1 = t_{1'} = t} , \quad (38)$$

Where K_1 denotes the kinetic energy and the $Tr_{(1)}$ is taken over coordinate space only. By using the Dyson equations we can express the observables defined in Eqs. (37,38) directly in terms of one-particle Green's functions and selfenergies[9].

THE KADANOFF-BAYM EQUATION

The Wigner transformation and the gradient approximation

The Dyson equations are very general but highly impractical. As they are, they take the full non-local behaviour in space and time into account. If we are in a situation where the system has a smooth behaviour in space and time, for instance at not to high densities, we can expand around a local space-time point up to first order in the gradients. This can be achieved by working in the Wigner-representation. Let $h(1, 1')$ be a one-body function, such as the one-particle Green's function or selfenergy. Introduce new variables $\vec{R}, T$ and $\vec{r}, t$:

$$\vec{R} = \frac{1}{2}(\vec{r_1} + \vec{r_{1'}}) \qquad\qquad T = \frac{1}{2}(t_1 + t_{1'}) \quad (39)$$

$$\vec{r} = (\vec{r_1} - \vec{r_{1'}}) \qquad\qquad t = (t_1 - t_{1'}) \quad . \quad (40)$$

The Wigner-transform h_W of h is defined by Fourier transforming only the relative variables. Using a 4-vector notation $R = \vec{R}, T$, $r = \vec{r}, t$ and $p = \vec{p}, \omega$, we have:

$$h_W(p, R) = \int d^4r e^{\frac{ipr}{\hbar}} h(R + \frac{1}{2}r, R - \frac{1}{2}r) , \quad (41)$$

and the 'classical' Wigner distribution function $f(\vec{p}, \vec{R}, T)$ is defined as :

$$f(\vec{p}, \vec{R}, T) = -i\hbar \int \frac{d\omega}{2\pi} g^<(p, R) . \tag{42}$$

If we would calculate $g^<(p, R)$ in an approximation where all quantum effects are neglected then the Wigner distribution function $f(\vec{p}, \vec{R}, T)$ corresponds to the classical distribution function and obeys the Boltzmann equation. We will demonstrate later that this is the case.

In the following, $\vec{R}$ and T are considered as the local variables around which we will expand. In the Dyson equations Eqs. (34,36), we observe that products of Green's functions and selfenergies appear. Wigner-transforming these products and expanding up to first order around R the following result is obtained;

$$H(x, x') = \int dx'' \Sigma(x, x'') g(x'', x') , \tag{43}$$

gives

$$\begin{aligned} H_W(p, R) &= \Sigma_W(p, R) g_W(p, R) + \\ &\quad \frac{1}{2} i\hbar [\partial_p \Sigma_W(p, R) \cdot \partial_R g_W(p, R) - \partial_R \Sigma_W(p, R) \cdot \partial_p g_W(p, R)] . \end{aligned} \tag{44}$$

The second term on the right-hand side is written as a four-vector Poisson bracket:

$$\{\Sigma, g\} = [\partial_p \Sigma \cdot \partial_R g - \partial_R \Sigma \cdot \partial_p g] . \tag{45}$$

The dot in Eqs. (44,45) implies a four-vector inproduct. In the following we will ommit the subscript W.

The Kadanoff-Baym equation

It is now straightforward to obtain the quantum transport equation in the case where the gradient approximation can be applied. From the Dyson equations we construct $s(1)g^<(1, 1') - s^*(1')g^<(1, 1')$, apply the Wigner-transformation and use the gradient approximation Eq. (44). This leads to the result:

$$i\hbar \partial_T g^<(p, R) + i\hbar \frac{\vec{p}}{m} . \partial_R g^<(p, R) - i\hbar \{\mathrm{Re}\Sigma^{(+}(p, R), g^<(p, R)\}$$
$$-i\hbar \{\Sigma^<(p, R), \mathrm{Re} g^{(+)}(p, R)\} = \Sigma^>(p, R) g^<(p, R) - \Sigma^<(p, R) g^>(p, R) \tag{46}$$

and an analogous equation[9] for $g^>(p, R)$. This equation is called the Kadanoff-Baym equation, since it has been proposed first by these authors in their famous textbook[13]. The right-hand side of the equation corresponds to the collision term.

The spectral properties of the system are determined by the spectral function $a(p, R)$ which is defined equivalently as:

$$\begin{aligned} a(p, R) &= i\hbar [g^>(p, R) - g^<(p, R)] \\ &= i\hbar [g^{(+)}(p, R) - g^{(-)}(p, R)] . \end{aligned} \tag{47}$$

The spectral function a can be calculated in the gradient approximation, either based on the first line (using Eq. (46) and the corresponding one for $g^>$), or based on the second line, which lead to apparently different expressions. However these can be shown to be identical[13]. The Wigner-transformed equation corresponding to Eq. (36) for $g^{(+)}$ and $g^{(-)}$ has the simple solution:

$$g^{(\pm)}(p, R) = [\omega - \frac{\vec{p}^2}{2m} - \Sigma^{(\pm)}(p, R) \pm i\eta]^{-1} , \tag{48}$$

which enables us to write down an explicit expression for the spectral function:

$$a(p, R) = \frac{-2\hbar \text{Im}\Sigma^{(+)}(p, R)}{\left[\omega - \frac{p^2}{2m} - \text{Re}\Sigma^{(+)}(p, R)\right]^2 + \left[\text{Im}\Sigma^{(+)}(p, R)\right]^2} \ . \tag{49}$$

The spectral function $a(p, R)$ obeys a very important sum-rule which is a consequence of the commutation relations for the field equations. In fact it can be shown that:

$$\lim_{t' \to t} a(r, r') = \delta^3(\vec{r} - \vec{r}') \ , \tag{50}$$

which in the Wigner-representation results in:

$$\int \frac{d\omega}{2\pi} a(p, R) = 1 \ . \tag{51}$$

Another important property is associated with the retarded and advanced character of any function $F^{(\pm)}$ defined as in Eq. (35). In Wigner-representation it implies that $F^{(\pm)}(\omega)$ is analytic in the upper, respectively lower half complex ω-plane. This can be most easily seen by using the integral representation in ω of the corresponding θ-functions in time. The analytic properties lead to dispersion relations between the real and imaginary parts of $F^{(\pm)}(\omega)$. For instance , the selfenergies obey the dispersion relation:

$$\text{Re}\Sigma^{(+)}(\omega) = \frac{P}{\pi} \int d\omega \frac{\text{Im}\Sigma^{(+)}(\omega)}{\omega' - \omega} \ . \tag{52}$$

where P stands for principal value. Under the assumption that we know how to calculate the different selfenergy contributions in the Eqs. (46,49), we have the full set of quantum transport equations available in the gradient approximation. The 'classical' distribution function f can be obtained through an integration, Eq. (42), using the solution of Eq. (46). However, we cannot construct a closed-form equation for the 'classical' distribution function itself since the integration $\int \frac{d\omega}{2\pi}$ cannot be transported directly onto $g^<$. This particular property is linked to the quantum aspects of the Kadanoff-Baym equations[15]. Therefore the Boltzmann equation, being an equation for the classical distribution function, must involve additional approximations.

The quasi-particle approximation and the Boltzmann equation

When discussing the Green's functions in the non-interacting case we remarked that the spectral function reduces to a δ-function. Our result for the spectral function in the quantum case, Eq. (49), shows a Lorentzian behaviour. In the limit that $\text{Im}\Sigma^{(+)}(\omega) \to 0$, we recover the δ-function. This limit is called the quasi-particle limit because it means that the 'life-time' for propagation is infinite. However this limit cannot be applied as such on the quantum transport equation as a whole because the collision term depends on $\text{Im}\Sigma^{(+)}$. The limit is applied only on the left-hand side of the equation based on the observation that, compared to the right-hand side, it is of higher order in the gradient expansion. This procedure implies that the $\{\Sigma^<, \text{Re}g^{(+)}\}$ term is zero as well, in order to achieve consistency in the two equivalent definitions of the spectral function a Eq. (47). Finally, it means that the ω-dependent part of $\text{Re}\Sigma^{(+)}(\omega)$ is zero in order to obey the dispersion relation Eq. (52). We will use the notation U for the ω-independent part of $\text{Re}\Sigma^{(+)}(\omega)$.

Applying the quasi-particle limit as outlined above we can now perform the integration $\int \frac{d\omega}{2\pi}$ on the transport equation Eq. (46), and obtain the Boltzmann equation for the classical distribution function $f(\vec{p}, \vec{R}, T)$ in the form:

$$[\frac{\partial}{\partial T} f(\vec{p}, \vec{R}, T) + \vec{\nabla}_R \epsilon(\vec{p}, \vec{R}, T) . \vec{\nabla}_p f(\vec{p}, \vec{R}, T) - \vec{\nabla}_p \epsilon(\vec{p}, \vec{R}, T) . \vec{\nabla}_R f(\vec{p}, \vec{R}, T)] =$$

$$- \int \frac{d\omega}{2\pi} (\Sigma^>(p, R) g^<(p, R) - \Sigma^<(p, R) g^>(p, R)) \; ;$$

$$\epsilon(\vec{p}, \vec{R}, T) = \frac{\vec{p}^2}{2m} + U(\vec{p}, \vec{R}, T) \; . \tag{53}$$

While the quasi-particle limit leads to the classical Boltzmann equation, some of its input might still be calculated quantummechanically.

In order for the quasi-particle limit to be valid, the limit $\text{Im}\Sigma^{(+)}(\omega) \to 0$ should be applicable. This is in fact determined by the behaviour of the spectral function a Eq. (47). If $\text{Im}\Sigma^{(+)}(\omega)$ is small compared to $\frac{\vec{p}^2}{2m} + \text{Re}\Sigma^{(+)}(p, R)$ it is indeed the case. How does this translates into a physical criterium? First of all, since $\text{Im}\Sigma^{(+)}(\omega)$ is related to the life time of a particular occupation of the momentum phase-space element p, it can be expressed in terms of the mean free path l. Secondly, the momentum is related to the wavelength λ, or if we include $\text{Re}\Sigma^{(+)}(p, R)$, to the effective medium-corrected wavelength λ_m. Then our criterium for validity of the quasi-particle approximation is expressed as:

$$\lambda_m < l \; . \tag{54}$$

This criterium is violated for systems where resonances occur (leading to short mean free paths as compared to the wavelength) or for quantum systems (where the wavelength is large compared to the mean free path). The first situation can occur even in classical systems, where the wavelength is relatively short. The second situation can occur also if the mean free path is long compared to interparticle distances, but of the order of the wavelength. This is called the 'mesoscopic' regime, with $\lambda_m \simeq l$ (which is sometimes called the Joffe-Regel criterium[14]).

Beyond the quasi-particle approximation

All local space-time observables can be obtained from the correlation Green's function by integrating the appropriate expressions over d^4p. Physically, the integrated correlation functions contain two ingredients. They can be considered as the integrated sum of complex amplitudes originating from different sources, each having its own frequency and momentum . The spectral function, associated with the first aspect, describes the distribution and weight of the different sources. Secondly there are the characteristics of each source which changes in space-time. These two ingredients can be exhibited explicitly in the correlation Green's functions by using the following substitutions:

$$\begin{aligned} -i\hbar g^<(p, R) &= a(p, R) F(p, R) \\ i\hbar g^>(p, R) &= a(p, R)(1 - F(p, R)) \; . \end{aligned} \tag{55}$$

The distribution function $F(p, R)$ describes the characteristics of each source and $a(p, R)$ is the spectral function or the spectral distribution of the different sources. Both functions a and F depend on $(\vec{p}, \omega)$ and $(\vec{R}, T)$. The spectral function can be obtained directly from Eq. (49) which is expressed solely in terms of the retarded self-energy $\Sigma^{(+)}$ which plays the role of a complex mean field. What about the explicit equation for $F(p, R)$? This equation must be contained within the Kadanoff-Baym transport equation, and can be extracted[9,15] by making two other substitutions for the correlation selfenergies $\Sigma^<$ and $\Sigma^>$:

$$
\begin{aligned}
i\hbar\Sigma^<(p, R) &= 2\hbar\mathrm{Im}\Sigma^{(+)}(p, R)F(p, R) + i\hbar X(p, R) \ , \\
i\hbar\Sigma^>(p, R) &= -2\hbar\mathrm{Im}\Sigma^{(+)}(p, R)(1 - F(p, R)) + i\hbar X(p, R) \ , \\
a(p, R)X(p, R) &\equiv -i\hbar(\Sigma^>(p, R)g^<(p, R) - \Sigma^<(p, R)g^>(p, R)) \ .
\end{aligned}
\tag{56}
$$

Using the substitutions Eqs. (55,56) in the quantum transport equation Eq. (46) we obtain[9,15] after some manipulations:

$$
a[(1 - \mathrm{Re}\alpha)\{\mathrm{Re}(g^{(+)})^{-1}, F\} + \mathrm{Im}\alpha\{\mathrm{Im}(g^{(+)})^{-1}, F\}] = \frac{aX}{i\hbar} + i\hbar\{X, \mathrm{Re}g^{(+)}\} \ , \tag{57}
$$

with

$$
1 - \alpha = 2i\mathrm{Im}(g^{(+)})^{-1}(g^{(+)})^* \ , \qquad (\mathrm{Re}\alpha)^2 + (\mathrm{Im}\alpha)^2 = 1 \ . \tag{58}
$$

The explicit expressions for $1 - \mathrm{Re}\alpha$ and $\mathrm{Im}\alpha$ are:

$$
\begin{aligned}
1 - \mathrm{Re}\alpha &= \frac{2[\Sigma^{(+)}(p, R)]^2}{\left[\omega - \frac{p^2}{2m} - \mathrm{Re}\Sigma^{(+)}(p, R)\right]^2 + [\mathrm{Im}\Sigma^{(+)}(p, R)]^2} \ , \\
\mathrm{Im}\alpha &= \frac{2\Sigma^{(+)}(p, R)[\omega - \frac{p^2}{2m} - \mathrm{Re}\Sigma^{(+)}(p, R)]}{\left[\omega - \frac{p^2}{2m} - \mathrm{Re}\Sigma^{(+)}(p, R)\right]^2 + [\mathrm{Im}\Sigma^{(+)}(p, R)]^2} \ .
\end{aligned}
\tag{59}
$$

The weight functions $1 - \mathrm{Re}\alpha$ and $\mathrm{Im}\alpha$ describe the response delay of the mean-field propagation as incorporated in the Poisson brackets on the left-hand side of the Eq. (57). This is a familiar effect in electromagnetic wave propagation in dielectrics, conductors and plasmas, where the response of an applied field is not synchronous but lags behind and energy is absorbed because of correlations. It is called dielectric relaxation and the analogous weight functions are the Debeye functions[16] for the frequency dependent complex dielectric constant. The collision term (right-hand side of Eq. (57)) is non-local and non-Markovian. It contains memory effects due to the finite duration of the collision, as expressed by the Poisson bracket $\{X, \mathrm{Re}g^{(+)}\}$. Equilibrium corresponds to $X(R, p) = 0$. The quantum-transport equation Eq. (57) for $F(p, R)$ is a new and quite interesting result.

Quantum diffusion

We now discuss, as an illustration, the diffusion process based on the quantum transport equation as formulated in the last subsection, and compare with the classical treatment. The procedure to calculate the diffusion constant follows closely the classical one and uses directly its definition Eq. (9) which again implies the Markovian limit[16]. In view of the earlier remarks on this limit this means that we take $\{X, \mathrm{Re}g^{(+)}\} \to 0$ in the quantum transport Eq. (57). The Markovian approximation does not affect conservation laws or causality requirements. It does not imply the quasi-particle limit and still respects important relations like $\Sigma^{(+)} - \Sigma^{(-)} = \Sigma^> - \Sigma^<$. In the non-Markovian situation one has to modify the definition of D, to take into account the non-localities in space and time[16]. The next simplification follows the SCDA-approach[2] where it is assumed that all retarded and advanced quantities, $\Sigma^{(\pm)}$, $g^{(\pm)}$ are independent of $\vec{R}, T$. In our case the Dyson equations for $g^{(\pm)}$ are local in $\vec{R}, T$ and do not contain gradients. On the other hand all correlation quantities $g^<, g^>, \Sigma^<, \Sigma^>$ do depend on $\vec{R}, T$. The dependence on $\vec{p}, \omega$ is fully incorporated in both categories.

The defining relation for D, Eq. (9), should be expressed explicitly into the correlation function $g^<(p,R)$ such that the macroscopic limit $T \to +\infty$, $\vec{R} \to +\infty$ can be performed. Since by definition $\vec{J}(R) = \int d^4p(-i\hbar)\frac{\vec{p}}{m}g^<(p,R)$ and $\vec{\nabla}_R n(R) = \int d^4p(-i\hbar)\vec{\nabla}_R g^<(p,R)$, we have to evaluate the latter gradient using the transport Eq. (57). The explicit expression for D can be found easily[17]. Fourier transforming the time variable T, and denoting the new variable s (now $s \to 0$ in the macroscopic limit), the result is ($\tilde{R} = \vec{R}, s$; $p = \vec{p}, \omega$):

$$D(\tilde{R}) = \frac{\int d^4p \left(\hat{\tilde{R}} \cdot \vec{p}\right)(-i\hbar)g^<(p,\tilde{R})}{\int d^4p \frac{M(p,\tilde{R})}{\hat{\tilde{R}}\cdot\vec{v}_{\vec{p}}}(-i\hbar)g^<(p,\tilde{R})} \ , \tag{60}$$

with

$$\begin{aligned}
M(p,\tilde{R}) &= -iv_\omega s + \frac{1}{i\hbar g^<}(\Sigma^> g^< - \Sigma^< g^>) \\
&= -iv_\omega s + \left(\frac{2}{\hbar}\text{Im}\Sigma^+ + \frac{\Sigma^< a}{\hbar^2 g^<}\right)
\end{aligned} \tag{61}$$

and

$$\begin{aligned}
\vec{v}_{\vec{p}}(p) &\equiv (1 - \text{Re}\alpha)\left(\frac{\vec{p}}{m} + \vec{\nabla}_{\vec{p}}\text{Re}\Sigma^+\right) - \text{Im}\alpha\vec{\nabla}_{\vec{p}}\text{Im}\Sigma^+ \ , \\
v_\omega(p) &\equiv (1 - \text{Re}\alpha)\left(1 - \frac{\partial}{\partial\omega}\text{Re}\Sigma^+\right) - \text{Im}\alpha\frac{\partial}{\partial\omega}\text{Im}\Sigma^+ \ ,
\end{aligned} \tag{62}$$

where $\hat{\tilde{R}}$ denotes the unit-vector in the $\vec{R}$-direction which is arbitrary, because of the ensemble-averaging. The expression for the diffusion constant D is the most general one within the approximations mentioned. The two velocity functions $\vec{v}_{\vec{p}}$ and v_ω can be associated with the group velocity and mass renormalization respectively[18], the latter being dimensionless.

Compared to the classical case, two new effects emerge if we disregard the subtleties involved in the differences between $f(\vec{R},\vec{p},T)$ and $g^<(p,R,T)$, which can only be assessed in a actual calculation. Consequently, our discussion here can only be very qualitative since we do not perform any detailed calculation. The new effects are related to the differences between the group velocities $\vec{v}_{gr}$ and $\vec{v}_{\vec{p}}$, as well as between the inverse relaxation time $\frac{1}{\tau}$ and memory kernel M. This becomes clear if we compare both the classical, Eq. (12) and the quantum expressions, Eq. (60) for the diffusion constant D in the macroscopic limit. There will be absence of diffusion, either if the group velocity becomes zero or if the memory kernel becomes infinite. The first phenomenon may occur in systems where resonances occur, classically corresponding to trapping in the mean-field potential. The second mechanism implies that the 'return' probability for a particle leaving a particular space-time point is enhanced. This can only be achieved if the phase information of the different paths, leading away from the space-time point and back, is conserved and not destroyed in between. Then, constructive interference between outgoing and ingoing paths can take place which will be maximal if both type of paths are related because of time-reversal invariance. In this way the phases are correlated and lead to the desired result. In the classical case, no phase information is present. In the quantum case, for purely elastic scattering, we have correlated phases since only inelastic scattering or time-dependent processes can destroy time-reversal invariance. The enhanced backward scattering because of interference is called weak localization and has been observed experimentally. It does not lead to absence of diffusion but it will reduce diffusion. In order to have absence of diffusion we need to have $D = 0$.

It has been shown by Vollhardt and Wölfle[2] that when the selfenergies occuring in the expression for M, Eq. (61), are evaluated for a particular type of diagrams (so-called maximally-crossed diagrams) the expression for $\Sigma^<$ reduces to a diverging series in two and three spatial dimensions. This is considered to be the equivalent of Anderson localization. The mechanism involved is again backward scattering, where the outgoing and ingoing scattering events are correlated. The ensemble average over the random scatterers is performed after the correlations have been included, otherwise the correlation between the two paths would be lost and they would contribute independently. The correlation in the case of the maximally-crossed diagrams involves the amplitude and the phase of the corresponding correlation Green's function. Performing then a selfconsistent summation of all these diagrams (SCDA, or selfconsistent diagrammatic approximation) produces a divergence in $\Sigma^<$ when the final momentum is opposite to the incoming one. This leads to $D = 0$ and hence localization. We refer to the literature for more details[2,3,4]. A nice introduction can be found in a recent textbook written by Datta[19].

CONCLUSION

In these lectures we have presented elements of Quantum transport theory as formulated in terms of Green's functions and we have discussed diffusion and localization phenomena as an example. Although no actual calculations have been performed, we have been able to show the importance of phase-information and interference effects in transport phenomena. The situation of strongly scattering media is especially interesting since there we expect a very non-classical behaviour. The new expression for the group velocity is our most significant result.

In order to get more quantitative results, the expressions for the selfenergies have to be evaluated. This is not simple, especially since we have to construct appropriate approximations and these should not violate any of the well-established conservation laws or causality requirements[20]. The result will be a highly non-linear set of selfconsistent equations. We know how to construct these in the so-called 'ladder-approximation'[9] and also for the 'maximally-crossed diagrams'[2]. The fact that the latter type introduces anomalies, as enhanced backward scattering, is very interesting and should be explored further. They belong to a class of diagrams which contribute quite differently in the classical or the quantum case due to their correlations. Also, there exists no BUU-type analogue which includes these diagrams. Since correlations are intimately linked to fluctuations, they may be of importance in describing a number of nuclear physics issues such as multi-fragmentation and sub-threshold particle production.

Acknowledgements
This work was performed as part of the research program of the 'Stichting voor Fundamenteel Onderzoek der Materie' (FOM), which is financially supported by the NWO.

REFERENCES

1. P.W. Anderson, Phys.Rev. **109**, 1492 (1958).
2. D. Vollhardt and P. Wölfle, Phys. Rev. B **22**, 4666 (1980).
3. For a recent review see, *Scattering and localization of Classical Waves in Random Media*, ed. Ping Sheng, (World-scientific, Singapore, 1990) and the contributions of : C. A. Condat and T. R. Kirkpatrick; Zhao-Qing Zhang and Ping Sheng.
4. M. P. van Albada, B. A. van Tiggelen, A. Lagendijk and A. Tip, Phys. Rev. Lett. **66**, 3132 (1991); A. Lagendijk and B. A. van Tiggelen, Phys. Rep. **270**, 143 (1996).
5. G.F. Bertsch and S. Das Gupta, Phys. Rep. **160**, 189 (1988). General reference: H.J. Kreuzer, *Nonequilibrium Thermodynamics and its Statistical Foundations*, (Clarendon, Oxford, 1981).
6. J. Schwinger, J. Math. Phys. **2**, 407, (1961).
7. L. V. Keldysh, Zh. Eksp. Teor. Fiz. **47**, 1515 (1964).
8. R.A. Craig, J. Math. Phys. **9**, 605, (1968).
9. W. Botermans and R. Malfliet, Phys. Rep. **198**, 115 (1990).
10. A.A. Abrikosov, L.P. Gorkov and I.E. Dzyaloshinski, *Methods of Quantum Field Theory in Statistical Physics*, (Dover, New York, 1963).
11. P. Danielewicz, Ann. Phys.**152**, 239, (1984).
12. J. E. Davies and R. J. Perry, Phys. Rev. C **43**, 1893 (1991).
13. L. P. Kadanoff and G. Baym, *Quantum Statistical Mechanics*, (Benjamin, New York, 1962).
14. A.F. Ioffe and A.R. Regel, Progress in Semiconductors **4**, 237 (1960).
15. R. Malfliet, Nucl. Phys. **A545**, 3 (1992).
16. R. Kubo, M. Toda and N. Hashitsume, *Statistical Physics*, Vol. 2, (Springer, Berlin, 1991).
17. R. Malfliet, submitted to Phys.Rev.Lett. (1996).
18. J. W. Negele and H. Orland, *Quantum Many-Particle Systems*, (Addison-Wesley, New York, 1988).
19. S. Datta, *Electronic Transport in Mesoscopic Systems*, (Cambridge University Press, Cambridge, 1995).
20. G. Baym and L. P. Kadanoff, Phys. Rev. **127**, 1391 (1962);
    G. Baym, Phys. Rev. **127**, 1391 (1962).

MULTIFRAGMENTATION
IN RELATIVISTIC HEAVY-ION REACTIONS

W. Trautmann

Gesellschaft für Schwerionenforschung mbH
D-64291 Darmstadt
Germany

ABSTRACT

Multifragmentation is the dominant decay mode of heavy nuclear systems with excitation energies in the vicinity of their binding energies. It explores the partition space associated with the number of nucleonic constituents and it is characterized by a multiple production of nuclear fragments with intermediate mass.

Reactions at relativistic bombarding energies, exceeding several hundreds of MeV per nucleon, have been found very efficient in creating such highly-excited systems. Peripheral collisions of heavy symmetric systems or more central collisions of mass asymmetric systems produce spectator nuclei with properties indicating a high degree of equilibration. The observed decay patterns are well described by statistical multifragmentation models.

The present experimental and theoretical studies are particularly motivated by the fact that multifragmentation is being considered a possible manifestation of the liquid-gas phase transition in finite nuclear systems. From the simultaneous measurement of the temperature and of the energy content of excited-spectator systems a caloric curve of nuclei has been obtained. The characteristic S-shaped behavior resembles that of ordinary liquids.

Signatures of critical phenomena in finite nuclear systems are searched for in multifragmentation data. These studies, supported by the success of percolation in reproducing the experimental mass or charge correlations, concentrate on the fluctuations observed in these observables. Attempts have been made to deduce critical-point exponents associated with multifragmentation.

INTRODUCTION

The hope to establish a link to the liquid-gas phase transition in nuclear matter has been a major motivation for the search for and the study of multifragment decays of heavy nuclei in recent years [1, 2]. Multifragmentation was predicted to be the dominant decay mode at excitation energies near the binding energy of nuclei of about 8 MeV per nucleon and at densities below the saturation density of nuclear matter [3, 4]. These conditions of high excitation and low density coincide with the liquid-gas coexistence region as predicted for nuclear matter from the Van-der-Waals type range dependence of the nuclear forces [5, 6]. It was also suggested early on that this region may be explored during the later stages of energetic nuclear reactions [7].

The experimental study of the nuclear liquid-gas phase transition in finite nuclei faces several serious difficulties related to the fact that excited nuclei are composed of a small number of constituents, that they are charged, and that there is no external pressure to counteract the internal pressure of the system at a given equilibrium condition [8]. Finite pressures may be maintained only dynamically and for very short periods of time during the disintegration process. There is also no heat bath available which would allow to predetermine the temperature of the system in order to measure its response to it.

In spite of these difficulties, stimulating new results have been presented very recently. They suggest that signals of the nuclear liquid-gas phase transition may be revealed by studying reactions of finite nuclei. From the simultaneous measurement of the temperature and the excitation energy for excited-projectile spectators in ^{197}Au + ^{197}Au collisions at 600 MeV per nucleon a caloric curve of nuclei has been obtained [9]. It exhibits a typical S-shaped behavior, reminiscent of first-order phase transitions in macroscopic systems. For the ^{197}Au on C reaction at 1.0 GeV per nucleon, the EOS collaboration has reported values of critical-point exponents which were derived from the correlations and fluctuations of the fragment sizes [10]. In both cases, the data providing the basis for the analysis were obtained from the fragmentation of heavy projectiles at relativistic energies in the range of up to about 1 GeV per nucleon. The decay properties of spectator nuclei produced in these reactions indicate that a high degree of equilibrium has been reached. This is a prerequisite for the study of the thermodynamic behavior of highly-excited nuclear matter and makes these reactions rather attractive for this purpose.

In the following notes, some main features of multifragment decays following heavy-ion reactions in the relativistic regime of bombarding energies will be summarized. The experimental material will be mostly taken from the work of the ALADIN collaboration, performed at the heavy-ion synchrotron SIS of the GSI in Darmstadt [11, 12]. The techniques used to determine the observables related to the liquid-gas phase transition will be briefly described. It is not intended, however, to give a complete account of the present discussion initiated by these results. Besides the references cited, the reader is referred to the proceedings of recent conferences or workshops [13-16] during which the topic has been discussed within a wider perspective. Further references on the subject of multifragmentation in general may be found in [17].

EXPERIMENTAL STUDY OF PROJECTILE DECAY

The first observations of multifragment decays of heavy projectiles have been made by exposing nuclear emulsions to the heavy-ion beams [18, 19]. This technique is still being

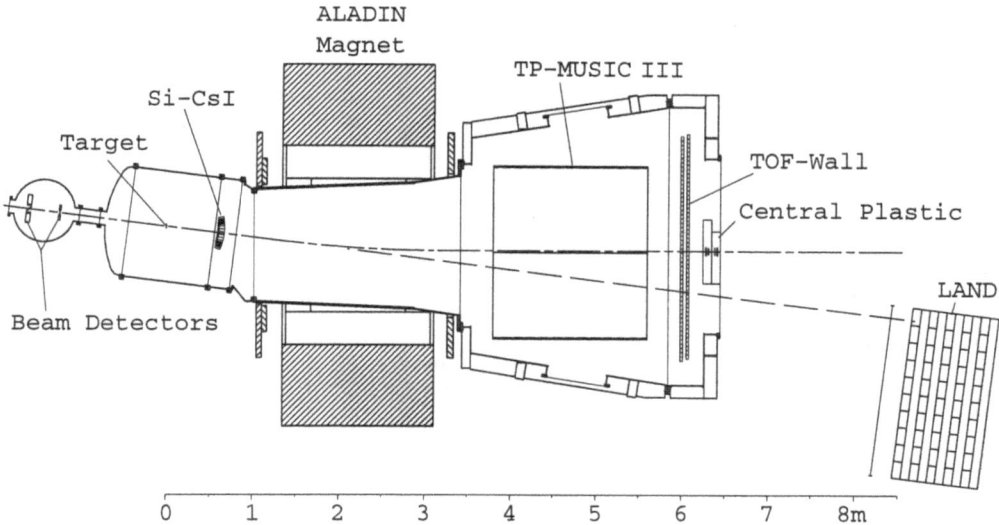

Figure 1. Cross-sectional view of the ALADIN facility in the configuration of the 1993 experiment. The beam enters from the left and is monitored by two beam detectors before reaching the target. Projectile fragments entering into the acceptance of the magnet are tracked and identified in the TP-MUSIC III detector and in the time-of-flight (TOF) wall. The Central Plastic detector covers the hole in the TOF wall at the exit for the beam. Fragments and particles emitted in forward directions outside the magnet acceptance and up to $\theta_{lab} = 16°$ are detected in the Si-CsI array. Neutrons emitted in directions close to $\theta_{lab} = 0°$ are detected with the large-area neutron detector (LAND). The dashed line indicates the direction of the incident beam. The dash-dotted line represents the trajectory of beam particles after they were deflected by an angle of 7.3° (from Ref.[11]).

used and, just recently, has produced first results on the fragmentation of gold nuclei at the energy of 10.6 GeV per nucleon, available from the AGS in Brookhaven[20, 21]. Another technique, employed in several studies[22], is based on plastic nuclear track detectors which have high charge resolution for atomic numbers $Z \geq 6$.

More detailed investigations are possible with electronic detection devices such as the ALADIN spectrometer at SIS[11, 23]. This detector system is built around A Large Acceptance DIpole magNet (ALADIN) with the target in front of the magnetic-field gap and the main detector systems behind it. The configuration used in the 1993 experiment, a systematic study covering a wide range of beams and targets, is shown in Fig. 1. In this experiment, complete acceptance within the kinematic region of projectile decay with good resolution was achieved. The solid angle adjacent to the acceptance of the ALADIN magnet was covered with 84 Si-CsI(Tl) telescopes in closely packed geometry. Behind the magnet, the MUltiple Sampling Ionization Chamber (MUSIC) served as the tracking detector. The high charge resolution of this detector permitted the identification of individual elements above a threshold atomic number $Z \geq 8$ (Fig. 2, bottom panel). Lighter fragments were tracked and identified by collecting and amplifying their ionization charges with proportional counters mounted in three positions at the anode plane of the drift volume.

The two-layered time-of-flight (TOF) wall extended over 2.4 m in the horizontal and 1.0 m in the vertical directions. Fragments with $Z \leq 15$ were elementally resolved with the TOF wall detectors (Fig. 2, top panel). The resolution of the time-of-flight measurement with respect to the beam detectors positioned upstream was between 200 ps and 400 ps (FWHM), depending on the fragment Z. It permitted the determination

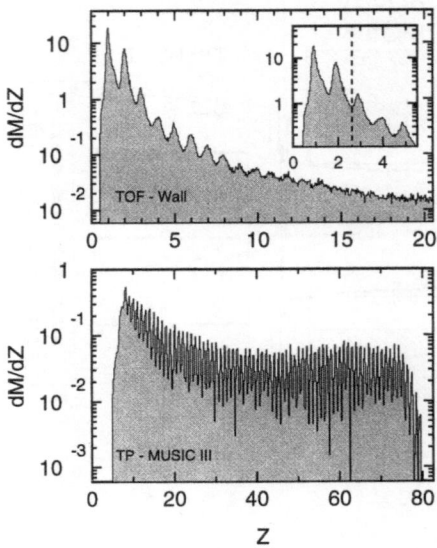

Figure 2. Z-identification spectra measured with the TOF wall (top) and the TP-MUSIC (bottom) for the reaction ^{197}Au on ^{197}Au at $E/A = 600$ MeV. The Z information from the TP-MUSIC was used to calibrate the response of the TOF wall in the region $Z > 15$. Note that the element yield at $Z > 65$ is affected by the experimental trigger.

Insert: Low-Z part of the TOF-wall spectrum. The dashed line indicates the equivalent sharp cut which was used for selecting fragments with $Z \geq 3$ (from Ref.[11]).

of the individual masses of the lighter products with Z up to about 12 from the momenta given by the tracking analysis. In addition to the charged projectile fragments, neutrons emitted in directions close to $\Theta_{lab} = 0°$ were measured with the Large-Area Neutron Detector (LAND) which was operated in a calorimetric mode.

The fragments from the decay of excited projectile spectators are well localized in rapidity. This is illustrated in Fig. 3 where rapidity spectra of light fragments from the reaction ^{197}Au on ^{197}Au at $E/A = 1000$ MeV are shown. The distributions are concentrated around a rapidity value very close to that of the projectile, y_P, and become increasingly narrower with increasing mass of the fragment. For the lighter fragments, the distributions extend into the mid-rapidity region. The widths and shapes of the distributions also depend on the impact parameter, as demonstrated for helium fragments in the two lower panels of Fig. 3. The bump in the peripheral He spectrum, located at a rapidity y between 0.8 and 0.9, and similar bumps observed for light fragments up to $Z \approx 4$ (Fig. 3, top) originate from mid-rapidity emission.

Based on these observations and on model studies, limits for the kinematic region of the projectile source were chosen for the off-line analysis. The condition $y \geq 0.75 \cdot y_P$ was adopted for the fragment rapidities (cf. Fig. 3), and upper limits in the laboratory angle were set which took the observed invariance of the transverse fragment momenta with bombarding energy into account. These definitions permitted a comparison of data measured at different bombarding energies on a quantitative level [11].

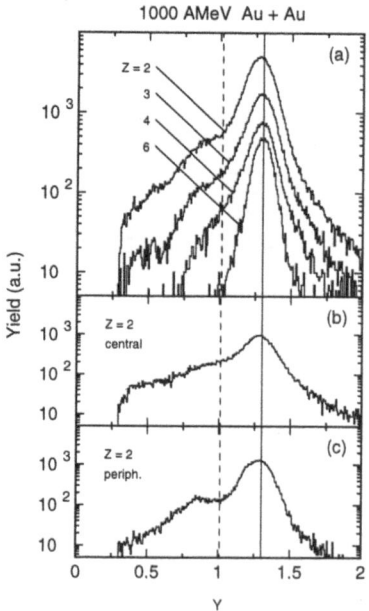

1000 AMeV Au + Au

Figure 3. (a): Rapidity spectra measured in the reaction ^{197}Au on ^{197}Au at $E/A = 1000$ MeV for fragments with $Z = 2$, 3, 4, and 6. The solid and dashed lines indicate the measured most probable rapidity $y = 1.32$ of the light fragments and the condition $y \geq 0.75 \cdot y_P$ adopted for fragments from the projectile spectator, respectively.
(b): Rapidity spectra of helium fragments, measured in central collisions ($Z_{bound} \leq 30$) for the same reaction.
(c): Same as (b) for peripheral collisions ($Z_{bound} \geq 50$) (from Ref.[11]).

UNIVERSALITY OF SPECTATOR DECAY

The decay of excited spectators exhibits universal features which become apparent in the observed Z_{bound} scaling of the measured charge correlations. The quantity Z_{bound} is defined as the sum of the atomic numbers Z_i of all projectile fragments with $Z_i \geq 2$. It represents the charge of the original spectator system reduced by the number of hydrogen isotopes emitted during its decay.

Scatter plots of two charge observables, the multiplicity M_{IMF} of intermediate-mass fragments (IMFs, $3 \leq Z \leq 30$) and the maximum fragment charge Z_{max} within the event, are shown in Fig. 4 as a function of Z_{bound} for ^{197}Au projectiles at 400 MeV per nucleon incident energy. The four rows of panels correspond to the results obtained with the four targets C, Al, Cu, and Au. It follows from the geometric properties of heavy-ion reactions at these energies that the mass (or charge) of the spectator, and therefore also Z_{bound}, is closely related to the impact parameter. At large Z_{bound}, the number of fragments is small and mainly one heavy residue nucleus with $Z_{max} \approx Z_{bound}$ is produced. With decreasing Z_{bound} the number of fragments increases and, correspondingly, Z_{max} is considerably smaller than Z_{bound}. Multifragment production dominates for impact parameters corresponding to $Z_{bound} \approx 40$. In central collisions with the heavier targets, the region of small Z_{bound} is strongly populated. Here, both M_{IMF} and Z_{max} decrease, reflecting the smaller size of the spectators produced in these collisions. This behavior was termed the rise and fall of multifragment emission[25]. On the side of the rise, at large Z_{bound}, the fragment production is governed by the amount

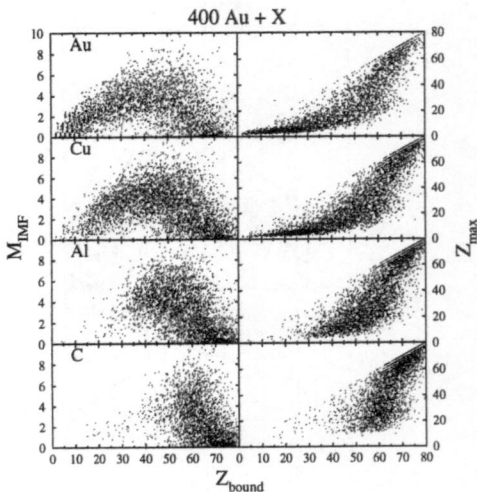

Figure 4. Multiplicity of intermediate-mass fragments (left-hand side) and atomic number of the largest fragment (right-hand side) as a function of Z_{bound} for the reaction ^{197}Au on ^{197}Au at $E/A = 400$ MeV. Random numbers, taken from the interval [-0.5,0.5], were added to the integer values M_{IMF} in order to preserve the intensity information in the scatter plot (from Ref.[24]).

of deposited energy, whereas in the fall region the limit of unconditional partitioning is approached [26].

The almost identical behavior of the observed charge correlations for different reactions, already suggested by the scatter plots, is best appreciated when looking at the mean values of these observables. In Fig. 5 the mean number of intermediate-mass fragments is shown as a function of Z_{bound} for the reaction of ^{197}Au on ^{197}Au at four bombarding energies. The rise and fall of fragment production is seen to be independent of the projectile energy within the experimental accuracy. This invariance also holds for other charge correlations that have been found useful to characterize the population of the partition space in the fragmentation process[11, 27].

The target invariance of the M_{IMF} versus Z_{bound} correlation was first observed for collisions of ^{197}Au projectiles with C, Al, Cu, and Pb targets at 600 MeV per nucleon [23,25-27]. In Fig. 6 (top) the universal nature of this correlation is demonstrated for ^{238}U projectiles at 1000 MeV per nucleon and for a set of seven targets, ranging from Be to U. The data for the lighter targets extend only over parts of the Z_{bound} range. This is more clearly seen in the bottom part of the figure where the differential cross sections $d\sigma/dZ_{bound}$ for four out of the seven targets are shown. From the cross sections, by assuming a monotonic relation between Z_{bound} and the impact parameter, an empirical impact parameter scale was obtained. Central collisions correspond to the smallest values of Z_{bound} reached with a given target, and given regions of Z_{bound}, in collisions with different targets, correspond to different impact parameters. The cross sections were found to depend somewhat on the bombarding energy. The range of Z_{bound} covered with, e.g., the Be or C targets increases with increasing bombarding energy.

The $\langle M_{IMF} \rangle$ versus Z_{bound} correlation depends on the mass of the projectile. The results obtained with the three projectiles ^{129}Xe, ^{197}Au, and ^{238}U at 600 MeV per nucleon show that, on the absolute scale, more fragments are produced in the decay of heavier projectiles (Fig. 7, left-hand side). However, a normalization with respect to

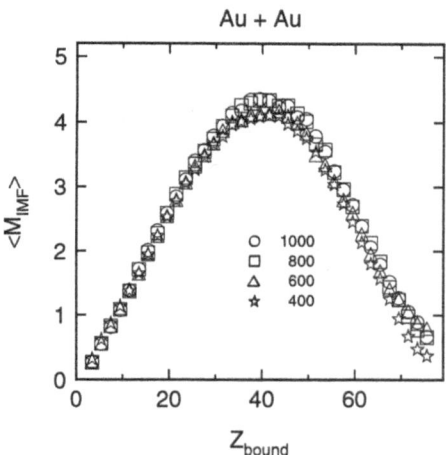

Figure 5. Mean multiplicity of intermediate-mass fragments $\langle M_{IMF} \rangle$ as a function of Z_{bound} for the reaction ^{197}Au on ^{197}Au at $E/A = 400, 600, 800,$ and 1000 MeV (from Ref.[11]).

the atomic number Z_P of the projectile reduces the three curves to a single universal relation (Fig. 7, right-hand side).

The observed Z_{bound} scaling thus comprises the dependences on the projectile and target mass and on the bombarding energy. The data obtained at the AGS with beams of 10.6 GeV per nucleon, in fact, indicate that it should be valid up to very-high bombarding energies[20, 21]. The reasons underlying this property of spectator decay may be sought in the mechanism of spectator excitation. Calculations with the intranuclear-cascade model[28] suggest that the relation between the excitation energy and the residual mass of the spectator should be universal[11, 29]. In this case, if the subsequent decay proceeds statistically, the final fragmentation patterns will only depend on Z_{bound} and not on the particular entrance channel of the reaction.

THE EQUILIBRATED SPECTATOR SOURCE

The observed universality of the spectator decay suggests that a high degree of equilibrium is reached in the initial stages of the reaction. This is confirmed by the analysis of the kinetic variables in the moving frame of the spectator.

The times and positions measured with the TOF wall were used to calculate the components of the fragment velocities in the reference frame of the original projectile. Results obtained for light fragments from the reaction ^{197}Au on ^{197}Au at $E/A = 800$ MeV are shown in Fig. 8. In the y direction, the acceptance of the magnet limits the observable range of velocities. Apart from this, the distributions are seen to be isotropic to a very good approximation. Gaussian widths fitted to the measured distributions of the velocity components in the moving frame confirm this isotropy quantitatively.

The intrinsic velocities have also been used to determine the fragment kinetic energies in the frame of the decaying spectator. Results for the reaction ^{197}Au on ^{197}Au at 600 MeV per nucleon, integrated over $20 \leq Z_{bound} \leq 60$, are shown in Fig. 9. It was assumed that either the longitudinal (squares) or the vertical transverse (circles) degrees of freedom represent one third of the total kinetic energies in the moving frame. The agreement between the two sets of results reflects the isotropy of the kinetic degrees of freedom. The intrinsic kinetic energies do not depend on the bombarding energy and

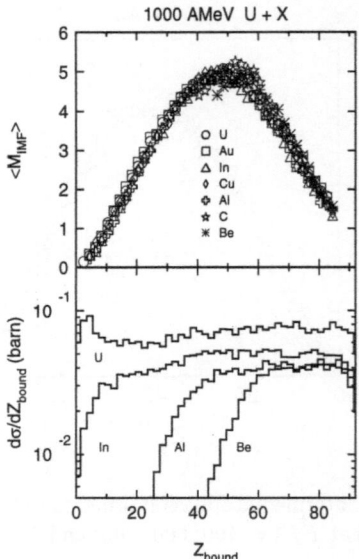

Figure 6. Top: Mean multiplicity of intermediate-mass fragments $\langle M_{IMF} \rangle$ as a function of Z_{bound} for the reactions of ^{238}U projectiles at $E/A = 1000$ MeV with the seven targets Be, C, Al, Cu, In, Au, and U.

Bottom: Measured cross sections $d\sigma/dZ_{bound}$ for the reactions of ^{238}U projectiles at $E/A = 1000$ MeV with the four targets Be, Al, In, and U. Note that the experimental trigger, for the case of uranium beams, affected the cross sections for $Z_{bound} \geq 70$ (from Ref.[11]).

thus are representative for the whole energy range over which the universal spectator decay prevails.

The mean kinetic energies per unit fragment mass $\langle E_{kin}/A \rangle$ decrease rapidly with atomic number Z. In the limit of purely thermal contributions to the kinetic energies, $\langle E_{kin}/A \rangle$ is expected to have a $1/A$ dependence which is approximately observed. However, on the order of one half of the kinetic energies in the rest frame of the decaying system may originate from Coulomb repulsion and sequential decays of excited fragments[31]. With this assumption the magnitude of the kinetic temperature $T = 2/3 \cdot 1/2 \cdot \langle E_{kin} \rangle$ assumes a value of approximately 15 MeV. This exceeds considerably the emission temperatures $T \approx 5$ MeV derived from the relative isotopic abundances (see below) or from relative yields of particle-unbound states[32] which represents a well known but up to now not fully-resolved problem [33-36].

TEMPERATURES AT BREAKUP

Several techniques have been developed for the measurement of temperatures of excited nuclear systems[37]. In the work leading to the caloric curve of nuclei the method suggested by Albergo *et al.*[38] has been used. It is based on the assumption of chemical equilibrium and requires the measurement of double ratios of isotopic yields.

In the limit of thermal and chemical equilibrium, the double ratio R built from the yields Y_i of two pairs of nuclides with the same differences in neutron and proton numbers is given by

$$R = \frac{Y_1/Y_2}{Y_3/Y_4} = a \cdot exp(((B_1 - B_2) - (B_3 - B_4))/T) \tag{1}$$

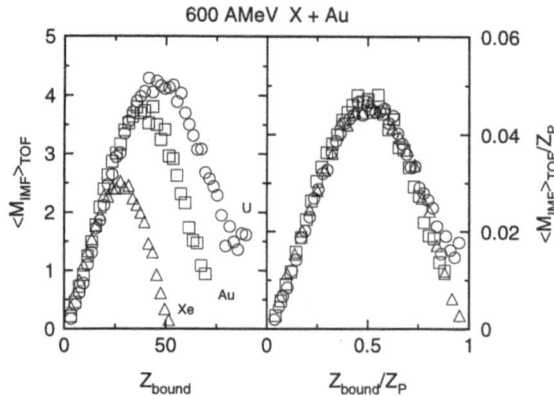

Figure 7. Left panel: Mean multiplicity of intermediate-mass fragments $\langle M_{IMF}\rangle_{TOF}$, observed with the TOF wall, as a function of Z_{bound} for the reactions ^{238}U on ^{197}Au (circles), ^{197}Au on ^{197}Au (squares), and ^{129}Xe on ^{197}Au (triangles) at $E/A = 600$ MeV. Note that also in Z_{bound} only fragments detected with the TOF wall are included.

Right panel: The same data, as shown in the left panel, after normalizing both quantities with respect to the atomic number Z_P of the projectile (from Ref.[11]).

where B_i denotes the binding energy of particle species i and the constant a contains their ground-state spins and mass numbers. In order to make the ratios sufficiently sensitive to the temperature T the double difference of binding energies should be larger than the typical temperature to be measured. For this reason, ^{3}He and ^{4}He are a useful choice for forming one of the two ratios because the difference in binding energy is 20.6 MeV. It may be combined with, e.g., the lithium yield ratio ^{6}Li/^{7}Li or with the hydrogen yield ratios p/d or d/t. Mass spectra obtained for the four isotopes ^{3}He, ^{4}He, ^{6}Li, and ^{7}Li from the tracking analysis are shown in Fig. 10. The ^{3}He yields reflect the sensitivity of this less strongly bound nuclide to the variation of the temperature with impact parameter.

Solving Eq. (1) with respect to T yields, for the case of He and Li isotopes, the following expression:

$$T_{HeLi,0} = 13.3 MeV/ln(2.2\frac{Y_{^6Li}/Y_{^7Li}}{Y_{^3He}/Y_{^4He}}). \qquad (2)$$

The subscript 0 of $T_{HeLi,0}$ refers to the fact that Eq. (1) is strictly valid only for the ground-state population of the considered isotopes at the breakup stage which later may be modified by feeding from the decay of excited states. The temperatures $T_{HeLi,0}$, therefore, will no longer be identical to the breakup temperature. The expected magnitude of this effect was investigated by performing calculations with the quantum-statistical model[39]. Results for an assumed density $\rho/\rho_0 = 0.3$ are shown in Fig. 11 (ρ_0 denotes the saturation density of nuclei). The relation between $T_{HeLi,0}$ and the input temperature of the model was found to be almost linear which is also the case for, e.g., $T_{Hedt,0}$. Other temperature probes, as illustrated for $T_{Hepd,0}$ in the figure, may be more strongly affected by sequential decays. Variations of the input density within reasonable limits or calculations with other decay models suggested that the accuracy of these estimates may lie within $\pm$ 15%[40]. Based on these findings, the constant correction $T_{HeLi} = 1.2 \cdot T_{HeLi,0}$, corresponding to the dotted line in Fig. 11, was adopted[9].

Temperatures T_{HeLi}, as deduced with this method from data measured in three

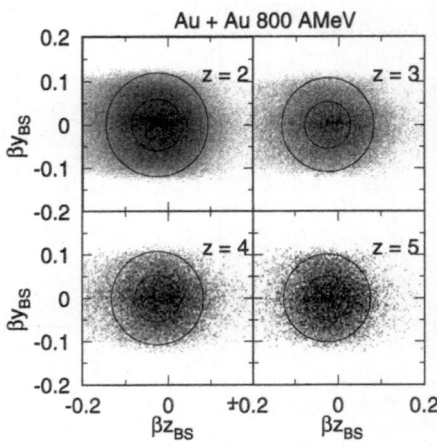

Figure 8. Projections of the fragment velocity β_{BS} in the moving frame (BS = beam system) into the y-z-plane for products with atomic number $Z = 2$ to 5 from the reaction ^{197}Au on ^{197}Au at $E/A = 800$ MeV (y and z denote the directions parallel to the magnetic field and along the beam direction, respectively). The component βy_{BS} is limited by the vertical acceptance of the magnet; the circles are meant to guide the eye (from Ref.[30]).

experiments, are shown in Fig. 12 as a function of Z_{bound}. Besides the results obtained for the projectile decay in ^{197}Au on ^{197}Au collisions at 600 MeV and 1000 MeV per nucleon[9, 40], also temperatures from a more recent study of the target decay in the same reaction at 1000 MeV per nucleon are given[41]. In the latter experiment, Z_{bound} was simultaneously measured for the coincident projectile decay. The temperatures increase slowly with decreasing Z_{bound} in the range $20 \leq Z_{bound} \leq 80$ but then increase more quickly at small Z_{bound} values.

The agreement between the temperatures for the projectile and the target spectators at 1000 MeV per nucleon is expected from the symmetry of the reaction and illustrates the accuracy of the measurements. The invariance with the bombarding energy, here established over the range 600 to 1000 MeV per nucleon, confirms the statistical interpretation of the observed universality of the spectator decay.

The lines shown in Fig. 12 represent results obtained with the statistical multifragmentation model[4]. The excitation energy and mass of the ensemble of excited spectator nuclei, required as input for the calculations, were chosen in such a way that the correlation between the mean multiplicity of intermediate-mass fragments with Z_{bound} (Fig. 5) was well reproduced. Within the given experimental and methodical uncertainties, the resulting mean value of the breakup temperature (full line) is in excellent agreement with the data. This means that the description of the fragmentation as a statistical process is internally consistent in that the temperatures needed to reproduce the observed partition patterns are equal to those measured. The dashed line gives the uncorrected isotopic temperature $T_{HeLi,0}$ deduced from the calculated isotope yields. The difference to the breakup temperature represents the correction for secondary decay according to the statistical multifragmentation model. It is in good qualitative agreement with the adopted correction factor of 1.2.

The consequences of sidefeeding from higher-lying states are presently investigated by several groups with different methods [42-47]. The results differ considerably in magnitude as well as in the sign of the required correction and, in some cases, exceed

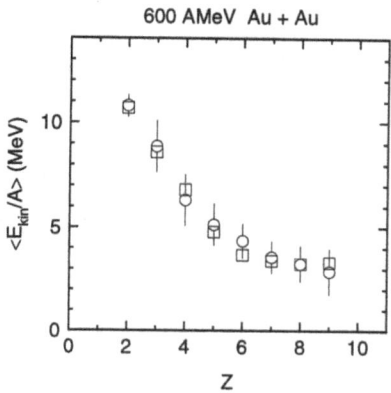

600 AMeV Au + Au

Figure 9. Mean kinetic energies per nucleon in the moving frame, deduced from the transverse (circles) and longitudinal (squares) momentum widths, for fragments from the reaction ^{197}Au on ^{197}Au at $E/A = 600$ MeV and for $20 \leq Z_{bound} \leq 60$ (from Ref.[11]).

the $\pm 15\%$ margin quoted above. Furthermore, temperatures for central collisions at lower bombarding energies obtained from isotopic yield ratios and from the population of particle-unstable resonances were found to deviate in a systematic fashion from each other[12]. These questions will have to be answered, eventually, in order to maintain a quantitative level in the investigation of thermodynamic properties of excited nuclear systems.

ENERGY DEPOSITION

Rather small fractions of the initial bombarding energy are imparted to the spectator nuclei in relativistic collisions. The actual amounts of energy deposition can only be reconstructed from the exit-channel configuration which requires a complete knowledge of all decay products, including their atomic numbers, masses, and kinetic energies.

A method to determine the excitation energy from the experimental data along this line was first presented by Campi *et al.*[48] and applied to the earlier ^{197}Au + Cu data[27]. The yields of hydrogen isotopes were determined by extrapolating to $Z = 1$ from the measured abundances for $Z \geq 2$, and the multiplicities of neutrons were inferred from a mass balance. The obtained asymptotic value of $E_x/A = 23$ MeV at $Z_{bound} = 0$ is the sum of the binding energy of 8 MeV and the kinetic energy of 15 MeV assigned to nucleons.

In the same type of analysis with the more recent data for ^{197}Au + ^{197}Au at 600 MeV per nucleon, the data on neutron production measured with LAND were taken into account[9]. Since the hydrogen isotopes were not detected assumptions concerning the overall N/Z ratio of the spectator, the intensity ratio of protons, deuterons, and tritons, and the kinetic energies of hydrogen isotopes had to be made. In addition, the EPAX parameterization[49, 50] was used for deriving masses from the atomic numbers of the detected fragments. The uncertainties resulting from the variation of these quantities within reasonable limits were included in the errors assigned to the results.

It is found that light particles and, in particular, the neutrons contribute considerably to the total spectator energy E_0. The balance of binding energies, i.e. the Q value associated with the fragmentation process, amounts to about 40% of it, fairly independent of Z_{bound}. The results for the mass A_0 and for the specific excitation en-

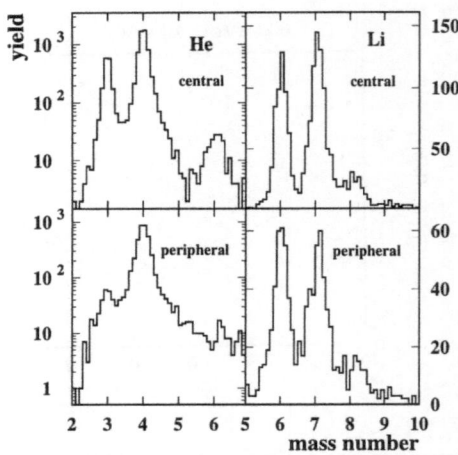

Figure 10. Mass spectra of He fragments (left panel) and Li fragments (right panel) for the reaction ^{197}Au on ^{197}Au at $E/A = 600$ MeV. The upper and lower panels correspond to central and peripheral collisions, respectively (from Ref.[40]).

ergy E_0/A_0 are given in Fig. 13. The data points represent the results for 10-unit-wide bins in a Z_{max}-versus-Z_{bound} representation (cf. Fig. 4). The mean mass A_0 decreases with decreasing Z_{bound}, seems to be independent of Z_{max}, and is in good agreement with the expectations from the geometric participant-spectator model[51]. The smallest mean spectator mass in the bin of $Z_{bound} \leq 10$ is $\langle A_0 \rangle \approx 50$. The excitation energy E_0 appears to be a function of both Z_{bound} and Z_{max}; the higher values correspond to the smaller Z_{max} values, i.e. to more complete disintegrations of a system of given mass. The maximum number of fragments, observed at $Z_{bound} \approx 40$, is associated with initial excitation energies of $\langle E_0 \rangle / \langle A_0 \rangle \approx 8$ MeV. With decreasing Z_{bound} the deduced excitation energies reach up to $\langle E_0 \rangle / \langle A_0 \rangle \approx 16$ MeV.

The experimentally-determined energies fall in between the higher predictions for the deposited energy of the intranuclear-cascade model and the much lower values obtained from analyses of the final partitions with the statistical multifragmentation model (Ref.[11] and references therein). A sequence of energies with this ordering is not unreasonable in that the formation of the equilibrated spectator in the initial reaction stages and its evolution towards the final breakup stage may be accompanied by the emission of fast light particles and, therefore, by a loss of excitation energy. On the experimental side, it is presently investigated to what extent the unexpectedly-high kinetic energies of protons, preequilibrium emission, and collective phenomena, in particular the bounce-off of the spectator systems, are influencing the deduced-energy deposits[30]. The result reported most recently[11] is about 15% higher than that shown in Fig. 13 which entered the caloric curve discussed in the next section.

THE CALORIC CURVE

The pairwise correlation of the temperatures and excitation energies, deduced as described in the last two sections, results in the caloric curve shown in Fig. 14. In addition to the data from projectile decays following ^{197}Au + ^{197}Au collisions at 600 MeV per nucleon, results from earlier experiments with ^{197}Au targets at intermediate energies of 30 to 84 MeV per nucleon and for compound nuclei produced in the ^{22}Ne

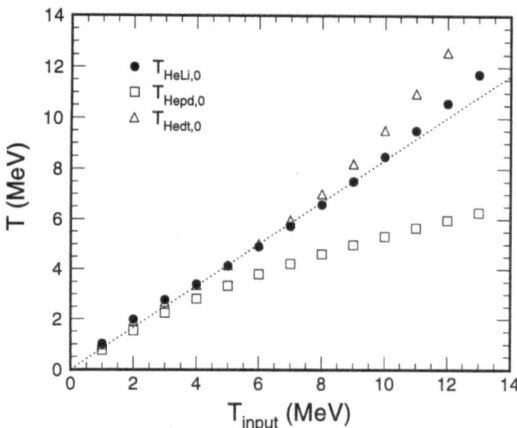

Figure 11. Temperatures $T_{HeLi,0}$, $T_{Hepd,0}$, and $T_{Hedt,0}$, according to the quantum-statistical model, as a function of the input temperature T_{input}. A breakup density $\rho/\rho_0 = 0.3$ is assumed. The dotted line represents the linear relation $T_{input}/1.2$ (from Ref.[41]).

+ ^{181}Ta reaction are included[52, 53]. All temperatures were deduced following the same method. For the reactions at intermediate energies, the excitation energy of the target residues was obtained from an energy balance based on moving-source analyses while, in the compound case, it is given by the collision energy.

One may first notice the consistency of the data obtained from different types of reactions, suggesting that the smooth S-shaped curve may represent a more general property of excited nuclei. In fact, at low energies, the deduced temperatures T_{HeLi} follow the low-temperature approximation for a Fermi-liquid, confirmed by several studies in the fusion evaporation regime [54, 55]. The full line depicts this behavior for a level-density parameter $a = A/10$ MeV^{-1}. At the high excitation energies, the rise of the temperature appears to be linear with the excitation energy, with the slope of 2/3 of a classical gas. In the limit of a free nucleon gas, the offset should be ≈ 8 MeV, corresponding to the mean binding energy of nuclei. A smaller offset may be caused by a freeze-out at a finite density and by the finite fraction of bound clusters and fragments of intermediate mass that are present even at these high excitation energies. The offset of 2 MeV is consistent with a breakup density ρ/ρ_0 between 0.15 and 0.3[9]. A final assessment, however, will have to await the completion of the ongoing analysis of the energy deposition.

Within the range of $\langle E_0 \rangle / \langle A_0 \rangle$ from 3 MeV to 10 MeV an almost constant value for T_{HeLi} of about 4.5 to 5 MeV is observed. This plateau may be related to the previous finding of almost constant emission temperatures over a broad range of incident energies which were deduced from the population of particle-unstable levels in He and Li fragments[32]. The plateau is suggestive of a first-order phase transition with a substantial latent heat (see also[56]). This is supported by the fact that the increase in excitation energy is associated with a disintegration into a larger number of fragments of smaller size as apparent from Fig. 4 ($\langle E_0 \rangle / \langle A_0 \rangle$ has to be translated back into the corresponding Z_{bound} with the help of Fig. 13). The surface energy needed for the formation of smaller constituents limits the rise of the temperature. This interpretation is consistent with the good description of the plateau temperatures by the statistical multifragmentation model which is based on the droplet model of nuclei (Fig. 12).

Alternative interpretations, in particular for the hitherto unobserved temperature

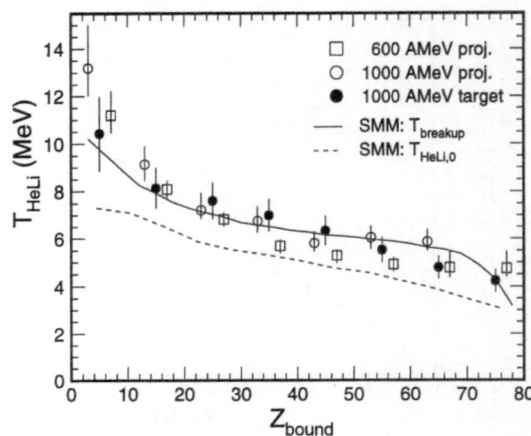

Figure 12. Temperatures T_{HeLi} for target ($E/A = 1000$ MeV) and projectile spectators ($E/A = 600$ and 1000 MeV) as a function of Z_{bound}. The data symbols represent averages over bins of 10-units width. The full and dashed lines represent the internal temperature $T_{breakup}$ and the uncorrected isotopic temperature $T_{HeLi,0}$ as given by the calculations with the statistical multifragmentation model (from Ref.[41]).

rise at the highest energies, have been presented by several groups. The interpretation of Natowitz et al.[57] relates the observed variation of the temperature to the variation of the system mass (Fig. 13) and to the mass dependence of the limiting temperatures obtained from theoretical descriptions of excited nuclei at the limit of their stability[58]. In the expansion scenario modeled by Papp and Nörenberg[59], the temperatures in the plateau region are found to be consistent with a spinodal decomposition in the dynamically unstable region of the temperature-versus-density plane. The upbend at high excitation energies, however, indicates a minimum breakup density rather than entry into the vapor phase. The results were found to be sensitive to the equation of state governing the expansion process which presents a further motivation for aiming at a high level of accuracy in these measurements.

CRITICAL FEATURES OF MULTIFRAGMENTATION

The apparent signatures of a first-order phase transition in nuclei, discussed in the last section, do not rule out the possibility that critical phenomena may be observed. In finite systems, a second-order phase transition is no longer characterized by a singular point, the associated fluctuations are rather spread over a finite interval in temperature[60]. Typical features of a first-order phase transition, like a latent heat, and signals indicating the proximity of the critical point, like diverging moments, are therefore not necessarily inconsistent.

The observation of a power law dependence of the fragment mass yields in reactions of high energy protons with Kr and Xe targets[61, 62], and its association with Fisher's prediction for droplet formation at the critical point[63] has initiated the intensive search for signatures of criticality [2, 16, 64]. The systematic investigations showed that the power law exponent τ approaches a value of ≈ 2.5 at high bombarding energies[65]. This is consistent with the limits $2.0 \leq \tau \leq 3.0$ given by the theory of critical phenomena[66]. While the interpretation of inclusive mass spectra was criticized[67], it was shown in exclusive measurements that the mass or charge distributions may approach the pure

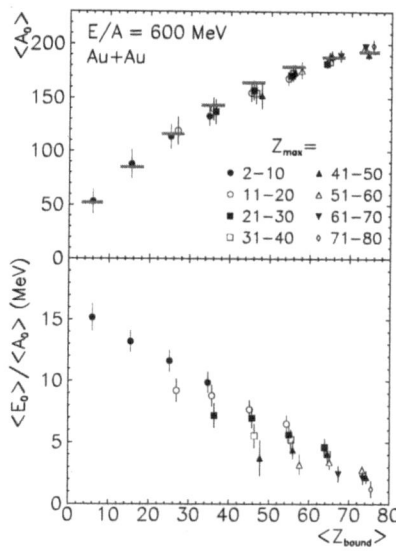

Figure 13. Reconstructed average mass $\langle A_0 \rangle$ (top) and excitation energy $\langle E_0 \rangle / \langle A_0 \rangle$ (bottom) of the decaying spectator system as functions of Z_{bound} (abscissa) and of Z_{max} (different data symbols). The horizontal bars represent the masses according to the participant-spectator model at the empirical impact parameter deduced from $d\sigma/dZ_{bound}$ (from Ref.[9]).

power law for certain values of the chosen sorting variable, e.g. Z_{bound}. In this case, the sorting variable may serve as the parameter controlling the distance to the critical point or critical region. For the fragmentation of ^{197}Au projectiles at 1 GeV per nucleon, this value is $Z_{bound} \approx 35$[68] which is close to the maximum fragment production (Fig. 5). Examples of Z_{bound}-gated spectra are shown in Ref.[27].

A more recent, equally stimulating, observation was the similarity of the charge fluctuations in fragmentation with those given by three-dimensional percolation on a small lattice[69]. Percolation is a mathematical model exhibiting a second-order phase transition in the limit of infinite lattice size[66]. Campi, therefore, concluded that 'nuclei break up like finite systems that show a clean phase transition in infinite size'. In the meanwhile, it has been demonstrated that nearly perfect descriptions of the partitions observed in multifragmentation reactions may be obtained with the percolation model[27].

The capability of percolation to describe the partition space is not trivial but it is also not unique to percolation. The fragment-charge correlations from the first experiments have been reproduced to high accuracy with the statistical multifragmentation model, but also with a variety of other models, such as classical-cluster formation [70], fragmentation-inactivation binary[71], and restructured-aggregation models[72]. Some of these models exhibit critical behavior in the limit of an infinite number of constituents. The apparent universality, inherent to disordered systems of interacting nucleons and to the models describing their fragmenting so well, remains to be understood[73].

A further step was taken by the EOS collaboration who have reported values for critical-point exponents from the charge correlations measured for the ^{197}Au on C reaction at 1 GeV per nucleon [10, 74, 75]. The analysis takes extensive recourse to results obtained for percolation. Below the critical point, the largest cluster of the partition is associated with the percolating cluster which corresponds to the liquid phase in a liquid-gas transition. It grows in proportion to t^β with the relative distance t from

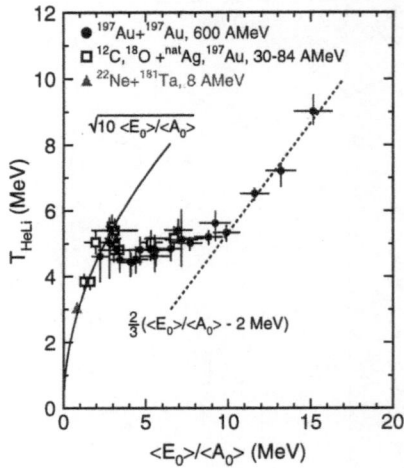

Figure 14. Caloric curve of nuclei as constituted by the temperature T_{HeLi} as a function of the excitation energy per nucleon. The lines are explained in the text (from Ref.[9]).

the critical point. The role of the susceptibility in magnetic systems or the isothermal compressibility in a liquid, i.e. the response to the external field, is assumed by the second moment M_2 of the fragment size distributions. It diverges with the power of $-\gamma$ as the critical point is approached (both β and γ are positive numbers).

With the guidance provided by percolation studies on small lattices, recipes were developed on how to extract critical exponents from the data[76]. The reported results $\beta = 0.29 \pm 0.02$ and $\gamma = 1.4 \pm 0.1$ are close to those of a liquid-gas system and significantly different from those of percolation or mean-field theory[10]. This conclusion relies on the correct assessment of the systematic errors inherent to the procedure which, however, has been questioned by other authors [77, 78]. A critical discussion may also be found in the contributions to Refs.[14, 16] by Müller *et al.* who, besides percolation, have studied Ising models on small lattices.

A critical-exponent analysis has been carried out with the ALADIN data for ^{197}Au and ^{238}U fragmentations[68]. The same numerical results were obtained if the procedure of Ref.[10] was followed in detail. However, also the same difficulties arising from the finite system size were encountered. As an example, the fits used to extract the exponent γ are shown in Fig. 15. The figure is virtually identical to the corresponding figure constructed from the EOS data[79]. Potential divergences at the critical point that will appear in the infinite system are smoothened over finite regions of the chosen control parameter, the charged-particle multiplicity m. The choice of the fit regions where the data are assumed to reflect the critical behavior is crucial. It has to rely on methods obtained from the study of other systems which, at this time, are not yet proven unambiguous. A consistent method was found for determining the exponent τ of the power law obeyed by the element distribution[68]. This may be related to the fact that τ describes a behavior *at* the critical point, and not the behavior how the critical point is approached. The finite size of the system may be less crucial in this case.

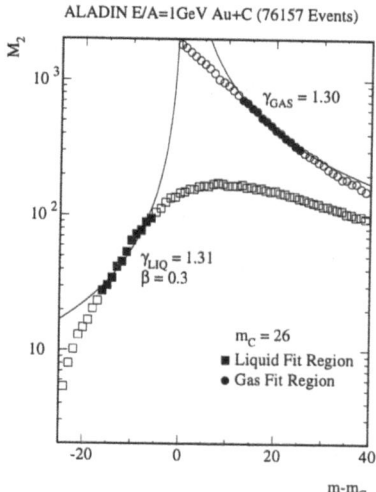

Figure 15. Second moment M_2 of the fragment-charge distribution, with (circles) and without (squares) including the largest fragment, as a function of the charged-particle multiplicity m for ^{197}Au + C at $E/A = 1$ GeV. The assumed critical multiplicity is $m_c = 26$, the full symbols indicate the selected fit regions (from Ref.[68]).

CONCLUSIONS AND PERSPECTIVES

The systematic set of data now available reveals the universal nature of multi-fragment decays of excited spectator nuclei at relativistic energies. It suggests that the correlation of excitation energy and mass of the produced spectator systems and the statistical nature of their decay are independent of the specific entrance channel. The emulsion data as well as the insight provided by calculations indicate that it should prevail up to very high bombarding energies with virtually-unchanged features. There are lower limits of the bombarding energy below which the spectator excitation does not suffice to induce a complete disassembly with maximum fragment production. These threshold energies depend on the collision partner, as shown for the case of gold nuclei in Fig. 16. In addition to data measured with the ALADIN spectrometer in inverse kinematics, also data for ^{197}Au on ^{197}Au at lower energies[80], for ^{84}Kr on ^{197}Au [81], and for ^{4}He on ^{197}Au[82] are included. The hatched line indicates these threshold energies as a function of the target mass. If the bombarding energy is raised above the threshold, the maximum fragment production will shift to more peripheral collisions. Below the threshold, the production of heavy residues will be the dominating process.

As one moves along the hatched line towards the central collisions of heavy systems, dynamical phenomena become important. The largest fragment multiplicities measured so far were observed in central ^{197}Au on ^{197}Au collisions at a bombarding energy of 100 MeV per nucleon[80]. The analysis of the kinetic-energy spectra in these reactions has revealed a considerable collective outward motion (radial flow), superimposed on the random motion of the constituents at the breakup stage[83]. The associated collective energy constitutes up to one-half of the incident kinetic energy in the center-of-mass frame and increases approximately linearly over the range of bombarding energies up to 1000 MeV per nucleon[84]. Furthermore, the fragment formation was found to be sensitive to the flow dynamics. The measured element yields are systematically correlated with the magnitude of the observed radial flow[84, 85].

This excursion to central collisions, not covered in the main part of these notes,

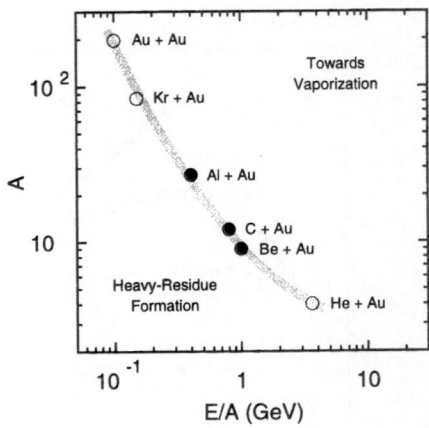

Figure 16. Mass number A of the collision partner versus bombarding energy for reactions of ^{197}Au nuclei for which maximum fragment production has been observed in central collisions (from Ref.[11]).

is meant to demonstrate that heavy-ion reactions at relativistic energies offer wide possibilities to study the response of excited nuclear matter under different conditions. The interplay of dynamical and statistical effects in these violent central collisions is of high current interest[86]. The measurement of breakup temperatures for these reactions, where equilibrium may be established only locally, has already been started and promises to produce stimulating new results[12].

Large dynamical effects are not expected in light-particle induced reactions with heavy targets, i.e. at the lower end of the hatched line in Fig. 16. Here, the main problem arises from the fact that the cross sections for large energy transfers to the target with maximum fragment production are comparatively small [82, 87, 88]. New experimental results including exclusive measurements with beams of antiprotons[89] and of protons with energies up to 14 GeV [90] have been obtained very recently, and a comprehensive picture of the light-ion reactions should emerge rather soon.

The phenomenology of multifragment decays, in particular the partitioning of the decaying systems, has been established with high accuracy, as shown in the first part of these notes. The critical discussion of the new methods for measuring thermodynamic properties of excited nuclear systems is important. Here, the systematic errors will have to be further reduced, which seems feasible with improved experiments and model calculations. The search for critical phenomena has to take the small number of constituents into account, and concepts developed for macroscopic systems have to be applied with caution. The finite system size, on the other hand, offers wide possibilities for model studies based on different approaches, aiming at an understanding of the universal properties of nuclear fragmentation and of fragmentation and clustering phenomena in other fields.

I am highly indebted to my colleagues of the ALADIN collaboration for valuable discussions and support during the preparation of this manuscript.

REFERENCES

1. P.J. Siemens, *Nature (London)* 305 (1983) 410.

2. J. Hüfner, *Phys. Rep.* 125 (1985) 129.

3. D.H.E. Gross, *Rep. Prog. Phys.* 53 (1990) 605.

4. J.P. Bondorf *et al.*, *Phys Rep.* 257 (1995) 133.

5. H. Jaqaman *et al.*, *Phys. Rev.* C 27 (1983) 2782 .

6. M. Brack *et al.*, *Phys. Rep.* 123 (1985) 275.

7. G. Bertsch and P.J. Siemens, *Phys. Lett.* 126B (1983) 9 .

8. W. Stocker, *Phys. Lett.* 142B (1984) 319 .

9. J. Pochodzalla *et al.*, *Phys. Rev. Lett.* 75 (1995) 1040 .

10. M.L. Gilkes *et al.*, *Phys. Rev. Lett.* 73 (1994) 1590 .

11. A. Schüttauf *et al.*, *Nucl. Phys.* A607 (1996) 457 .

12. for the most recent reports of the collaboration see, e.g.,
G. Immé *et al.*, to appear in *Proceedings of the 1st Catania Relativistic Ion Studies: Critical Phenomena and Collective Observables*, Acicastello, Italy, 1996;
W.F.J. Müller *et al.*, ibid.;
J. Pochodzalla *et al.*, ibid.

13. *Proceedings of the International Workshop XXII*, Hirschegg, 1994, edited by H. Feldmeier and W. Nörenberg (GSI, Darmstadt, 1994).

14. *Proceedings of the XXXIII International Winter Meeting on Nuclear Physics*, Bormio, 1995, edited by I. Iori (Ricerca Scientifica ed Educazione Permanente, Milano, 1995).

15. *Proceedings of the 12th Winter Workshop on Nuclear Dynamics*, Snowbird, Utah, USA (1996).

16. *Proceedings of the 1st Catania Relativistic Ion Studies: Critical Phenomena and Collective Observables*, Acicastello, Italy (1996).

17. For a recent review see L.G. Moretto and G.J. Wozniak, *Ann. Rev. Nucl. Part. Sci.* 43 (1993) 379.

18. B. Jakobsson *et al.*, *Z. Phys.* A 307 (1982) 293 .

19. E.M. Friedlander *et al.*, *Phys. Rev.* C 27 (1983) 2436 .

20. P.L. Jain *et al.*, *Phys. Rev.* C 50 (1994) 1085 .

21. M.L. Cherry *et al.*, *Phys. Rev.* C 52 (1995) 2652 .

22. G. Rusch *et al.*, *Phys. Rev.* C 49 (1994) 901 .

23. J. Hubele *et al.*, *Z. Phys.* A 340 (1991) 263 .

24. W.D. Kunze, PhD thesis, Universität Frankfurt (1996), unpublished.

25. C.A. Ogilvie *et al.*, *Phys. Rev. Lett.* 67 (1991) 1214 .

26. J. Hubele *et al.*, *Phys. Rev.* C 46 (1992) R1577 .

27. P. Kreutz *et al.*, *Nucl. Phys.* A556 (1993) 672 .

28. Y. Yariv and Z. Fraenkel, *Phys. Rev.* C 20 (1979) 2227 ; *Phys. Rev.* C 24 (1981) 488 .

29. V.S. Barashenkov and V.D. Toneev, Interactions of high-energy particles and atomic nuclei with nuclei, Moscow, Atomizdat, 1972 (in Russian);
V.D. Toneev, private communication (1994).

30. A. Schüttauf, PhD thesis, Universität Frankfurt (1996), unpublished.

31. V. Lindenstruth, PhD thesis, Universität Frankfurt (1993), report GSI-93-18.

32. G.J. Kunde *et al.*, *Phys. Lett.* B272 (1991) 202 .

33. H.W. Barz *et al.*, *Phys. Lett.* B217 (1989) 397 .

34. D.H. Boal *et al.*, *Phys. Rev. Lett.* 62 (1989) 737 .

35. H.W. Barz *et al.*, *Phys. Lett.* B228 (1989) 453 .

36. W. Bauer, *Phys. Rev.* C 51 (1995) 803 .

37. D. Morrissey *et al.*, *Ann. Rev. Nucl. Part. Sci.* 44 (1994) 27.

38. S. Albergo *et al.*, *Il Nuovo Cimento* 89 A (1985) 1.

39. J. Konopka *et al.*, *Phys. Rev.* C 50 (1994) 2085 .

40. T. Möhlenkamp, PhD thesis, Universität Dresden (1996), unpublished.

41. Hongfei Xi *et al.*, to be published.

42. M.B. Tsang *et al.*, *Phys. Rev.* C 53 (1996) R1057 .

43. A. Kolomiets *et al.*, *Phys. Rev.* C 54 (1996) R472 .

44. X. Campi *et al.*, *Phys. Lett.* B, in print.

45. M.B. Tsang *et al.*, preprint MSUCL-1035 (1996).

46. Hongfei Xi *et al.*, preprint MSUCL-1040 (1996).

47. Z. Majka *et al.*, preprint 96-03, Texas A&M University (1996).

48. X. Campi *et al.*, *Phys. Rev.* C 50 (1994) R2680 .

49. K. Sümmerer *et al.*, *Phys. Rev.* C 42 (1990) 2546 .

50. A.S. Botvina *et al.*, *Nucl. Phys.* A584 (1995) 737 .

51. J. Gosset *et al.*, *Phys. Rev.* C 16 (1977) 629 .

52. R. Trockel *et al.*, *Phys. Rev.* C 39 (1989) 729 ; R. Trockel, PhD thesis, Universität Heidelberg (1988), unpublished.

53. C. Borcea *et al.*, *Nucl. Phys.* A415 (1984) 169 .

54. G. Nebbia *et al.*, *Phys. Lett.* B176 (1986) 20 .

55. D. Fabris *et al.*, *Phys. Lett.* B196 (1987) 429 .

56. L.G. Moretto *et al.*, *Phys. Rev. Lett.* 76 (1996) 2822 .

57. J.B. Natowitz *et al.*, *Phys. Rev.* C 52 (1995) R2322 .

58. P. Bonche *et al.*, *Nucl. Phys.* A436 (1986) 265 .

59. G. Papp and W. Nörenberg, preprint GSI-95-30 (1995).

60. A.L. Goodman *et al.*, *Phys. Rev.* C 30 (1984) 851 .

61. J.E. Finn *et al.*, *Phys. Rev. Lett.* 49 (1982) 1321 .

62. A.S. Hirsch *et al.*, *Phys. Rev.* C 29 (1984) 508 .

63. M.E. Fisher, *Physics (N.Y.)* 3 (1967) 255.

64. J. Richert, *Int. J. Mod. Phys.* E2 (1993) 679.

65. W. Trautmann *et al.*, *Z. Phys.* A 344 (1993) 447 .

66. D. Stauffer and A. Aharony, *Introduction to Percolation Theory*, (Taylor & Francis, London, 1992).

67. J. Aichelin *et al.*, *Phys. Rev.* C 37 (1988) 2451 .

68. A. Wörner, PhD thesis, Universität Frankfurt (1995), unpublished.

69. X. Campi, *Phys. Lett.* B208 (1988) 351 .

70. J.B. Garcia and C. Cerruti, *Nucl. Phys.* A578 (1994) 597 .

71. R. Botet and M. Ploszajczak, *Phys. Lett.* B312 (1993) 30 ; *Acta Physica Polonica* B25 (1994) 353.

72. S. Leray and S. Souza, *Proceedings of Second European Biennial Conference on Nuclear Physics*, Megève (1993), edited by D. Guinet (World Scientific, Singapore, 1995) p. 81.

73. B. Elattari *et al.*, *Phys. Lett.* B356 (1995) 181 ; *Nucl. Phys.* A592 (1995) 385 .

74. J.A. Hauger *et al.*, *Phys. Rev. Lett.* 77 (1996) 235 .

75. J.B. Elliott *et al.*, *Phys. Lett.* B381 (1996) 35 .

76. J.B. Elliott *et al.*, *Phys. Rev.* C 49 (1994) 3185 .

77. W. Bauer and W.A. Friedman, *Phys. Rev. Lett.* 75 (1995) 767 .

78. W. Bauer and A.S. Botvina, *Phys. Rev.* C 52 (1995) R1760 .

79. H.G. Ritter *et al.*, *Nucl. Phys.* A583 (1995) 491c .

80. M.B. Tsang *et al.*, *Phys. Rev. Lett.* 71 (1993) 1502 .

81. G.F. Peaslee *et al.*, *Phys. Rev.* C 49 (1994) R2271 .

82. V. Lips *et al.*, *Phys. Rev. Lett.* 72 (1994) 1604 .

83. W.C. Hsi *et al.*, *Phys. Rev. Lett.* 73 (1994) 3367 .

84. W. Reisdorf *et al.*, in Ref.[13], p. 93.

85. G.J. Kunde *et al.*, *Phys. Rev. Lett.* 74 (1995) 38 .

86. J. Bondorf *et al.*, *Phys. Rev. Lett.* 73 (1994) 628 .

87. K. Kwiatkowski *et al.*, *Phys. Rev. Lett.* 74 (1995) 3756 .

88. L. Pienkowski *et al.*, *Phys. Lett.* B336 (1994) 147 .

89. F. Goldenbaum *et al.*, *Phys. Rev. Lett.* 77 (1996) 1230 .

90. V.E. Viola *et al.*, *Proceedings of Third International Conference on Nuclear Physics at Storage Rings*, Bernkastel-Kues, Germany (1996).

HANBURY-BROWN/TWISS INTERFEROMETRY FOR RELATIVISTIC HEAVY-ION COLLISIONS: THEORETICAL ASPECTS

Ulrich Heinz [1]

Institut für Theoretische Physik, Universität Regensburg,
D-93040 Regensburg, Germany
Email: Ulrich.Heinz@physik.uni-regensburg.de
[1] Work supported by BMBF, DFG and GSI.

INTRODUCTION

Relativistic heavy-ion collisions at center-of-mass energies in the range of many to many hundreds of GeV per nucleon are performed with the ultimate goal of generating a new state of matter, the quark-gluon plasma (QGP). To make a QGP, one must heat and compress nuclear matter to such high energy densities that the hadrons overlap, creating a homogeneous piece of matter consisting only of the stuff usually hidden inside hadrons. In the region thus created the quarks and gluons, which are usually imprisoned inside the hadrons, become "deconfined", i.e. they are able to travel around freely over regions which are large compared to the usual confinement length scale of about 1 fm (the typical radius of a hadron). Such matter existed in and filled all of the volume of the Early Universe during the first microsecond of its life but, as far as we know, it has not been recreated in the laborarory anywhere since.

How do we know our heavy-ion collision experiment has been successful in creating QGP? The answer to this question is surprisingly complex. The main reason for this is that quarks and gluons cannot travel over large distances outside the hot and dense QGP region and thus cannot be detected directly. Everything we can measure in the experiment provides therefore only indirect knowledge about the hoped-for QGP state. This is in particular true for the bulk of secondary particles produced in the collision which are hadrons: they can only be formed towards the end of the collision process when the reaction zone, by expansion into the surrounding vacuum, has cooled down sufficiently to allow the quarks and gluons to re-hadronize.

One approach to "prove" the making of QGP (which I stress must still be complemented by other experimental tests to check for consistency)tries to reach a complete understanding of the space-time structure and dynamical state of the reaction zone at the "freeze-out point" where the measured hadrons decouple. One hopes and by now, from the experiments performed at the Brookhaven AGS and CERN SPS during the last decade, has accumulated a reasonably convincing body of evidence that at decoupling the matter has reached a state of local thermal and perhaps even chemical equilibrium. It also features strong longitudinal and transverse collective (hydrodynamic) expansion. While some of the longitudinal expansion may be "primordial", i.e. due to incomplete stopping of the two nuclei in the collision region, all of the transverse expansion must have been generated dynamically in the reaction. If it is possible to determine the energy density of the state at decoupling and simultaneously its expansion velocity, then one can try to extrapolate this state backwards in time to a point of vanishing transverse expansion and check whether there the energy density was above the critical value of about 1 GeV/fm^3 where one expects the phase transition to the QGP to occur.

But how does one measure the energy density at decoupling? To measure the total energy of the state is not too difficult: one measures the momenta of all emitted particles and their masses and adds them up. But to compute the energy density we must also know the volume of the reaction zone. And there is no known way to measure this volume directly. The collision fireball just doesn't hang around long enough so that we could shine light on it (or a suitable other species of particles) and measure its size by diffraction.

The only known way to obtain indirect experimental information on the space-time structure of the particle-emitting source created in a relativistic nuclear collision is through two-particle intensity (Hanbury-Brown–Twiss (HBT)) interferometry[1]. The goal of this method is to extract the *space-time* structure of the source from *momentum spectra* which are the only measurable quantities, making use of the quantum statistical correlations between pairs of identical particles. The basic idea, presented in a very naive and oversimplified way which to correct I will spend three hours of lectures during this Advanced Summer Institute, is as follows: Consider a source (for simplicity spherically symmetric) with radius R, and the emission of a pair of identical particles from point $\boldsymbol{x}_1$ in the source with momentum $\boldsymbol{p}_1$ and point $\boldsymbol{x}_2$ with momentum $\boldsymbol{p}_2$. If these two points are well separated in phase-space, i.e. they satisfy

$$(x_1^i - x_2^i)(p_1^i - p_2^i) \gg 2\pi\hbar\,, \qquad i = 1 \text{ or } i = 2 \text{ or } i = 3\,, \tag{1}$$

this process can be treated classically. If on the other hand,

$$(x_1^i - x_2^i)(p_1^i - p_2^i) \lesssim 2\pi\hbar\,, \qquad i = 1,\ 2 \text{ and } 3\,, \tag{2}$$

the two particles sit close in phase-space, and quantum mechanics can no longer be ignored. The most important quantum mechanical correction to be taken into account is the (anti-)symmetrization of the two-particle wave function: it ensures vanishing probability for two identical fermions to originate from the same phase-space point, and for bosons it leads to an enhanced probability to find them at the same point in phase-space compared to the classical expectation (bosons are "social subjects").

Since the distance in coordinate space $(x_1^i - x_2^i)$ is limited by the finite diameter $2R$ of the source, we can force the system into the quantum domain by measuring particle pairs with smaller and smaller relative momentum $\boldsymbol{p}_1 - \boldsymbol{p}_2$. Once $q_i \equiv p_1^i - p_2^i$ becomes smaller than $\pi\hbar/R$, the two particles can no longer avoid quantum mechanics by escaping to larger relative distances in coordinate space. Their emission probability will be affected by wave function symmetrization, leading in the case of bosons (fermions) to an enhanced (reduced) pair emission probability compared to the classical expectation (which would be simply the product of the individual single-particle emission probabilities). The two-particle *correlation function* is thus expected to begin to appreciably deviate from unity for relative momenta $q < q^* \simeq (\hbar/R)$. The critical value q^* at which this effect sets in (conveniently one chooses for q^* the value where the correlation is half way between it maximum (minimum) value and 1) is thus a measure for the geometric radius R of the source.

So far the naive picture of how two-particle HBT interferometry works. Unfortunately, the only situation where it applies more or less directly is for photon interferometry of stars for which the method was invented. I will spend my time here in Dronten to explain to you why this naive picture is generically *wrong* and how to substitute it correctly. The basic reason why the above simple cartoon ceases to work in high-energy nuclear and particle physics is that the sources created in hadronic or heavy-ion collisions live only for very short time periods and feature inhomogeneous temperature profiles and strong collective dynamical expansion. I will show you that for such sources the HBT radius parameters (half widths of the correlation function) generally don't measure the full source size, but only so-called "space-time regions of homogeneity" inside which the momentum distribution varies sufficiently little so that the particles can actually show the quantum-statistical correlations. They also mix the spatial and temporal structure of the source, and we will learn tricks (in particular a new way of parametrizing the correlation functions) which are designed unfold these different aspects from the data.

The size of the just mentioned homogeneity regions varies with the momentum of the emitted particles, causing an important dependence of the HBT parameters on the pair momentum. I will show you how this momentum dependence can be used to extract the strength of the collective flow of the source at decoupling. To do so in a quantitative way requires the detailed consideration of many physical features of the particle-emission process. One such feature which I will discuss here in some detail is the fact that often some of the particles one uses in constructing the pair-correlation function don't come directly from the source but are created well after decoupling by the decay of unstable resonances. Resonance decays not only affect the size and momentum dependence of some of the HBT radius parameters but also the overall strength of the correlation as described by the so-called "chaoticity parameter" λ.

My lectures are structured in the following way: In the First Lecture, I discuss the general connection of the measured one- and two-particle spectra with the phase-space distribution of particles in the source. In particular I discuss the important aspects of "chaotic" versus "coherent" particle emission and how to implement them in a formal way. This establishes the formal background which I exploit in the Second Lecture to extract in a more quantitative way general ("model-independent") analytic relationships between the geometric and dynamic space-time structure of the source and certain features of the pair correlation function. In the Third Lecture I analyze these relationships quantitatively in the framework of a general class of model sources with finite geometric extension and three-dimensional collective expansion. The results should give you a good qualitative and even semi-quantitative feeling of the expected behaviour of the pair correlation function in relativistic heavy-ion collisions.

The Fourth Lecture should have presented a comparison of the calculations with the data, but fortunately the organizers gave me only three hours time so that I was spared the embarrassment of having to admit that at the present moment no quantitative such comparison exists: the theoretical analysis is still so new that most of the diagrams you will see are less than a few months old, and the new high-quality heavy-ion data which allow for the detailed multidimensional analysis advocated here are still so hot ("preliminary") that the experimentalists wouldn't give me permission to show them anyhow. Please look into the Proceedings of the *Quark Matter '96* Conference in

Heidelberg, 20-24 May 1996, if you want to catch a first glimpse of the data.

LECTURE 1: SPECTRA AND EMISSION FUNCTION

One- and two-particle spectra

The covariant single- and two-particle distributions are defined by

$$P_1(\boldsymbol{p}) \;=\; E\,\frac{dN}{d^3p} = E\,\langle \hat{a}_{\boldsymbol{p}}^+ \hat{a}_{\boldsymbol{p}} \rangle \,, \tag{3}$$

$$P_2(\boldsymbol{p}_a, \boldsymbol{p}_b) \;=\; E_a\,E_b\,\frac{dN}{d^3p_a\,d^3p_b} = E_a\,E_b\,\langle \hat{a}_{\boldsymbol{p}_a}^+ \hat{a}_{\boldsymbol{p}_b}^+ \hat{a}_{\boldsymbol{p}_b} \hat{a}_{\boldsymbol{p}_a} \rangle \,, \tag{4}$$

where $\hat{a}_{\boldsymbol{p}}^+$ ($\hat{a}_{\boldsymbol{p}}$) creates (destroys) a particle with momentum $\boldsymbol{p}$. The angular brackets denote an ensemble average,

$$\langle \hat{O} \rangle = \mathrm{tr}\,(\hat{\rho}\hat{O}) \,, \tag{5}$$

where $\hat{\rho}$ is the density operator associated with the ensemble. (When talking about an ensemble we may think of either a single large, thermalized source, or a large number of similar, but not necessarily thermalized collision events.) The single-particle spectrum is normalized to the average number of particles, $\langle N \rangle$, per collision,

$$\int \frac{d^3p}{E}\,P_1(\boldsymbol{p}) = \langle N \rangle \,, \tag{6}$$

while the two-particle distribution is normalized to the number of possible pairs, $\langle N(N-1) \rangle$, per event:

$$\int \frac{d^3p_a}{E_a}\,\frac{d^3p_b}{E_b}\,P_2(\boldsymbol{p}_a, \boldsymbol{p}_b) = \langle N(N-1) \rangle \,. \tag{7}$$

The two-particle correlation function is defined as[3]

$$C(\boldsymbol{p}_a, \boldsymbol{p}_b) = \frac{\langle N \rangle^2}{\langle N(N-1) \rangle}\,\frac{P_2(\boldsymbol{p}_a, \boldsymbol{p}_b)}{P_1(\boldsymbol{p}_a) P_1(\boldsymbol{p}_b)} \,. \tag{8}$$

If the two particles are emitted independently and final-state interactions are neglected I will show that it is possible to prove a generalized Wick theorem,

$$P_2(\boldsymbol{p}_a, \boldsymbol{p}_b) = \frac{\langle N(N-1) \rangle}{\langle N \rangle^2} \left(P_1(\boldsymbol{p}_a) P_1(\boldsymbol{p}_b) \pm |\bar{S}(\boldsymbol{p}_a, \boldsymbol{p}_b)|^2 \right) \,, \tag{9}$$

where the plus (minus) sign refers to bosons (fermions), and we have defined the following covariant quantity:

$$\bar{S}(\boldsymbol{p}_a, \boldsymbol{p}_b) = \sqrt{E_a E_b}\,\langle \hat{a}_{\boldsymbol{p}_a}^+ \hat{a}_{\boldsymbol{p}_b} \rangle \,. \tag{10}$$

If the ensemble corresponds to a thermalized source in global thermodynamic equilibrium, i.e. $\hat{\rho}$ is the (grand) canonical density operator, this is just the well-known "thermal Wick theorem" [4]. But even in non-thermal sources emission may be "chaotic", i.e. the emission of particle a at point x_a may be completely independent from the emission of particle b at x_b. This can be described by assigning the wave function of the emitted particle a random phase ϕ which is averaged over in the ensemble average. I will show that this is also sufficient to guarantee Wick's theorem (10). – The opposite to independent, "chaotic" particle emission is emission from a coherent state (e.g. in a laser) where the phases of all emitted particles are fully correlated with each other; in that case the second term in (10) is completely missing. The general situation can be parametrized by a superposition of density operators,

$$\hat{\rho} = \alpha \, \hat{\rho}_{\text{chaotic}} + (1 - \alpha) \, \hat{\rho}_{\text{coherent}} \,, \tag{11}$$

with $0 \leq \alpha \leq 1$, where only the chaotic first part contributes to the second term in (9):

$$\begin{aligned}
\langle \hat{a}_a^+ \hat{a}_b^+ \hat{a}_b \hat{a}_a \rangle &= \langle \hat{a}_a^+ \hat{a}_a \rangle \langle \hat{a}_b^+ \hat{a}_b \rangle \pm \alpha |\langle \hat{a}_a^+ \hat{a}_b \rangle_{\text{ch}}|^2 \\
&+ \alpha(1 - \alpha) \left[\left(\langle \hat{a}_a^+ \hat{a}_a \rangle_{\text{ch}} - \langle \hat{a}_a^+ \hat{a}_a \rangle_{\text{coh}} \right) \left(\langle \hat{a}_b^+ \hat{a}_b \rangle_{\text{ch}} - \langle \hat{a}_b^+ \hat{a}_b \rangle_{\text{coh}} \right) \right] \,. \tag{12}
\end{aligned}$$

The last term contributes only if the chaotic and coherent parts of the density operator generate different single-particle spectra. One easily checks that for $\alpha < 1$ in such an ensemble the correlations are "incomplete", i.e. at $\boldsymbol{p}_a = \boldsymbol{p}_b$ the correlation function (8) approaches a value less than 2 for bosons and larger than 0 for fermions. – Since so far the heavy ion data don't appear to require a coherent component in the particle-production process, I will for the remainder of these lectures assume $\alpha = 1$ and refer the interested reader for the possible effects from partial coherence to the literature[5].

Assuming the generalized Wick theorem (9) the correlation function (8) can be written as

$$C(\boldsymbol{p}_a, \boldsymbol{p}_b) = 1 \pm \frac{|\langle \hat{a}_{\boldsymbol{p}_a}^+ \hat{a}_{\boldsymbol{p}_b} \rangle|^2}{\langle \hat{a}_{\boldsymbol{p}_a}^+ \hat{a}_{\boldsymbol{p}_a} \rangle \langle \hat{a}_{\boldsymbol{p}_b}^+ \hat{a}_{\boldsymbol{p}_b} \rangle} \,. \tag{13}$$

Note that the second term is positive definite, i.e. the correlation function cannot, for example, oscillate around unity. [If you see such a behaviour in the literature[6] (and the authors did not include final-state interactions) it is wrong.]

From now on I will assume that the emitted particles are bosons, and for convenience I will call them pions, although nearly everything in the first two lectures applies equally well to other bosonic particles (except where explicitly stated). In the last Lecture I will be a little more specific and distinguish between 2-pion and 2-kaon correlations for comparison.

The generalized Wick theorem

Pions are created in heavy-ion collisions throughout the history of the collision process, but we are only interested in those pions which reach the detector as free, non-interacting particles. (Unfortunately, pion interferometry can only be practically performed with charged pions, because neutral pions decay too rapidly. But charged pions have a long-range final-state interaction, the Coulomb repulsion from the other pion in the pair with equal charge, as well as the attractive or repulsive Coulomb interaction with the hundreds of other produced charged particles, including the charge of the protons in the fireball. A proper treatment of this many-body Coulomb problem is a difficult task[7]. Since I don't know yet how to do it properly, I will simply neglect the Coulomb interaction in the final state – assuming that somehow the experimentalists know how to approximately correct their measured correlations for it)

We assume that the last interaction of the pion, in which the finally observed pion is created in its free asymptotic state, can be parametrized by a classical source amplitude $J(x)$. The solution of the free Klein-Gordon equation for pions generated by such a classical current,

$$\left(\Box + m^2 \right) \hat{\phi}(x) = J(x) \,, \tag{14}$$

with outgoing boundary conditions is given[3] by a classical ("coherent") state

$$|J\rangle = e^{-\bar{n}/2} \exp \left(i \int d^3 p \, \tilde{J}(\boldsymbol{p}) \, \hat{a}_{\boldsymbol{p}}^+ \right) |0\rangle \tag{15}$$

where

$$\tilde{J}(\boldsymbol{p}) = \int \frac{d^4x}{\sqrt{(2\pi)^3 2E_p}} \exp[i(E_p t - \boldsymbol{p}\cdot\boldsymbol{x})] \, J(x) \qquad (16)$$

is the on-shell Fourier transform of the source $J(x)$, and the normalization of the state is given by

$$\bar{n} = \int d^3p \, |\tilde{J}(\boldsymbol{p})|^2 \,. \qquad (17)$$

The state (15) is an eigenstate of the destruction operator:

$$\hat{a}_{\boldsymbol{p}}|J\rangle = i\tilde{J}(\boldsymbol{p})|J\rangle \,. \qquad (18)$$

Emission from a single coherent state

If there is only a single classical source $J(x)$, the corresponding density operator of the "ensemble" is just the projection operator on the coherent state, $\hat{\rho}_{\text{coherent}} = |J\rangle\langle J|$, and, using (18), the single- and two-particle spectra (3,4) are easily evaluated:

$$E\frac{dN}{d^3p} = E\langle J|\hat{a}_{\boldsymbol{p}}^+ \hat{a}_{\boldsymbol{p}}|J\rangle = E|\tilde{J}(\boldsymbol{p})|^2 \,, \qquad (19)$$

$$E_a E_b \frac{dN}{d^3p_a \, d^3p_b} = E_a E_b \langle J|\hat{a}_{\boldsymbol{p}_a}^+ \hat{a}_{\boldsymbol{p}_b}^+ \hat{a}_{\boldsymbol{p}_b} \hat{a}_{\boldsymbol{p}_a}|J\rangle = E_a|\tilde{J}(\boldsymbol{p}_a)|^2 \cdot E_b|\tilde{J}(\boldsymbol{p}_b)|^2 \,. \qquad (20)$$

Obviously, there is no exchange term in the two-particle spectrum which is simply given by the product of the single-particle spectra. *A coherent state thus has no Bose-Einstein correlations.*

Emission by a chaotic superposition of classical sources

This changes if we consider a superposition of classical source amplitudes each of which emits free pions independently, i.e. with a random phase[3]:

$$J(x) = \sum_{i=1}^{N} e^{i\phi_i} \, e^{-ip_i \cdot (x-x_i)} \, J_0(x - x_i) \,. \qquad (21)$$

The construction rule[8,9] for this source is obvious: we take N sources $J_0(x)$ with identical internal structure, give each of them a boost with 4-momentum p_i, then translate them to different positions x_i in the fireball and supply them with a random phase ϕ_i. This allows for arbitrary x-p correlations [8] (i.e. correlations between the momentum spectrum of the emitted particles and the point from where they are emitted). The momenta p_i of the sources can, but need not be on the pion mass-shell; for example, the source could be a decaying Δ-resonance with 3-momentum $\boldsymbol{p}_i$. The on-shell Fourier transform of (21) is

$$\tilde{J}(\boldsymbol{p}) = \sum_{i=1}^{N} e^{i\phi_i} \, e^{ip\cdot x_i} \, \tilde{J}_0(p - p_i) \,, \qquad (22)$$

where

$$\tilde{J}_0(p - p_i) = \int \frac{d^4x}{\sqrt{(2\pi)^3 2E_p}} \, e^{i(p-p_i)\cdot x} \, J_0(x) \qquad (23)$$

is the (regular) Fourier transform of $J_0(x)$, and p is on-shell while p_i may be off-shell. The state $|J\rangle$ which is defined by inserting (22) into (15) now depends on the parameters $\{x_i, p_i, \phi_i; i = 1, \dots, N\}$:

$$|J\rangle \equiv |J[N; \{x, p, \phi\}]\rangle \,. \qquad (24)$$

The ensemble of sources can be defined in terms of a density operator $\hat{\rho}$ which fixes the distribution of these parameters. We assume that the number of sources N is distributed with a probability distribution P_N, the phases ϕ are distributed randomly between 0 and 2π, and the source positions x_i and momenta p_i are distributed with a phase-space density $\rho(x,p)$, with normalizations

$$\sum_{N=0}^{\infty} P_N = 1, \qquad \sum_{N=0}^{\infty} N P_N = \langle N \rangle, \qquad \int d^4x \, d^4p \, \rho(x,p) = 1. \tag{25}$$

The corresponding ensemble average is given by

$$\text{tr}(\hat{\rho}\hat{O}) = \sum_{N=0}^{\infty} P_N \prod_{i=1}^{N} \int d^4x_i \, d^4p_i \, \rho(x_i,p_i) \int_0^{2\pi} \frac{d\phi_i}{2\pi} \, \langle J[N;\{x,p,\phi\}] | \hat{O} | J[N;\{x,p,\phi\}] \rangle. \tag{26}$$

The calculation of the single-particle spectrum is straightforward:

$$\langle \hat{a}_{\boldsymbol{p}}^+ \hat{a}_{\boldsymbol{p}} \rangle = \sum_{N=0}^{\infty} P_N \prod_{i=1}^{N} \int d^4x_i \, d^4p_i \, \rho(x_i,p_i) \int_0^{2\pi} \frac{d\phi_i}{2\pi}$$
$$\times \sum_{n,n'=1}^{N} e^{i(\phi_n - \phi_{n'})} e^{ip \cdot (x_n - x_{n'})} \tilde{J}_0^*(p - p_{n'}) \tilde{J}_0(p - p_n). \tag{27}$$

After performing the integrations over the phases ϕ_i in the double sum over n and n', only the diagonal terms with $n = n'$ survive. For each term in the remaining single sum over n the integrations over x_i and p_i, $i \neq n$, can be done using the normalization condition (25). After suitably relabelling the dummy integration variables for the one remaining x- and p-integration we end up with N identical terms under the sum over n. This allows to perform the sum over N, and we simply get

$$P_1(\boldsymbol{p}) = E_p \langle |\tilde{J}(\boldsymbol{p})|^2 \rangle = \langle N \rangle E_p \int d^4x' \, d^4p' \, \rho(x',p') \, |\tilde{J}_0(p - p')|^2 \tag{28}$$

$$= \langle N \rangle E_p \int d^4p' \, \tilde{\rho}(p') \, |\tilde{J}_0(p - p')|^2. \tag{29}$$

The single-particle spectrum is thus obtained by folding the intrinsic momentum spectrum $|\tilde{J}_0(p)|^2$ of the individual source currents J_0 with the 4-momentum distribution of the sources, $\tilde{\rho}(p) = \int d^4x \, \rho(x,p)$.

The algebra for the two-particle spectrum is a little more involved. It is useful to first compute

$$\langle \hat{a}_{\boldsymbol{p}_a}^+ \hat{a}_{\boldsymbol{p}_b} \rangle = \sum_{N=0}^{\infty} P_N \prod_{i=1}^{N} \int d^4x_i \, d^4p_i \, \rho(x_i,p_i) \int_0^{2\pi} \frac{d\phi_i}{2\pi}$$
$$\times \sum_{n,n'=1}^{N} e^{i(\phi_n - \phi_{n'})} e^{i(p_b \cdot x_n - p_a \cdot x_{n'})} \tilde{J}_0^*(p_a - p_{n'}) \tilde{J}_0(p_b - p_n). \tag{30}$$

Again only the terms $n = n'$ survive the phase average, and after doing the dummy integrations over x_i, p_i, $i \neq n$, one finds that the remaining sum over n contains again N identical terms, such that the sum over N can be performed:

$$\langle \hat{a}_{\boldsymbol{p}_a}^+ \hat{a}_{\boldsymbol{p}_b} \rangle = \langle N \rangle \int d^4x' \, d^4p' \, \rho(x',p') \, e^{i(p_b - p_a) \cdot x'} \, \tilde{J}_0^*(p_a - p') \tilde{J}_0(p_b - p'). \tag{31}$$

With this auxiliary result at hand we can now attack the two-particle spectrum. From the definitions one finds

$$\langle \hat{a}_{\boldsymbol{p}_a}^+ \hat{a}_{\boldsymbol{p}_b}^+ \hat{a}_{\boldsymbol{p}_b} \hat{a}_{\boldsymbol{p}_a} \rangle = \sum_{N=0}^{\infty} P_N \prod_{i=1}^{N} \int d^4 x_i \, d^4 p_i \, \rho(x_i, p_i) \int_0^{2\pi} \frac{d\phi_i}{2\pi}$$

$$\times \sum_{n,n',m,m'=1}^{N} e^{i(\phi_n + \phi_m - \phi_{n'} - \phi_{m'})} e^{ip_a \cdot (x_n - x_{n'})} e^{ip_b \cdot (x_m - x_{m'})}$$

$$\times \tilde{J}_0^*(p_a - p_{n'}) \tilde{J}_0^*(p_b - p_{m'}) \tilde{J}_0(p_b - p_m) \tilde{J}_0(p_a - p_n). \quad (32)$$

The integration over the phases ϕ_i now yields two types of non-vanishing contributions: $n = n', m = m'$ and $n = m', m = n'$. The term where all four summation indices are equal, $n = m = n' = m'$, should be omitted[3]: it corresponds to emission of both particles from the same elementary source, and if one carefully first puts the whole system in a finite volume V, performs the calculation there and lets $V \to \infty$ in the end, then this term is suppressed relative to the others by a factor $1/V$. We thus get

$$\langle \hat{a}_{\boldsymbol{p}_a}^+ \hat{a}_{\boldsymbol{p}_b}^+ \hat{a}_{\boldsymbol{p}_b} \hat{a}_{\boldsymbol{p}_a} \rangle = \sum_{N=0}^{\infty} P_N \prod_{i=1}^{N} \int d^4 x_i \, d^4 p_i \, \rho(x_i, p_i) \sum_{n \neq m}^{N} \Big[|\tilde{J}_0(p_a - p_n)|^2 \, |\tilde{J}_0(p_b - p_m)|^2$$

$$+ \; e^{i(p_a - p_b) \cdot (x_n - x_m)} \, \tilde{J}_0^*(p_a - p_m) \tilde{J}_0^*(p_b - p_n) \tilde{J}_0(p_b - p_m) \tilde{J}_0(p_a - p_n) \Big]$$

$$= \sum_{N=0}^{\infty} P_N \prod_{i=1}^{N} \int d^4 x_i \, d^4 p_i \, \rho(x_i, p_i)$$

$$\times \left[\left(\sum_{n \neq m}^{N} |\tilde{J}_0(p_a - p_n)|^2 \right) \left(\sum_{m=1}^{N} |\tilde{J}_0(p_b - p_m)|^2 \right) \right.$$

$$+ \left(\sum_{n \neq m}^{N} e^{i(p_a - p_b) \cdot x_n} \, \tilde{J}_0^*(p_b - p_n) \tilde{J}_0(p_a - p_n) \right)$$

$$\left. \times \left(\sum_{m=1}^{N} e^{-i(p_a - p_b) \cdot x_m} \, \tilde{J}_0(p_b - p_m) \tilde{J}_0^*(p_a - p_m) \right) \right]. \quad (33)$$

After again doing the dummy integrations over x_i, p_i, $i \neq n, m$, one realizes that each of the two terms in the square bracket contains $N(N-1)$ identical terms, yielding a factor $\langle N(N-1) \rangle$ after performing the sum over N. Up to this factor, the first term is just a product of two terms of the type (28), i.e. a product of single-particle spectra, while the second term is recognized as a product of (31) and its complex conjugate. We thus have

$$P_2(\boldsymbol{p}_a, \boldsymbol{p}_b) = \frac{\langle N(N-1) \rangle}{\langle N \rangle^2} E_a E_b \Big[\langle \hat{a}_{\boldsymbol{p}_a}^+ \hat{a}_{\boldsymbol{p}_a} \rangle \langle \hat{a}_{\boldsymbol{p}_b}^+ \hat{a}_{\boldsymbol{p}_b} \rangle + |\langle \hat{a}_{\boldsymbol{p}_a}^+ \hat{a}_{\boldsymbol{p}_b} \rangle|^2 \Big] \quad (34)$$

$$= \frac{\langle N(N-1) \rangle}{\langle N \rangle^2} E_a E_b \Big[\langle |\tilde{J}(\mathbf{p}_a)|^2 \rangle \langle |\tilde{J}(\mathbf{p}_b)|^2 \rangle + |\langle \tilde{J}^*(\mathbf{p}_a) \tilde{J}(\mathbf{p}_b) \rangle|^2 \Big], \quad (35)$$

which proves the generalized Wick theorem (9).

Source Wigner function and spectra

These expressions can be rewritten in a very nice and suggestive way by introducing the so-called "emission function" $S(x, K)$ [10,11,5]:

$$S(x, K) = \int \frac{d^4 y}{2(2\pi)^3} e^{-iK \cdot y} \left\langle J^*(x + \tfrac{1}{2}y) J(x - \tfrac{1}{2}y) \right\rangle. \quad (36)$$

144

It is the Wigner transform of the density matrix associated with the classical source amplitudes $J(x)$. This Wigner density is a quantum-mechanical object defined in phase-space (x, K); in general it is neither positive definite nor real. But, when integrated over x or K it yields the classical (positive definite and real) source density in momentum or coordinate space, respectively, in exactly the same way as a classical phase-space density would behave. Furthermore, textbooks on Wigner functions show that their non-reality and non-positivity are genuine quantum effects resulting from the uncertainty relation and are concentrated at short phase-space distances; when the Wigner function is averaged over phase-space volumes which are large compared to the volume $(2\pi\hbar)^3$ of an elementary phase-space cell, the result is real and positive definite and behaves exactly like a classical phase-space density.

The emission function $S(x, K)$ is thus the quantum-mechanical analogue of the classical phase-space distribution which gives the probability of finding at point x a source which emits free pions with momentum K. Please note that K in $S(x, K)$ can be off-shell. Also, it is defined in terms of a 4-dimensional Wigner transform of the source density matrix[10], in contrast to the 3-dimensional expression suggested by Pratt[11] which neglects retardation and off-shell effects.

Using Eq. (16) it is easy to establish the following relationship:

$$
\begin{aligned}
\tilde{J}^*(\boldsymbol{p}_a)\,\tilde{J}(\boldsymbol{p}_b) &= \int \frac{d^4x_1\,d^4x_2}{(2\pi)^3\,2\sqrt{E_aE_b}}\,\exp(-ip_a\cdot x_1 + ip_b\cdot x_2)J^*(x_1)J(x_2) \\
&= \int \frac{d^4x\,d^4y}{(2\pi)^3\,2\sqrt{E_aE_b}}\,\exp(-iq\cdot x - iK\cdot y)J^*(x + \tfrac{1}{2}y)J(x - \tfrac{1}{2}y)\,, \quad (37)
\end{aligned}
$$

where $x = \tfrac{1}{2}(x_1 x_2)$ and $y = x_1 - x_2$. Inserting this into Eqs. (28) and (35) one finds the fundamental relations:

$$
E_K \frac{dN}{d^3K} = \int d^4x\, S(x, K)\,, \tag{38}
$$

$$
C(\boldsymbol{q}, \boldsymbol{K}) = 1 + \frac{|\int d^4x\, S(x, K)\,e^{iq\cdot x}|^2}{\int d^4x\, S(x, K + \tfrac{1}{2}q)\,\int d^4x\, S(x, K - \tfrac{1}{2}q)}\,. \tag{39}
$$

For the single-particle spectrum (38), the Wigner function $S(x, K)$ on the r.h.s. must be evaluated on-shell, i.e. at $K^0 = E_K = \sqrt{m^2 + \boldsymbol{K}^2}$. For the correlator (39) we have defined the relative momentum $\boldsymbol{q} = \boldsymbol{p}_a - \boldsymbol{p}_b$, $q^0 = E_a - E_b$ between the two particles in the pair, and the total momentum of the pair $\boldsymbol{K} = (\boldsymbol{p}_a + \boldsymbol{p}_b)/2$, $K^0 = (E_a + E_b)/2$. Of course, since the 4-momenta $p_{a,b}$ of the two measured particles are on-shell, $p_i^0 = E_i = \sqrt{m^2 + \boldsymbol{p}_i^2}$, the 4-momenta q and K are in general off-shell. They satisfy the orthogonality relation

$$
q \cdot K = 0\,. \tag{40}
$$

Thus, the Wigner function on the r.h.s. of Eq. (39) is *not* evaluated at the on-shell point $K^0 = E_K$. This implies that for the correlator, in principle, we need to know the off-shell behaviour of the emission function, i.e. the quantum-mechanical structure of the source. Obviously, this makes the problem appear rather untractable!

Fortunately, nature is nice to us: the interesting behaviour of the correlator (its deviation from unity) is concentrated at small values of $|\boldsymbol{q}|$. Expanding $K^0 = (E_a + E_b)/2$ for small q one finds

$$
K^0 = E_K \left(1 + \frac{\boldsymbol{q}^2}{8E_K^2} + \mathcal{O}\left(\frac{\boldsymbol{q}^4}{E_K^4}\right)\right) \approx E_K\,. \tag{41}
$$

Since the relevant range of q is given by the inverse size of the source (more properly: the inverse size of the regions of homogeneity in the source – see Lecture 2), the validity of this approximation is ensured in practice as long as the Compton wavelength of the particles is small compared to this "source size". For the case of pion, kaon, or proton interferometry for heavy-ion collisions this is true automatically due to the rest mass of the particles: even for pions at rest, the Compton wavelength of 1.4 fm is comfortably smaller than any typical nuclear source size. This is of enormous practical importance because it allows you essentially to replace the source Wigner density by a classical phase-space distribution function for on-shell particles. This provides a necessary theoretical foundation for the calculation of HBT correlations from classical hydrodynamic or kinetic (e.g. cascade or molecular dynamics) simulations of the collision.

In photon interferometry there is no rest mass available to help you: for photons, the approximation $K^0 \approx E_K$ can only be justified if they escape from the source with high momentum, and in HBT interferometry with soft photons the quantum-mechanical nature of the emission function needs to be explicitly considered. In practice this means that one must study the photon-production processes microscopically and quantum mechanically.

If the single-particle spectrum is an exponential function of the energy then it is easy to prove[12] that one can replace the product of single-particle distributions in the denominator of (39) by the square of the single-particle spectrum evaluated at the average momentum K:

$$C(\boldsymbol{q}, \boldsymbol{K}) \approx 1 + \left| \frac{\int d^4x \, e^{iq \cdot x} \, S(x, K)}{\int d^4x \, S(x, K)} \right|^2 \equiv 1 + \left| \langle e^{iq \cdot x} \rangle \right|^2. \tag{42}$$

The deviations from this approximation are proportional to the curvature of the single-particle distribution in logarithmic representation[12]. They are small in practice because the measured single-particle spectra are usually more or less exponential. In the second equality of (42) we defined $\langle \ldots \rangle$ as the average taken with the emission function; due to the K-dependence of $S(x, K)$ this average is a function of K. This notation will be used extensively in Lecture 2.

The ensemble average on the r.h.s. of (36) is defined in the sense of Eq. (26) and can be evaluated with the help of the definition (21). One finds

$$S(x, K) = \langle N \rangle \int d^4z \, d^4q \, \rho(x - z, q) \, S_0(z, K - q), \tag{43}$$

where

$$S_0(x, p) = \int \frac{d^4y}{2(2\pi)^3} \, e^{-ip \cdot y} J_0^*(x + \tfrac{1}{2}y) J_0(x - \tfrac{1}{2}y) \tag{44}$$

is the Wigner function associated with an individual source J_0. This establishes a similar folding relation for the Wigner function itself as we have already obtained in (29) for the single-particle spectrum: the emission function of the complete source is obtained by folding the Wigner function for an individual pion source J_0 with the Wigner distribution ρ of these sources. Eq. (43) is useful for the calculation of quantum-statistical correlations from classical Monte Carlo event generators for heavy-ion collisions: $\langle N \rangle \rho(x, p)$ can be considered as the distribution of the classical phase-space coordinates of the pion emitters (decaying resonances or 2-body collision systems), and $S_0(x, p)$ as the Wigner function of the free pions emitted at these points (for example, a Gaussian in Quantum-Molecular-Dynamics calculations[13]). Replacing the former by a sum of δ-functions describing the space-time locations of the last interactions and the pion momenta just afterwards, and the latter by a product of two Gaussians with momentum spread Δp and coordinate spread Δx such that $\Delta x \Delta p \geq \hbar/2$, we recover the expressions derived in[14].

The fundamental relations (38) and (39) resp. (42) show that *both the single-particle spectrum and the two-particle correlation function can be expressed as simple integrals over the emission function.* The emission function thus is the crucial ingredient in the theory of HBT interferometry: if it is known, the calculation of one- and two-particle spectra is straightforward (even if the evaluation of the integrals may in some cases be technically involved); more interestingly, measurements of the one- and two-particle spectra provide access to the emission function and thus to the space-time structure of the source. This latter aspect is, of course, the motivation for exploiting HBT in practice. In my second and third Lecture I will concentrate on the question to what extent this access to the space-time structure from only momentum-space data really works, whether it is complete, and (since we will find it is not and HBT analyses will thus be necessarily model dependent) what can be reliably said about the extension and dynamical space-time structure of the source anyhow, based on a minimal set of intuitive and highly-suggestive model assumptions.

LECTURE 2: MODEL-INDEPENDENT DISCUSSION OF HBT CORRE-LATION FUNCTIONS

In this lecture I will discuss very general relations between the space-time structure of the source (as encoded in the x-dependence of the emission function $S(x, K)$) and the shape of the two-particle correlation function. These relations are valid for arbitrary emission functions, and in this sense the discussion is *model-independent*. It nevertheless provides important insight into the physical features of HBT interferometry, in particular for short-lived dynamical sources, and it clarifies what HBT can achieve and what not.

The mass-shell constraint

Expressions (39,42) show that the correlation function is related to the emission function by a Fourier transformation. At first sight this might suggest that one should easily be able to reconstruct the emission function from the measured correlation function by inverse Fourier transformation, the single particle spectrum (38) providing the normalization. This is, however, not correct. The reason is that, since the correlation function is constructed from the on-shell momenta of the measured particle pairs, not all four components of the relative momentum q occurring on the r.h.s. of (42) are independent. They are related by the "mass-shell constraint" (40) which can, for instance, be solved for q^0:

$$q^0 = \boldsymbol{\beta} \cdot \boldsymbol{q} \qquad \text{with} \qquad \boldsymbol{\beta} = \frac{\boldsymbol{K}}{K^0} \approx \frac{\boldsymbol{K}}{E_K} . \tag{45}$$

$\boldsymbol{\beta}$ is (approximately) the velocity of the c.m. of the particle pair. The Fourier transform in (42) is therefore not invertible, and the reconstruction of the space-time structure of the source from HBT measurements will thus always require additional model assumptions.

It is instructive to insert (45) into (42):

$$C(\boldsymbol{q}, \boldsymbol{K}) \approx 1 + \left| \frac{\int d^4x \, \exp(i\boldsymbol{q} \cdot (\boldsymbol{x} - \boldsymbol{\beta} \, t)) \, S(x, K)}{\int d^4x \, S(x, K)} \right|^2 . \tag{46}$$

This shows that the correlator $C(\boldsymbol{q}, \boldsymbol{K})$ actually mixes the spatial and temporal information on the source in a non-trivial way which depends on the pair velocity $\boldsymbol{\beta}$. Only for time-independent sources things seem to be simple: the correlator then just measures the Fourier transform of the spatial source distribution. Closer inspection shows, however, that it does so only in the directions *perpendicular* to $\boldsymbol{\beta}$ since the time integration leads to a δ-function $\delta(\boldsymbol{\beta} \cdot \boldsymbol{q})$:

$$\lim_{T \to \infty} \left| \frac{\int_{-T}^{T} dt \, \exp(-i \, \boldsymbol{q} \cdot \boldsymbol{\beta} \, t)}{\int_{-T}^{T} dt} \right|^2 = \lim_{T \to \infty} \frac{2\pi}{T} \delta(\boldsymbol{q} \cdot \boldsymbol{\beta}). \tag{47}$$

This implies that there are no correlations in the direction *parallel* to the pair velocity $\boldsymbol{\beta}$ (which will be called the "outward" direction below), i.e. $C = 1$ for $q_{\text{out}} \neq 0$. The width of the correlator in this direction vanishes! This should puzzle you: wouldn't you have thought that the width of the correlator in the "outward" direction is inversely related to the source size in that direction (which is, of course, perfectly finite)? As we will see in the next subsection this unexpected behaviour is just another consequence of the mixing of the spatial and temporal structure of the source in the correlator: The width parameter of the correlator in the "outward" direction receives also a contribution from the lifetime of the source which in this case diverges, leading to the vanishing width of the correlator.

K-dependence of the correlator

Eq. (42) shows that in general the correlator is a function of *both* $\boldsymbol{q}$ and $\boldsymbol{K}$. Only if the emission function factorizes in x and K, $S(x, K) = F(x) \, G(K)$, which means that every point x in the source emits particles with the same momentum spectrum $G(K)$ (no "x-K-correlations"), the K-dependence in $G(K)$ cancels between numerator and denominator of (42), and the correlator seems to be K-independent. However, not even this is really true: even after the cancellation of the explicit K-dependence $G(K)$, there remains an implicit K-dependence via the pair velocity $\boldsymbol{\beta} \approx \boldsymbol{K}/E_K$ in the exponent on the r.h.s. of Eq. (46)! Only if both conditions, factorization of the emission function in x and K *and* time-independence of the source, apply simultaneously, the correlation function is truely K-independent (because then the $\boldsymbol{\beta}$-dependence resides only in the δ-function (47)).

The only practical situation which I know where this occurs and a K-independent correlation function should thus be expected is in HBT interferometry of stars for which the method was invented[15]. It is hard to believe that this complication in the application of the original HBT idea to high-energy collisions went nearly unnoticed for more than 20 years and was stumbled upon more or less empirically by Scott Pratt in his pioneering work on HBT interferometry for heavy-ion collisions[11] only in 1984!

If one parametrises it by a Gaussian in q (see below) this means that in general the parameters ("HBT radii") depend on K. Typical sources of x-K correlations in the emission function are a collective expansion of the emitter and/or temperature gradients in the particle source: in both cases the momentum spectrum $\sim \exp[-p \cdot u(x)/T(x)]$ of the emitted particles (where $u^\mu(x)$ is the 4-velocity of the expansion flow) depends on the emission point. In the case of collective expansion, the spectra from different emission points are Doppler shifted relative to each other. If there are temperature gradients, e.g. a high temperature in the center and cooler matter at the edges, the source will look smaller for high-momentum particles (which come mostly from the hot center) than for low-momentum ones (which receive larger contributions also from the cooler outward regions).

We thus see that collective expansion of the source induces a K-dependence of the correlation function. But so do temperature gradients. The crucial question is: does a careful measurement of the correlation function, in particular of its K-dependence, permit a separation of such effects, i.e. can the collective dynamics of the source be quantitatively determined through HBT experiments? We will see that this is not an easy task; however, with sufficiently good data, it should be possible. In any case, the K-dependence of the correlator is a decisive feature which puts the HBT game into a completely new ball park. Even if it sounds exaggerated and may at first offend some of my experimentalist friends who are busy fighting the limited statistics of their data: two-particle correlation measurements which are not able to resolve the K-dependence of the HBT parameters are, in high-energy nuclear and particle physics, essentially useless. [Unfortunately, this applies to all the HBT data from pp and e^+e^- collisions which I am aware of. In my opinion, a renewed investigation of two-particle correlations from pp and e^+e^- collisions, using the powerful new tool of multidimensional HBT analysis, should be a high-priority project – as it is, we have practically nothing with which to compare our heavy-ion results in a meaningful way.]

The Gaussian approximation

As motivated in the Introduction, the most interesting feature of the two-particle correlation function is its half-width. Actually, since the relative momentum $\boldsymbol{q} = \boldsymbol{p_1} - \boldsymbol{p_2}$ has three Cartesian components, the fall-off of the correlator for increasing q is not described by a single half-width, but rather by a (symmetric) 3×3 tensor[16] which describes the curvature of the correlation function near $\boldsymbol{q} = 0$. We will see that in fact nearly all relevant information that can be extracted from the correlation function resides in the 6 independent components of this tensor[17]. This in turn implies that in order to compute the correlation function C it is sufficient to approximate the source function S by a Gaussian in x which contains only information on its space-time moments up to second order. [Gaussian approximations for the emission function have been used for the discussion of HBT correlation functions in many different variants [12,16--22], a perfect example how research proceeds by trial and error. Here I give the rigorous derivation first published in[21].]

Let us write the arbitrary emission function $S(x, K)$ in the following form:

$$S(x, K) = N(K)\, S(\bar{x}(K), K)\, e^{-\frac{1}{2}\tilde{x}^{\mu}(K) B_{\mu\nu}(K)\tilde{x}^{\nu}(K)} + \delta S(x, K), \qquad (48)$$

where we adjust the parameters $N(K)$, $\bar{x}^{\mu}(K)$, and $B_{\mu\nu}(K)$ of the Gaussian first term in such a way that the correction term δS has vanishing zeroth, first and second order space-time moments:

$$\int d^4x\, \delta S(x, K) = \int d^4x\, x^{\mu}\, \delta S(x, K) = \int d^4x\, x^{\mu}x^{\nu}\, \delta S(x, K) = 0. \qquad (49)$$

This is achieved by choosing

$$N(K) \;=\; E_K \frac{dN}{d^3K} \frac{\det B_{\mu\nu}(K)}{S(\bar{x}(K), K)}, \qquad (50)$$

$$\bar{x}^{\mu}(K) \;=\; \langle x^{\mu}\rangle, \qquad (51)$$

$$\left(B^{-1}\right)_{\mu\nu}(K) \;=\; \langle \tilde{x}_{\mu}\tilde{x}_{\nu}\rangle \equiv \langle (x-\bar{x})_{\mu}(x-\bar{x})_{\nu}\rangle. \qquad (52)$$

The (K-dependent) average over the source function $\langle\ldots\rangle$ has been defined in Eq. (42). The normalization factor (50) ensures that the Gaussian term in (48) gives the correct single-particle spectrum (38); it fixes the normalization on-shell, i.e. for $K^0 = E_K$, but as we discussed this is where we need the emission function also for the computation of the correlator. (Note that for photon interferometry this may not be true, and (50) should then be replaced by a suitable generalization.) $\bar{x}(K)$ in (51) is the centre of the emission function $S(x, K)$ and approximately equal to its "saddle point", i.e. the point of highest emissivity for particles with momentum K. The second equality in (52) defines $\tilde{x}$ as the space-time coordinate relative to the centre of the emission function; only this quantity enters the further discussion, since, due to the invariance of the momentum spectra under arbitrary translations of the source in coordinate space, the absolute position of the emission point is not measurable in experiments which determine only particle momenta. Since $\bar{x}(K)$ is not measurable, neither is the normalization $N(K)$ [21] as its definition (50) involves the emission function at $\bar{x}(K)$. Finally, Eq. (52) ensures that the Gaussian first term in (48) correctly reproduces the variances $\langle \tilde{x}_\mu \tilde{x}_\nu \rangle$ of the original emission function, in particular its r.m.s. widths in the various space-time directions.

Inserting the decomposition (48) into Eq. (42) we obtain for the correlation function

$$C(\boldsymbol{q}, \boldsymbol{K}) = 1 + \exp[-q^\mu q^\nu \langle \tilde{x}_\mu \tilde{x}_\nu \rangle (\boldsymbol{K})] + \delta C(\boldsymbol{q}, \boldsymbol{K}). \qquad (53)$$

The Gaussian in q results from the Fourier transform of the Gaussian contribution in (48); the last term δC receives contributions from the second term δS in (48) which contains information on the third and higher-order space-time moments of the emission function, like sharp edges, wiggles, secondary peaks, etc. in the source. It is at least of fourth order in q, i.e. the second derivative of the full correlator at $q = 0$ is given *exactly* by the Gaussian in (53). Please note that the exponent of the correlator contains no term linear in q; since the correlator must be symmetric under $\boldsymbol{q} \to -\boldsymbol{q}$ because it does not matter which of the two particles of the pair receives the label 1 or 2, a linear q-dependence could only arise in the form $\exp(-R|\boldsymbol{q}|)$. The only type of emission function yielding such a q-dependence of the correlator would be a spherically symmetric Lorentzian. Any emission function which at large x falls off faster than $1/x^2$ results in the leading Gaussian behaviour (53) instead. This settles, in my opinion, the old issue whether Gaussian or exponential fits of the correlation function should be preferred.

[In the past it has repeatedly been observed that the correlation data appear to be better fit by exponentials than by Gaussians. However, as far as I know, this happened always when one tried to fit the correlator as a function of the single Lorentz invariant variable $Q_{\text{inv}}^2 = (q^0)^2 - \boldsymbol{q}^2$. Contemplating the structure of Eq. (53) one realizes that such a fit does not make sense: the generic structure of the exponent, $-q^\mu q^\nu \langle \tilde{x}_\mu \tilde{x}_\nu \rangle$, tells us that the term $(q^0)^2$ should come with the time variance of the source while the spatial components $(q^i)^2$ should come with the spatial variances of the source. Since all variances are positive semidefinite by definition, it does not make sense to parametrize the correlation function by a variable in which $(q^0)^2$ and $\boldsymbol{q}^2$ appear with the opposite sign! Such a fit could only work if the time variance and all mixed variances would vanish identically, and all three spatial variances were equal. This is certainly not the general case in nature. The good exponential fits of the correlation functions from pp and e^+e^- collisions are thus, in my mind, purely accidental and an empirical curiosity without physical meaning. *The variable Q_{inv} should not be used for fitting HBT data.*]

Please note also that Eq. (53) has no factor $\frac{1}{2}$ in the exponent. If the measured correlator is fitted by a Gaussian as defined in (53), its q-width can be directly interpreted in terms of the r.m.s. widths of the source in coordinate space. Any remaining factors of $\sqrt{3}$, $\sqrt{3}$, or $\sqrt{5}$ (which you can sometimes find in the literature) are due to reexpressing the r.m.s. width of the source in terms of certain other width parameters chosen for the parametrization of the source in coordinate space. The confusion connected with such factors is easily avoided by always expressing the source parametrization directly in terms of r.m.s. widths.

Eqs. (48) and (53) would, of course, not be useful if the contributions from δS and δC were not somehow small enough to be neglected. This requires a numerical investigation. It was shown numerically in Ref.[17] that in typical (and even in some not so typical) situations δS *has a negligible influence on the half width of the correlation function*. It contributes only weak, essentially unmeasurable structures in $C(\boldsymbol{q}, \boldsymbol{K})$ at large values of $\boldsymbol{q}$. The reader can easily verify this analytically for an emission function with a sharp box profile; the results for the exact correlator and the one resulting from the Gaussian approximation (48) are given in [16] and differ by less than 5% in the half widths; the exact correlator has, as a function of q, secondary maxima with an amplitude below 5% of the value of the correlator at $q = 0$. We have checked that similar statements remain even true for a source with a doughnut structure, i.e. with a hole in the middle, which was obtained by rotating the superposition of two 1-dimensional Gaussians separated by twice their r.m.s. widths around their center. The only situation where these statements require qualification is if the correlator receives contributions from the decay of long-lived resonances; this will be discussed in Lecture 3.

From Eq. (53) we conclude that the two-particle correlation function measures the second order space-time variances of the emission function. That's it – finer features of its space-time structure (edges, wiggles, holes) cannot be measured with two-particle correlations. The variances $\langle \tilde{x}_\mu \tilde{x}_\nu \rangle$ are in general *not* identical with our naive intuitive notion of the "source radius": unless the source is stationary and has no x-K-correlations at all, the variances depend on the momentum $\boldsymbol{K}$ of the pair and cannot be interpreted in terms of simple overall source geometry. Their correct physical interpretation[23,19,12] is in terms of "lengths of homogeneity" which give, for each pair momentum $\boldsymbol{K}$, the size of the region around the point of maximal emissivity $\bar{x}(\boldsymbol{K})$ over which the emission function is sufficiently homogeneous to contribute to the correlation function. Thus HBT measures "regions of homogeneity" in the source and their variation with the momentum of the particle pairs. As we will see, the latter is the key to their physical interpretation.

Gaussian parametrizations for the correlation function

A full characterization of the source in terms of its second order space-time variances requires knowledge of the 10 parameters $\langle \tilde{x}_\mu \tilde{x}_\nu \rangle$. These quantities appear in the expression (42) for the correlation function but this expression still uses all four components of the relative momnetum q^μ. However, as already noted only three of the four components are independent, due to the mass-shell constraint (45). Thus only 6 linear combinations of the variances $\langle \tilde{x}_\mu \tilde{x}_\nu \rangle(K)$ are actually measurable [16].

If the source is azimuthally symmetric around the beam axis, this counting changes as follows: Even if the source is azimuthally symmetric in coordinate space, the emission function $S(x, K)$ in phase space is for finite $\boldsymbol{K}$ no longer azimuthally symmetric because the transverse components $\boldsymbol{K}_\perp$ of the pair momentum distinguish a direction transverse to the beam direction. There remains, however, a reflection symmetry with respect to the plane spanned by $\boldsymbol{K}$ and the beam axis. If we call the direction orthogonal to this plane y, all mixed variances which are linear in y must vanish due to this reflection symmetry, and the correlator must be symmetric under $q_y \to -q_y$. Thus only 7 non-vanishing variances $\langle \tilde{x}_\mu \tilde{x}_\nu \rangle$ survive in general, of which, due to the mass-shell constraint

(45) only 4 linear combinations are measurable.

Before the correlator (53) can be fit to experimental data, the redundant components of q must first be eliminated from the exponent of the Gaussian. At this point it is useful to introduce a cartesian coordinate system with z along the beam axis and K lying in the x-z-plane. Customarily one labels the z-component of a 3-vector by l (for *longitudinal*), the x-component by o (for *outward*) and the y-component by s (for *sideward*). The above choice of the orientation of the x and y axes is natural for azimuthally symmetric collision events or event samples because then the transverse components $K_\perp$ of the pair momentum distinguish a direction in the transverse plane, and it is convenient to orient one of the coordinate axes along this direction such that K has only one transverse component:

$$K = (K_x, K_y, K_z) = (K_\perp, 0, K_L). \tag{54}$$

For sources without azimuthal symmetry, e.g. from collisions at finite impact parameter which have been selected according to the orientation of the collision plane, it is probably more useful to orient the x axis along the collision plane; then K will be characterized by three parameters, K_L, $K_\perp$ and the azimuthal angle Φ of $K_\perp$ relative to the x-z collision plane.

A useful formalism for HBT interferometry of finite impact parameter collisions has not yet been developed. I will therefore limit my discussion to azimuthally symmetric event samples and exploit the symmetry to orient the x-axis along $K_\perp$. Then from (45) we see that $\beta_s = 0$ such that

$$q^0 = \beta_\perp q_o + \beta_l q_l \tag{55}$$

with $\beta_\perp = |K_\perp|/K^0$ being (approximately) the velocity of the particle pair transverse to the beam direction while β_l is its longitudinal component.

This constraint can now be used in various ways to eliminate the redundant q-components from the exponent of Eq. (53). But whichever choice one makes, all the K-dependent parameters ("HBT radii") in the resulting Gaussian function of q can be easily calculated from the variances $\langle \tilde{x}^\mu \tilde{x}^\nu \rangle$, i.e. by simple quadrature formulae, for arbitrary emission functions $S(x, K)$. The relation between the HBT parameters and the variances is *model-independent*, i.e. it does not depend on the form of the emission function $S(x, K)$.

Here I discuss two specific parametrizations: the standard Cartesian one (mostly for historic reasons[11,24,25,18]), and the more physically motivated Yano-Koonin-Podgoretskiĭ one [16,21,22].

Standard Cartesian parametrization

The standard form[12,18] for the parametrization of the correlation function is obtained by using (55) to eliminate q^0 from Eq. (53). One obtains

$$C(q, K) = 1 + \exp\left[- \sum_{i,j=s,o,l} R_{ij}^2(K) \, q_i \, q_j \right] \tag{56}$$

where the 6 HBT radius parameters R_{ij} are defined in terms of the following variances of the source function[12,18,26]:

$$R_{ij}^2(K) = \langle (\tilde{x}_i - \beta_i \tilde{t})(\tilde{x}_j - \beta_j \tilde{t}) \rangle, \quad i,j = s,o,l. \tag{57}$$

For an azimuthally symmetric sample of collision events $C(q, K)$ is symmetric with respect to $q_s \to -q_s$ [16]. Then $R_{os}^2 = R_{sl}^2 = 0$ and

$$C(q, K) = 1 + \exp\left[-R_s^2(K)q_s^2 - R_o^2(K)q_o^2 - R_l^2(K)q_l^2 - 2R_{ol}^2(K)q_o q_l \right], \tag{58}$$

152

with

$$R_s^2(\boldsymbol{K}) = \langle \tilde{y}^2 \rangle, \tag{59}$$

$$R_o^2(\boldsymbol{K}) = \langle (\tilde{x} - \beta_\perp \tilde{t})^2 \rangle, \tag{60}$$

$$R_l^2(\boldsymbol{K}) = \langle (\tilde{z} - \beta_l \tilde{t})^2 \rangle, \tag{61}$$

$$R_{ol}^2(\boldsymbol{K}) = \langle (\tilde{x} - \beta_\perp \tilde{t})(\tilde{z} - \beta_l \tilde{t}) \rangle. \tag{62}$$

The cross-term (62) was only recently discovered[18]. Clearly these HBT radius parameters mix spatial and temporal information on the source in a non-trivial way. Furthermore, since they multiply combinations of the components q^μ which are not invariant under longitudinal boosts of the measurement frame, their interpretation depends on the frame in which the particle momenta are specified. This complicates their physical interpretation. An extensive discussion of these parameters, in particular of the meaning of the generally non-vanishing cross-term R_{ol}^2, can be found in Refs.[12,16--18,27,28], where the expressions (59)-(62) were analyzed analytically and numerically for a large class of (azimuthally-symmetric) model source functions. Some of these results will be discussed in the Lecture 3.

An important observation resulting from these studies is that the difference

$$R_{\text{diff}}^2 \equiv R_o^2 - R_s^2 = \beta_\perp^2 \langle \tilde{t}^2 \rangle - 2\beta_\perp \langle \tilde{x}\tilde{t} \rangle + (\langle \tilde{x}^2 \rangle - \langle \tilde{y}^2 \rangle) \tag{63}$$

is generally dominated by the first term on the r.h.s. and thus provides access to the lifetime $\Delta t = \sqrt{\langle t^2 \rangle - \langle t \rangle^2}$ of the source (more exactly: the duration of the particle-emission process)[29]. However, in relativistic heavy-ion collisions, due to rapid expansion of the source one would not expect $\langle \tilde{t}^2 \rangle$ to be generically much larger than either $\langle \tilde{x}^2 \rangle$ or $\langle \tilde{y}^2 \rangle$ unless there is a phase transition to a quark-gluon plasma and the collision fireball is initiated within a certain range of energy densities above the critical energy density where the transition occurs[30]. In the standard fit one is not sensitive to small values of Δt since Eq. (63) then involves a small difference of two large numbers, each associated with standard experimental errors. The factor $\beta_\perp^2 \leq 1$ in front of $\langle \tilde{t}^2 \rangle$ further complicates its extraction, in particular at low $K_\perp$ where $\Delta t(\boldsymbol{K})$ is usually largest (see below). Successful attempts to determine the duration of particle emission from HBT measurements have been reported from low-energy heavy-ion collisions (using 2-proton correlations) where the measured lifetimes are very long: 25 ± 15 fm/c in Ar+Sc collisions at $E/A = 80$ MeV[31] and 1400 ± 300 fm/c in Xe+Al collisions at $E/A = 31$ MeV[32] (the latter is the typical evaporation time of a compound nucleus). Two-pion correlations at ultra-relativistic energies ($E/A = 200$ GeV) so far failed to yield positive evidence for a non-vanishing emission duration [33,34], except for the heaviest collision system Pb+Pb[35], but even there the effective lifetime is only a few fm/c.

Yano-Koonin-Podgoretskiǐ (YKP) parametrization

The Yano-Koonin-Podgoretskiǐ parametrization of the correlation function is the generalization to azimuthally symmetric systems of the Yano-Koonin parametrization[36]. It was first written down by M.I. Podgorestskiǐ in 1983 for moving azimuthally symmetric, but non-expanding sources[37], with K-independent parameters. Not knowing about this paper, we reinvented it in [16], with K-dependent parameters for expanding azimuthally symmetric systems. The YKP parametrization is based on an elimination in Eq. (53) of q_o and q_s in terms of $q_\perp = \sqrt{q_o^2 + q_s^2}$, q^0, and q_3, using Eq. (55):

$$C(\boldsymbol{q}, \boldsymbol{K}) = 1 + \exp\left[-R_\perp^2 q_\perp^2 - R_\parallel^2 \left(q_l^2 - (q^0)^2\right) - \left(R_0^2 + R_\parallel^2\right)(q \cdot U)^2\right]. \tag{64}$$

Like the standard Cartesian form (58) it has four K-dependent fit parameters, but now only three of them, $R_\perp(\boldsymbol{K})$, $R_\parallel(\boldsymbol{K})$, and $R_0(\boldsymbol{K})$, have dimensions of length while the fourth parameter, $U(\boldsymbol{K})$, is a 4-velocity with only a longitudinal spatial component:

$$U(\boldsymbol{K}) = \gamma(\boldsymbol{K})\,(1,0,0,v(\boldsymbol{K})), \quad \text{with} \quad \gamma = \frac{1}{\sqrt{1-v^2}}. \tag{65}$$

This parametrization has the advantage that the "YKP radii" $R_\perp$, $R_\parallel$, and R_0 extracted from such a fit do not depend on the longitudinal velocity of the observer system in which the correlation function is measured; they are invariant under longitudinal boosts. Their physical interpretation is easiest in terms of coordinates measured in the frame where $v(\boldsymbol{K})$ vanishes. There they are given by[16]

$$R_\perp^2(\boldsymbol{K}) = R_s^2(\boldsymbol{K}) = \langle \tilde{y}^2 \rangle, \tag{66}$$

$$R_\parallel^2(\boldsymbol{K}) = \left\langle (\tilde{z} - \beta_l \tilde{x}/\beta_\perp)^2 \right\rangle - \beta_l^2 \langle \tilde{y}^2 \rangle / \beta_\perp^2 \approx \langle \tilde{z}^2 \rangle, \tag{67}$$

$$R_0^2(\boldsymbol{K}) = \left\langle \left(\tilde{t} - \tilde{x}/\beta_\perp \right)^2 \right\rangle - \langle \tilde{y}^2 \rangle / \beta_\perp^2 \approx \langle \tilde{t}^2 \rangle, \tag{68}$$

where in the last two expressions the approximation consists of dropping terms which were found in[16] to be generically small. (A more quantitative discussion of this point will follow in Lecture 3.) The first expression (66) remains true in any longitudinally boosted frame.

Eq. (68) shows that the YKP parameter $R_0(\boldsymbol{K})$ essentially measures the time duration $\Delta t(\boldsymbol{K}) = \sqrt{\langle \tilde{t}^2 \rangle}$ during which particles of momentum $\boldsymbol{K}$ are emitted, in the frame were the YKP velocity $v(\boldsymbol{K}) = 0$. It enters as the leading contribution in R_0, is fitted directly and no longer obtained as the difference of two large fit parameters as in the standard Cartesian fit.

Eqs. (66)-(68) were written down in the special frame where $v(\boldsymbol{K}) = 0$ which we call the *Yano-Koonin (YK) frame*. In Refs.[16,22] it is shown that for a large class of models this frame essentially coincides with the longitudinal rest frame of the fluid cell around the point $\bar{x}(\boldsymbol{K})$ of maximum emissivity at momentum $\boldsymbol{K}$ (i.e. the *Longitudinal Saddle Point System* LSPS[20]). This is true also for sources which are not longitudinally boost-invariant and for which the LSPS and the LCMS (the *Longitudinally CoMoving System* in which the pion pair has $\beta_l = 0$[29]) do not coincide.

In general the measurement system will not coincide with the (K-dependent) YK-frame, and the YKP radii will be given by more complicated combinations of the space-time variances of the source expressed in the coordinates of the measurement frame. This is the simple result of a Lorentz-boost between the two frames. However, I stress that in any frame the YKP parameters can again be easily calculated from the second order moments of the source function $S(x, K)$, i.e. by simple quadrature. Introducing the notational shorthands

$$A = \left\langle \left(\tilde{t} - \tilde{\xi}/\beta_\perp \right)^2 \right\rangle, \tag{69}$$

$$B = \left\langle \left(\tilde{z} - \beta_l \tilde{\xi}/\beta_\perp \right)^2 \right\rangle, \tag{70}$$

$$C = \left\langle \left(\tilde{t} - \tilde{\xi}/\beta_\perp \right) \left(\tilde{z} - \beta_l \tilde{\xi}/\beta_\perp \right) \right\rangle, \tag{71}$$

where $\tilde{\xi} \equiv \tilde{x} + i\tilde{y}$ and $\langle \tilde{y} \rangle = \langle \tilde{x}\tilde{y} \rangle = 0$ for azimuthally-symmetric sources such that $\langle \tilde{\xi}^2 \rangle = \langle \tilde{x}^2 - \tilde{y}^2 \rangle$, one finds in an arbitrary longitudinal reference frame

$$v = \frac{A+B}{2C} \left(1 - \sqrt{1 - \left(\frac{2C}{A+B} \right)^2} \right), \tag{72}$$

$$R_{\parallel}^2 = \frac{1}{2}\left(\sqrt{(A+B)^2 - 4C^2} - A + B\right) = B - vC, \tag{73}$$

$$R_0^2 = \frac{1}{2}\left(\sqrt{(A+B)^2 - 4C^2} + A - B\right) = A - vC. \tag{74}$$

The Yano-Koonin velocity v is zero in the frame where the expression (71) for C vanishes[16]; this fixes also the sign in front of the square root in (72). For small values of C the Yano-Koonin velocity is given approximately by

$$v \approx \frac{C}{A+B} \approx \frac{\langle \tilde{z}\tilde{t}\rangle}{\langle \tilde{t}^2\rangle + \langle \tilde{z}^2\rangle}, \tag{75}$$

where in the second approximation we again neglected generically small terms[16] proportional to $\langle \tilde{z}\tilde{x}\rangle$, $\langle \tilde{x}\tilde{t}\rangle$, and $\langle \tilde{x}^2 - \tilde{y}^2\rangle$. The accuracy of the approximate expression (75) for $v(\boldsymbol{K})$ was tested numerically in[22] and found to be excellent in the situations discussed below. In the same limit the expressions for R_0^2 and $R_{\parallel}^2$ simplify to $R_0^2 \approx A$ and $R_{\parallel}^2 \approx B$, in agreement with (67) and (68).

Since the standard Cartesian and YKP parametrizations are mathematically equivalent (being simply based on a different choice of independent q-components), the resulting HBT parameters obey simple relations[21]:

$$R_s^2 = R_{\perp}^2, \tag{76}$$

$$R_{\text{diff}}^2 = R_o^2 - R_s^2 = \beta_{\perp}^2 \gamma^2 \left(R_0^2 + v^2 R_{\parallel}^2\right), \tag{77}$$

$$R_l^2 = \left(1 - \beta_l^2\right) R_{\parallel}^2 + \gamma^2 (\beta_l - v)^2 \left(R_0^2 + R_{\parallel}^2\right), \tag{78}$$

$$R_{ol}^2 = \beta_{\perp}\left(-\beta_l R_{\parallel}^2 + \gamma^2 (\beta_l - v)^2 \left(R_0^2 + R_{\parallel}^2\right)\right). \tag{79}$$

Although a mathematical triviality, these relations provide a powerful consistency check on the experimental fitting procedure, of similar value as the relation[16,17] $\lim_{K_{\perp}\to 0}(R_o(\boldsymbol{K}) - R_s(\boldsymbol{K})) = 0$ which results from azimuthal symmetry.

LECTURE 3: HBT CORRELATIONS FOR EXPANDING SOURCES

In this third lecture I will present a quantitative discussion of the HBT parameters, both in the standard Cartesian and in the YKP fits. The emphasis will be on their K-dependence because, as discussed in the previous two lectures, only a careful study of this pair-momentum dependence permits a separation of the geometric from the dynamic aspects of the source. You probably remember that at the beginning of Lecture 2 I stressed that a model-independent reconstruction of the emission function from HBT measurements is not possible. Therefore any quantitative discussion must necessarily occur within the framework of specific source models.

I will choose a relatively simple analytical parametrization of the emission function which was first suggested by T. Csörgő and B. Lörstad in 1994 in an unpublished preprint (see also [20,18,12,16]) and which incorporates many of the (as we believe) relevant physical features of the typical sources created in relativistic nuclear collisions: approximate thermalization at decoupling, finite transverse and longitudinal extension of the source, collective expansion in both the longitudinal and transverse directions, and a finite duration of the particle-emission process. All these features are controlled by parameters which can be freely tuned, thus allowing for extensive parameter studies [16,17,22,28] which have given us a good intuitive understanding of which properties of the source are important for the correlation function and which aren't, and how to look in the correlation function for specific space-time properties of the emitter.

A simple model for an expanding source

Let us consider the following model for the emission function [16,20]:

$$S(x, K) = \frac{2J + 1}{(2\pi)^3} K \cdot n(x) \exp\left[-\frac{K \cdot u(x) - \mu(x)}{T(x)}\right] H(x) \qquad (80)$$

with

$$H(x) = \frac{1}{\pi \Delta \tau} \exp\left[-\frac{r^2}{2R^2} - \frac{(\eta - \eta_0)^2}{2(\Delta\eta)^2} - \frac{(\tau - \tau_0)^2}{2(\Delta\tau)^2}\right] \qquad (81)$$

and

$$K \cdot n(x) = M_\perp \cosh(\eta - Y). \qquad (82)$$

The factor $2J + 1$ counts the spin degeneracy of the observed particle species and is included because the detectors usually do not identify the polarization of the measured particles. There is no such factor for isospin because of the requirement that the two particles in the pair be identical, i.e. have the, for example, the same electric charge. The Lorentz covariant Boltzmann factor $\exp[-(K\cdot u(x) - \mu(x))/T(x)]$ implements the assumption of local thermal equilibrium of the emitted particles at freeze-out, with local temperature $T(x)$ and chemical potential $\mu(x)$, and the collective expansion of the source with hydrodynamic flow 4-velocity $u_\mu(x)$. I will here take T and μ as constants and refer to Refs. [20,28] for an investigation of the effects of temperature gradients. The factor $H(x)$ describes the geometric properties of the source; it can be interpreted as a space-time modulation of the fugacity $\exp[\mu(x)/T(x)]$. Space-time is parametrized by *longitudinal proper time* $\tau = \sqrt{t^2 - z^2}$ and *space-time rapidity* $\eta = \frac{1}{2}\ln[(t + z)/(t - z)]$ in the longitudinal and temporal directions, and by $r = \sqrt{x^2 + y^2}$ and ϕ in the transverse directions. The volume element is then $d^4x = \tau \, d\tau \, d\eta \, r \, dr \, d\phi$. The Gaussian factors $\exp[-r^2/(2R^2)]$ and $\exp[-(\eta - \eta_0)^2/(2(\Delta\eta)^2)]$ provide smooth geometric limits for the source in the transverse and longitudinal directions, scaled by R and $L = \tau\Delta\eta$, respectively. The function $H(x)$ is normalized to the total comoving 3-volume according to

$$\int d^4x \, H(x) = \pi r_{\text{rms}}^2 \cdot 2\tau_0 \eta_{\text{rms}} \qquad (83)$$

where

$$r_{\text{rms}}^2 = 2R^2 = x_{\text{rms}}^2 + y_{\text{rms}}^2, \qquad \eta_{\text{rms}} = \Delta\eta \qquad (84)$$

are the r.m.s. expectation values for the transverse radius r and for η, respectively. (The r.m.s. widths for x and y are each given by R.) If the Gaussians in $H(x)$ were replaced by box functions, the equivalent box dimensions (with the same r.m.s. radii) would be $\tilde{R} = 2R$, $\tilde{\eta} = \sqrt{3}\Delta\eta$. (Here you see the "unneccesary" factors $\sqrt{2}$ and $\sqrt{3}$ mentioned before!)

$K\cdot n(x)$ is the flux factor through the freeze-out hypersurfaces whose normal direction is given by the unit vector $n_\mu(x)$. In our model these hypersurfaces are for simplicity assumed to be surfaces of constant longitudinal proper time τ, parametrized by surface coordinates $\Sigma_{(\tau)}(x) = (\tau \cosh\eta, r \cos\phi, r \sin\phi, \tau \sinh\eta)$. The last factor in $H(x)$ provides a Gaussian smearing of the proper time around the mean values τ_0 with dispersion $\Delta\tau$, thereby implementing particle emission over a finite time interval of order $\Delta\tau$ in the local comoving frame. With these assumptions the flux factor reduces [16] to the form given in Eq. (82). For $\Delta\tau \to 0$ the Gaussian in τ approaches a δ-function centered at τ_0, simulating instantaneous freeze-out at constant proper time as often implemented in hydrodynamical situations.

[Let me make some side remarks on hydrodynamical simulations here, because they are a very popular tool for computing the space-time dynamics of the reaction zone

156

in heavy-ion collisions. In hydrodynamics freeze-out is always assumed to occur on a sharp hypersurface (not a smeared one as here), and one writes the emission function in the form[38]

$$S(x, K) = \frac{2J + 1}{(2\pi)^3} \int_\Sigma \frac{K^\mu d^3\sigma_\mu(x')\,\delta^{(4)}(x - x')}{\exp[(K{\cdot}u(x') - \mu(x'))/T(x')] \pm 1}, \qquad (85)$$

where $d^3\sigma_\mu(x')$ is the normal vector on the freeze-out surface $\Sigma(x')$, and we have correctly accounted for the quantum-statistical effects in the local thermal distribution functions. The latter are unimportant for heavy particles but should, in a quantitative comparison with data, be included for pions – at least in the single-particle spectrum. Inserting the ansatz (85) into the expression (38) for the single-particle spectrum one obtains the well-known Cooper-Frye formula[39]

$$E_K \frac{dN}{d^3K} = \int_\Sigma K^\mu d^3\sigma_\mu(x)\, f(x, p) \qquad (86)$$

with the local equilibrium distribution function

$$f(x, p) = \frac{2J + 1}{(2\pi)^3} \frac{1}{\exp[(K{\cdot}u(x) - \mu(x))/T(x)] \pm 1}. \qquad (87)$$

For the numerator of the correlator in (39) one obtains

$$\int_\Sigma K^\mu d^3\sigma_\mu(x)\, K^\nu d^3\sigma_\nu(y)\, f(x, K)\, f(y, K)\, \exp[iq{\cdot}(x - y)], \qquad (88)$$

similar, but not identical with to the one given in [40]. In[40] each of the two distribution functions under the integral featured on-shell arguments p_a and p_b, respectively, instead of the common (off-shell) average argument K as in (88). This error in[40] can be traced back to an inaccurate transition from finite discrete volumes along the freeze-out surface Σ to the continuum limit [41]. Taking over this inaccuracy produces (in particular for very rapidly expanding sources) unphysical oscillations of the correlation function around unity at large values of q (see e.g.[6]) which are inconsistent with the manifestly positive definite nature of the exchange term noted in Lecture 1. Recently Aichelin[42] pointed out that, for technical reasons, the same erroneous assumption is made in all simulations of HBT correlations using Monte Carlo event enerators. This problem should certainly receive more attention in the future.]

Since according to Eq. (41) the time-component of the pair momentum may be approximated by the on-shell value $K_0 = E_K = \sqrt{m^2 + \mathbf{K}^2}$, the pair momentum K can be parametrised in terms of the momentum rapidity $Y = \frac{1}{2}\ln[(1 + \beta_l)/(1 - \beta_l)]$ and the transverse mass $M_\perp = \sqrt{m^2 + K_\perp^2}$,

$$K^\mu = (M_\perp \cosh Y, K_\perp, 0, M_\perp \sinh Y). \qquad (89)$$

We implement longitudinal and azimuthally symmetric transverse expansion of the source by parametrising the flow velocity in the form

$$u^\mu(x) = \left(\cosh \eta_l(\tau, \eta)\cosh \eta_t(r),\; \frac{x}{r}\sinh \eta_t(r),\; \frac{y}{r}\sinh \eta_t(r),\; \sinh \eta_l(\tau, \eta)\cosh \eta_t(r)\right). \qquad (90)$$

For the longitudinal flow rapidity we take $\eta_l(\tau, \eta) = \eta$ independent of τ, i.e. we assume a Bjorken scaling profile[43] $v_l = z/t$ in the longitudinal direction. The growth of the longitudinal flow velocity in the longitudinal direction is then automatically limited by the Gaussian in η in (81). For the transverse flow rapidity we take a linear profile of strength η_f:

$$\eta_t(r) = \eta_f \left(\frac{r}{R} \right) . \tag{91}$$

The scalar product in the exponent of the Boltzmann factor can then be written as

$$K \cdot u(x) = M_\perp \cosh(\eta - Y) \cosh \eta_t(r) - K_\perp \frac{x}{r} \sinh \eta_t(r) , \tag{92}$$

Please note that for non-zero transverse momentum $K_\perp$, a finite transverse flow breaks the azimuthal symmetry of the emission function via the second term in (92). For $\eta_f = 0$ the source has no explicit $K_\perp$-dependence, and $M_\perp$ is the only relevant scale. As will be discussed later this gives rise to perfect $M_\perp$-scaling of the YKP radius parameters in the absence of transverse flow, which is again broken for non-zero transverse flow [44].

Besides η_f, the model parameters are the freeze-out temperature T, the transverse geometric (Gaussian) radius R, the average freeze-out proper time τ_0 as well as the mean proper emission duration $\Delta\tau$, the centre of the source rapidity distribution η_0, and the (Gaussian) width of the space-time rapidity profile $\Delta\eta$. A rough spatial picture of the source at various fixed coordinate times can be gleaned from Figs. 1 and 2 in Ref. [45] to which I would like to refer those readers having trouble with visualizing Gaussians in η and τ in regular Cartesian coordinates. Although the source in Ref. [45] has sharp edges in space and time whereas ours is smoothed by Gaussian profiles, the essential space-time features of the sources are very similar.

The calculations presented below were done for pions ($m = m_{\pi^\pm} = 139$ MeV/c^2) and kaons ($m = m_{K^\pm} = 494$ MeV/c^2) and (except were noted otherwise) for the fixed set of parameters $T = 140$ MeV, $R = 3$ fm, $\tau_0 = 3$ fm/c, $\Delta\tau = 1$ fm/c, $\Delta\eta = 1.2$, and $\eta_0 = 0$ (i.e. all our rapidities will be given relative to the rapidity η_0 of the c.m. of the source). The strength η_f of the transverse flow will be varied systematically to investigate its effects on the HBT parameters. A detailed discussion of how the variation of some of the other parameters affects the correlation function can be found in Refs. [22,28].

K-dependence of the HBT radii

Analytical approximations – HBT radii as lengths of homogeneity

For a qualitative understanding of the physical origin of the pair-momentum dependence of the HBT parameters it is instructive to start from their model-independent expressions in terms of space-time variances and evaluate the latter approximately analytically. For the standard Cartesian fit parameters (59)-(62) this was done, at different levels of accuracy, in Refs. [12,17,19,20] by exploiting the method of saddle-point integration for the evaluation of the variances. This method was introduced by Makhlin and Sinyukov[23] in the context of infinitely long sources with boost-invariant longitudinal expansion where they used it to show that the longitudinal HBT radius R_l is finite and determined by the inverse of the longitudinal velocity gradient, noting for the first time that R_l has the property of a "longitudinal length of homogeneity" in the source rather than being related to the longitudinal geometric size of the entire source. It was later observed[12,17] that not all of the important qualitative features of the K-dependence of the HBT parameters can be obtained from the leading term in the saddle point approximation (see Fig. 1). In particular, in the presence of transverse flow the saddle point moves away from the beam axis $r = 0$, and this must be taken into account in order to obtain reasonable approximations[17]. Unfortunately, this renders the whole procedure

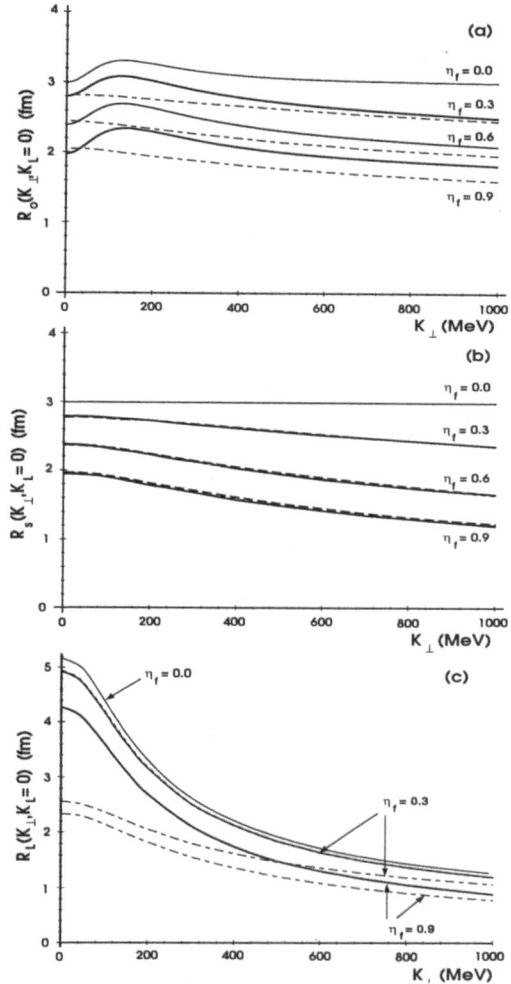

Figure 1. $K_\perp$-dependence of the standard Cartesian HBT radii R_o (top), R_s (middle), and R_l, evaluated in the LCMS (the frame where $K_L = 0$), for an infinitely long source ($\Delta\eta \to \infty$ in Eqs. (80,81)) with boost-invariant longitudinal expansion and a linear transverse flow rapidity profile. The strength of the tranverse flow η_f is varied between $\eta_f = 0$ and $\eta_f = 0.9$ as indicated. The duration parameter $\Delta\tau$ was set to 0 (locally instantaneous freeze-out). For such a source the cross-term R_{ol} vanishes identically in the LCMS. Solid lines represent results from a full numerical evaluation of the integrals in Eqs. (59-62). Dashed lines give the results from the leading-order saddle-point approximation to these integrals. One sees that the saddle-point integration misses the rise of R_o at small $K_\perp$ (i.e. the contribution from the finite duration of particle emission in the LCMS frame, see text) and gives a very bad approximation to R_l at small $K_\perp$. The solid line for R_l for $\eta_f = 0$ is given analytically by $R_l = \tau_0\sqrt{T/M_\perp}\sqrt{K_2(M_\perp/T)/K_1(M_\perp/T)}$ [26] while the saddle-point approximation (dashed) gives for $\eta_f = 0$ the Makhlin-Sinyukov formula $R_l = \tau_0\sqrt{T/M_\perp}$ [23]. (Figure taken from Ref. [17].)

rather cumbersome[17], and in the end, e.g. for a quantitative comparison with data, a full numerical evaluation of the integrals for the variances cannot be avoided.

In spite of their unreliability on a quantitative level, the analytical results from saddle-point integration are still very instructive on a qualitative level. I will here discuss the leading results for a longitudinally finite ($L = \tau_0 \Delta \eta$) source of type (80) in the limit $\Delta \tau = 0$, in the LCMS (i.e. in the Longitudinally CoMoving System where the pairs have $K_L = 0$). One finds[12]

$$R_s^2(M_\perp, K_L = 0) = R_*^2, \tag{93}$$

$$R_o^2(M_\perp, K_L = 0) = R_*^2 + \beta_\perp^2 (\Delta t_*)^2, \tag{94}$$

$$R_l^2(M_\perp, K_L = 0) = L_*^2, \tag{95}$$

$$R_{ol}^2(M_\perp, K_L = 0) = \frac{\beta_\perp Y}{\sqrt{2}} L_* \Delta t_* \left(\frac{L_*^2}{L^2} \right), \tag{96}$$

where R_*, L_*, and Δt_* are functions of $M_\perp$ defined by

$$\frac{1}{R_*^2} = \frac{1}{R^2} + \frac{1}{R_H^2}, \tag{97}$$

$$\frac{1}{L_*^2} = \frac{1}{L^2} + \frac{1}{L_H^2}, \tag{98}$$

$$\Delta t_* = \sqrt{2} \left(\sqrt{\tau_0^2 + L_*^2} - \tau_0 \right), \tag{99}$$

with the transverse and longitudinal "dynamical lengths of homogeneity"

$$R_H(M_\perp) = \frac{R}{\eta_f} \sqrt{\frac{T}{M_\perp}} = \frac{1}{\partial \eta_t(r)/\partial r} \sqrt{\frac{T}{M_\perp}}, \tag{100}$$

$$L_H(M_\perp) = \tau_0 \sqrt{\frac{T}{M_\perp}} = \frac{1}{\partial \cdot u_l} \sqrt{\frac{T}{M_\perp}}. \tag{101}$$

Strictly speaking, expression (100) is only valid for weak transverse flow $\eta f \ll 1$. In (101) we have defined the longitudinal flow 4-velocity $u_l = (\cosh \eta_l, 0, 0, \sinh \eta_l) = (\cosh \eta, 0, 0, \sinh \eta)$.

The physical interpretation of these results is quite interesting: in addition to geometry (implemented by the Gaussian cutoff factors in the function $H(x)$), dynamics affects the HBT radii through the dynamical homogeneity lenghts R_H, L_H. The latter are inversely proportional to the gradients of the expansion-velocity field in the respective direction, but smeared by a thermal smearing factor $\sqrt{T/M_\perp}$ resulting from the random thermal motion of the particles around the fluid velocity. The HBT radii are determined by the shorter of the two (geometric or dynamic) lengths scales. In the absence of random thermal motion ($T \to 0$) any velocity gradient in the system would lead to a vanishing dynamical length of homogeneity and consequently to vanishing HBT radii. At finite T, the dynamical smearing decreases with increasing transverse mass $M_\perp$, leading to a decrease of the HBT radii at large $M_\perp$. (It turns out that the $\sqrt{1/M_\perp}$-scaling of the HBT radii at large $M_\perp$ suggested by these analytical expressions is unreliable and a consequence of the saddle-point approximation[17]. A numerical analysis[44,22] shows that the power of $M_\perp$ itself, by which the HBT radii decrease for increasing $M_\perp$, is proportional to the expansion velocity gradient.)

The following intuitive picture results from these considerations: if the expansion velocity is small, i.e. all velocity gradients can be essentially neglected over the range where the geometric Gaussians in $H(x)$ are large, then HBT measures the geometric parameters R, L which tell you where the function $H(x)$ (and thus the whole emission function $S(x, K)$) gets cut off. If, on the other hand, the velocity gradients are large, they effectively cut off the emission function at a distance R_H resp. L_H from the saddle point, and the matter outside these homogeneity regions decouples from the correlator because it cannot contribute particles with sufficiently small relative momenta to see the effects of quantum statistics. This explains my statement above that HBT does in general not measure the geometry of the source, but rather the regions of homogeneity inside the source at a given wavelength $1/K$.

Of course, a space-time dependence of the temperature field $T(x)$ can induce additional gradients into the emission function and thus affect the size of the regions of homogeneity in the source and the HBT radii. This was investigated in some detail in Refs. [20,28] to which I refer the interested reader.

The last point to be discussed is the origin of the quantity Δt_* in Eq. (94) for R_o^2. Comparing with Eq. (63) we see that it has the meaning of an effective source lifetime. But where does it come from, since we set the width $\Delta \tau$ of the proper time distribution to zero? This apparent paradox has an interesting answer which is reflected in the mathematical structure of Eq. (99): since the correlator receives non-vanishing contributions from a longitudinal region of homogeneity of size $R_l = L_*$, it probes emission from different points z at different times $t = \sqrt{\tau_0^2 + z^2}$ along the proper-time hyperbola $\tau = \tau_0$, with maximal range $-L_* \lesssim z \lesssim L_*$. Thus even for sharp freeze-out at constant proper time the correlator sees a non-vanishing effective lifetime in the fixed observer frame (here the LCMS) which is in principle measurable via the difference $R_o^2 - R_s^2$. Since L_* is a decreasing function of $M_\perp$, so is Δt_*, and for large $M_\perp$ the difference $R_o^2 - R_s^2$ vanishes. (If $\Delta \tau$ were non-zero, the difference would at large $M_\perp$ approach the limit $(\Delta \tau)^2$.)

Cartesian HBT radii in the CMS and LCMS

In Fig. 2 I show the HBT radius parameters from the standard Cartesian fit (58) for pion pairs with c.m. rapidity $Y = 1.5$ where the fit of the correlator is done in the CMS[28]. The different thick curves correspond to different strengths η_f of the transverse flow. Without transverse flow R_s is $M_\perp$-independent because the we consider (80) for constant temperature and neglect possible transverse temperature gradients. As the transverse flow increases, R_s develops an increasing dependence on $M_\perp$. As shown in Fig. 6 below it can be approximated by an inverse power law, with the power increasing monotonously with η_f[17,22].

R_l features a very strong $M_\perp$-dependence even without transverse flow, due to the strong longitudinal expansion of the source. It can also be described by an inverse power law, with a larger power $\simeq 0.55$, in rough agreement with the approximate $\sqrt{T/M_\perp}$-scaling law suggested in[23] (see, however, [17,26] for a more quantitative discussion). The increase of R_o at small $M_\perp$ is due to the contribution (63) from the effective lifetime. As seen in Fig. 5 below, in the YK frame (source rest frame) the latter is of order 2.5 fm/c at small $M_\perp$; Fig. 2b shows that its effect on R_o compared to R_s in the CMS is much smaller (and thus more difficult to measure). Fig. 2d shows that the cross-term is small in the CMS but non-zero. It vanishes at $K_\perp = 0$ by symmetry and also becomes small again at large $K_\perp$.

The thin lines in Fig. 2 show for comparison approximate results for the HBT radii calculated from the approximate analytical results given in Ref.[20] which were derived by evaluating Eqs. (59-62) by saddle point integration. It is clear that this method fails here (see Ref.[17] for a quantitative discussion of this approximation), and that the analytical expressions should not be used for a quantitative analysis of HBT data.

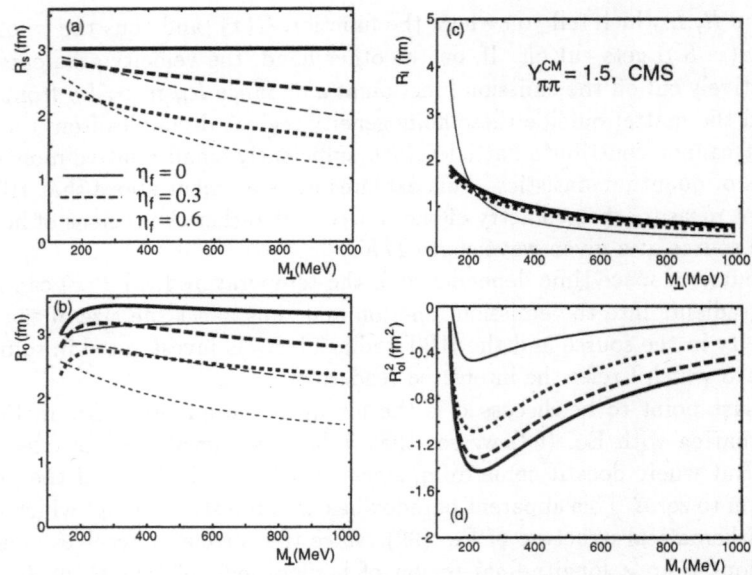

Figure 2. The standard Cartesian parameters R_s (a), R_o (b), R_l (c), and R_{ol}^2 (d) in the CMS for pion pairs with c.m. rapidity $Y = 1.5$, as functions of $M_\perp$ for 3 different values for the transverse flow η_f. The thick lines are exact numerical results from Eqs. (59-62), the thin lines are obtained from the analytical approximations given in Ref. [20]. (Figure taken from Ref. [28].)

Fig. 3 shows the same situation as Fig. 2, but now all HBT radii are evaluated in the LCMS (longitudinally comoving system[29]) which moves with the pair rapidity $Y = 1.5$ relative to the CMS. A comparison with Fig. 2 shows the strong reference frame dependence of the standard HBT radii. In particular, the cross-term changes sign and is now much larger. The analytical approximations from Ref.[20] work much better in the LCMS[20], but for R_o and R_{ol}^2 they are still not accurate enough (in particular in view of the delicate nature of the lifetime effects on R_o).

The Yano-Koonin velocity

Fig. 4 shows (for pion pairs) the dependence of the YK velocity on the pair momentum **K**. In Fig. 4a we show the YK rapidity $Y_{YK} = \frac{1}{2}\ln[(1+v)/(1-v)]$ as a function of the pair rapidity Y (both relative to the CMS) for different values of $K_\perp$, in Fig. 4b the same quantity as a function of $K_\perp$ for different Y. Solid lines are without transverse flow, dashed lines are for $\eta_f = 0.6$. For large $K_\perp$ pairs, the YK rest frame approaches the LCMS (which moves with the pair rapidity Y); in this limit all pairs are thus emitted from a small region in the source which moves with the same longitudinal velocity as the pair. For small $K_\perp$ the YK frame is considerably slower than the LCMS; this is due to the thermal smearing of the particle velocities in our source

around the local fluid velocity $u^\mu(x)$[22]. The linear relationship between the rapidity Y_{YK} of the Yano-Koonin frame and the pion pair rapidity Y is a direct reflection of the boost-invariant longitudinal expansion flow[21]. For a non-expanding source Y_{YK} would be independent of Y. Additional transverse flow is seen to have nearly no effect. The dependence of the YK velocity on the pair rapidity thus measures directly the longitudinal expansion of the source and cleanly separates it from its transverse dynamics. A detailed discussion of these features is given in Ref.[22] where it is also

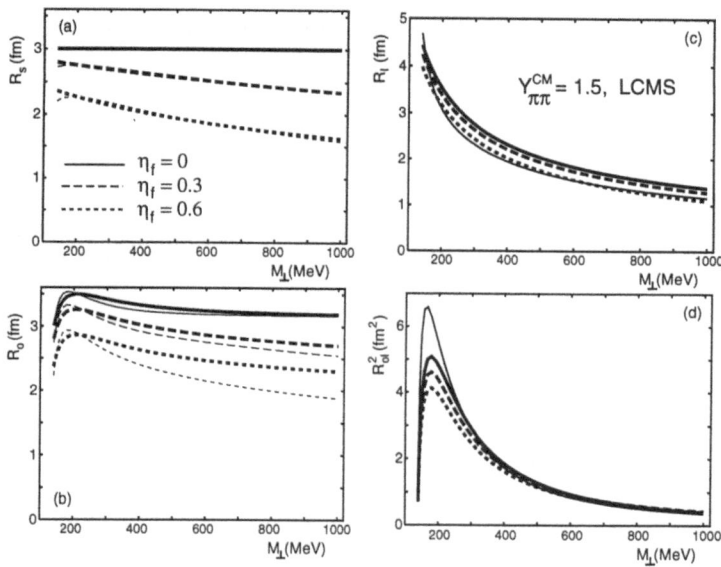

Figure 3. Same as Fig. 2, but now evaluated in the LCMS. Please note the change of sign and magnitude of the cross-term. (Figure taken from Ref. [28].)

shown that the YK velocity is always very close to the velocity of the Longitudinal Saddle Point System LSPS (i.e. to the longitudinal velocity of the fluid element around the point of maximal emissivity at momentum K). This last observation establishes the usefulness of the YK velocity (which can be directly extracted from an YKP fit to the data) as a measure for the longitudinal expansion velocity of the source.

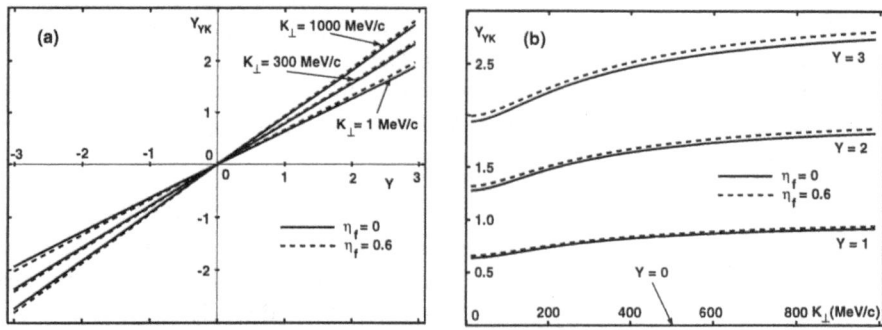

Figure 4. (a) The Yano-Koonin rapidity for pion pairs, as a function of the pair c.m. rapidity Y, for various values of $K_\perp$ and two values for the transverse flow η_f. (b) The same, but plotted against $K_\perp$ for various values of Y and η_f. (Figure taken from Ref. [21].)

YKP radii: $M_\perp$-scaling and transverse flow

In the absence of transverse flow, a thermal source like (80) depends on the particle rest mass and on the transverse momentum $K_\perp$ only through the combination $M_\perp^2 = m^2 + K_\perp^2$ (see Eq. (92)). Furthermore, the source is then azimuthally and $x \to -x$ reflection symmetric. Hence $\langle \tilde{x}\tilde{t} \rangle$, $\langle \tilde{x}\tilde{z} \rangle$, and $\langle \tilde{x}^2 - \tilde{y}^2 \rangle$ all vanish and the approximations in Eqs. (67,68) become exact. As a result, all three YKP radii (66)-(68) are only functions of $M_\perp$, too (as well as of Y, of course), i.e. they do not depend explicitly on the particle rest mass.

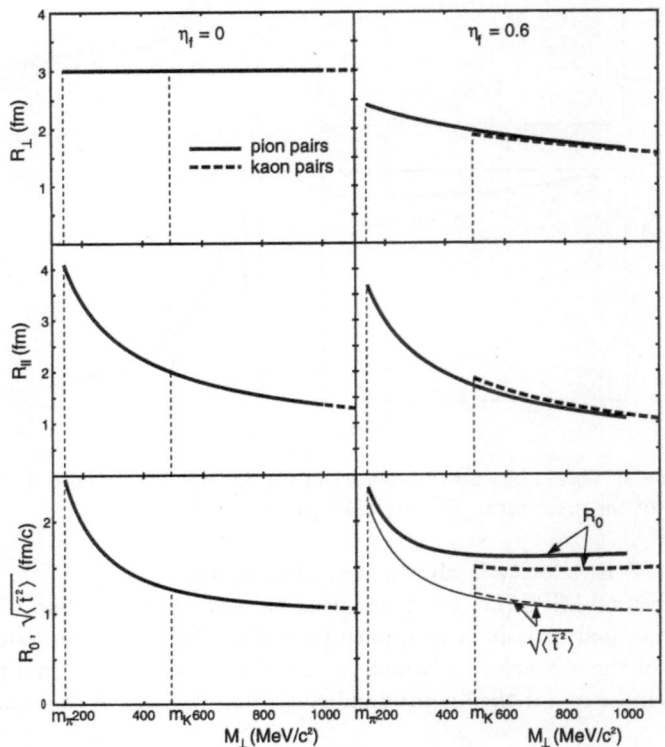

Figure 5. The YKP radii $R_\perp$, $R_\parallel$, and R_0 (from top to bottom) for vanishing transverse flow (left column) and for $\eta_f = 0.6$ (right column), as functions of $M_\perp$ for pairs at $Y_{cm} = 0$. Solid (dashed) lines are for pions (kaons). The breaking of the $M_\perp$-scaling by transverse flow is obvious in the right column. Also, as shown in the lower right panel, for non-zero transverse flow R_0 does not agree exactly with the effective source lifetime $\sqrt{\langle \tilde{t}^2 \rangle}$. (Figure taken from Ref. [22].)

This is seen in the left column of Fig. 5 where the three YKP radii are plotted for $Y_{cm} = 0$ pion and kaon pairs as functions of $M_\perp$; they agree perfectly. The transverse radius here shows no $M_\perp$-dependence due to the absence of transverse temperature gradients, but even with temperature gradients it would only depend on $M_\perp$. (Of course, this discussion neglects resonance decays which will be studied in Sec. 4.3.) The very strong $M_\perp$-dependence of the longitudinal radius parameter $R_\parallel$ is again due to the strong longitudinal expansion of the source. Note that $M_\perp$-scaling in the absence of transverse flow applies only to the YKP radius parameters: since the expressions (60)-(62) involve non-vanishing variances with $\beta_\perp$- or β_l-prefactors (which depend explicitly on the rest mass), the HBT radii from the standard Cartesian fit do not exhibit $M_\perp$-scaling.

For non-zero transverse flow $\eta_f \neq 0$ this $M_\perp$-scaling is broken by two effects: first, the second term in (92) destroys the $M_\perp$-scaling of the emission function itself, and second the β-dependent correction terms in (67,68) are now non-zero because the same term also breaks, for $K_\perp \neq 0$, the $x \to -x$ and $x \to y$ symmetries. The magnitude of the associated scale breaking due to the pion-kaon mass difference is seen in the right column of Fig. 5 for $\eta_f = 0.6$. The effects are small and require very accurate

164

experiments for their detection. However, the sign of the effect is opposite for $R_\parallel$ and for $R_\perp$, R_0 which may help to distinguish flow-induced effects from resonance decay contributions.

Since for $Y_{cm} = 0$ the YK and CMS frames coincide, $\beta_l = 0$ in the YK frame and the approximation in (67) remains exact even for non-zero transverse flow. The same is not true for the approximation in (68), and therefore I show in the lower right panel of Fig. 5 also the effective source lifetime $\sqrt{\langle \tilde{t}^2 \rangle}$ for comparison. The apparently rather large discrepancies between the YKP parameter R_0 and the effective source lifetime is due to a rather extreme choice of parameters: a large flow transverse flow and a small intrinsic source lifetime of $\Delta\tau = 1$ fm/c in (81). Since $\sqrt{\langle \tilde{t}^2 \rangle}$ approaches $\Delta\tau$ in the limit of large $M_\perp$ while the dominant[22] correction term $\langle \tilde{x}^2 - \tilde{y}^2 \rangle$ does not depend on $\Delta\tau$, the YKP parameter R_0 will track the effective source lifetime more accurately for larger values of $\Delta\tau$ (and for smaller values of η_f).

Why do $\sqrt{\langle \tilde{t}^2 \rangle}$ and R_0 increase at small $M_\perp$? Due to the rapid longitudinal expansion, the longitudinal region of homogeneity $R_\parallel$ is a decreasing function $M_\perp$. Since for different pair momenta R_0 measures the source lifetime in different YK reference frames, the freeze-out "hypersurface" will in general appear to have different shapes for pairs with different momenta. Only in our model, where freeze-out occurs at fixed proper time τ_0 (up to a Gaussian smearing with width $\Delta\tau$), is it frame-independent. It is thus generally unavoidable (and here, of course, true in any frame) that freeze-out at different points z in the source will occur at different times t in the YK frame. Since a z-region of size $R_\parallel$ contributes to the correlation function, $R_\parallel$ determines how large a domain of this freeze-out surface (and thus how large an interval of freeze-out times in the YK frame) is sampled by the correlator. This interval of freeze-out times combines with the intrinsic Gaussian width $\Delta\tau$ to yield the total effective duration of particle emission. It will be largest at small pair momenta where the homogeneity region $R_\parallel$ is biggest, and will reduce to just the variance of the Gaussian proper time distribution at large pair momenta where the longitudinal (and transverse) homogeneity regions shrink to zero. The rise of $\Delta t(\mathbf{K})$ at small $\mathbf{K}$ is thus generic.

While the strong $M_\perp$-dependence of the longitudinal radius parameter $R_\parallel$ arises from the strong longitudinal expansion, the weaker $M_\perp$-dependence of the transverse radius parameter reflects the weaker transverse expansion of our source. Following a suggestion by Th. Alber[46], this relation can be made quantitative: in Fig. 6 we plot in the left column the transverse and longitudinal YKP radii $R_\perp$ and $R_\parallel$ versus $M_\perp$ on a double-logarithmic scale. We see that both can be approximately represented by power laws. (The same is not true for R_0.) While in such a plot the slope of $R_\perp$ clearly increases with the strength η_f of the transverse flow, the slope of $R_\parallel$ appears to be insensitive to transverse flow. This can be seen quantitatively in the right column of Fig. 6 where we plot the powers $\alpha_\perp, \alpha_\parallel$ extracted from a fit

$$ R_\perp(M_\perp) \propto M_\perp^{-\alpha_\perp}, \qquad R_\parallel(M_\perp) \propto M_\perp^{-\alpha_\parallel} \tag{102} $$

as a function of η_f. As indicated in Fig. 6b the extracted power $\alpha_\perp$ for $R_\perp$ depends somewhat on the fit region because $R_\perp$ doesn't follow an exact power law; independent of the fit region it increases, however, monotonously and nearly linearly with the strength η_f of the transverse flow. Kaons "feel" the transverse flow more strongly than pions, as reflected by the somewhat larger powers $\alpha_\perp$ at fixed η_f. Note that even for a rather strong transverse flow $\eta_f = 0.6$ (heavy-ion data seem to require less flow) $\alpha_\perp$ remains below 0.25.

The power $\alpha_\parallel$ for $R_\parallel$ is shown in Fig. 6d. It has a much larger value of 0.55-0.56 even for $\eta_f = 0$, reflecting the strong boost-invariant longitudinal expansion. As η_f

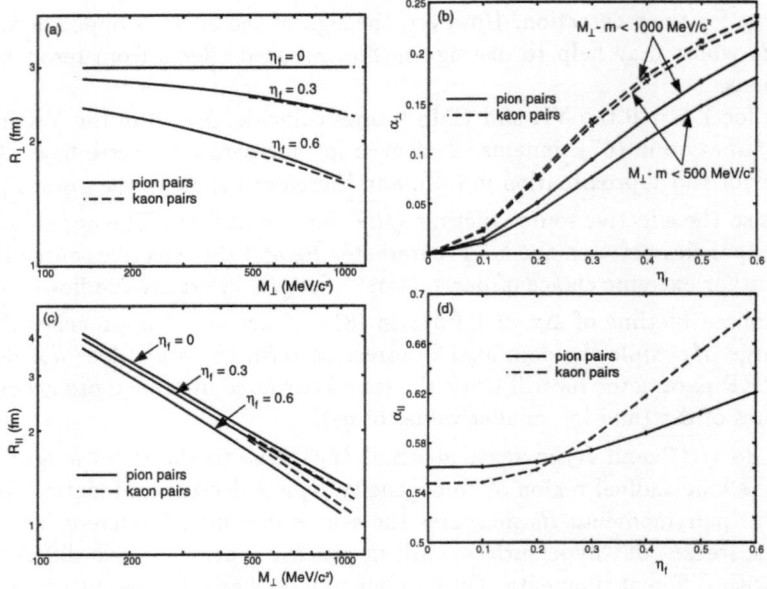

Figure 6. (a) $R_\perp$ as a function of $M_\perp$ at $Y_{CM} = 0$, for pions (solid) and kaons (dashed) and different transverse flow rapidities η_f. (b) The scaling coefficient $\alpha_\perp$ defined by $R_\perp \approx M_\perp^{-\alpha_\perp}$ for pions (solid) and kaons (dashed) as a function of the transverse flow rapidity η_f. The different results obtained by fitting in the regions $M_\perp - m < 500$ MeV/c^2 and $M_\perp - m < 1000$ MeV/c^2 are shown separately. (c) Same as (a), but for $R_\parallel$. (d) Same as (b), but for $\alpha_\parallel$. (Figure taken from Ref. [22].)

increases, $\alpha_\parallel$ also increases, but only by a few percent. (Note the suppressed zero in Fig. 6d!) Again, kaons are affected more strongly by the transverse flow than pions, but altogether the $M_\perp$-dependence of $R_\parallel$ is rather insensitive to transverse flow.

As observed by Th. Alber in his thesis[46] these features agree qualitatively with the heavy-ion data: for $\alpha_\parallel$ he found values of order 0.4–0.5, while $\alpha_\perp$ remained smaller, of order 0.1–0.2; in ^{32}S-induced collisions, the central values for both numbers showed a systematic tendency to increase with the mass of the target nucleus, indicating stronger collective flow in larger collision systems than in smaller ones. The error bars on the α's were, however, large, and one should wait for independent confirmation before firmly drawing such a conclusion.

The Y-dependence of the Yano-Koonin rapidity Y_{YK} and the $M_\perp$-dependence of $R_\parallel$ can thus be used as a quantitative measure for the longitudinal expansion of the source which is hardly affected at all by the presence and strength of transverse flow. On the other side, $R_\perp$ being boost-invariant, the $M_\perp$-dependence of $R_\perp$ is independent of the longitudinal expansion of the source and reflects only its transverse expansion. Together with the breaking of the $M_\perp$-scaling of the YKP radii, the $M_\perp$-dependence of $R_\perp$ can thus be used to extract quantitatively the transverse expansion velocity of the source [44,22].

Resonance decays

As already mentioned pions in particular have the problem that in high-energy collisions only a fraction of about 50% or less of all pions come directly from the decoupling source while the rest are produced after decoupling by the decay of unstable resonances. The above considerations presuppose that the resonance decays do not affect the $M_\perp$-dependence of $R_\perp$. This is not obvious, in particular since it is known that the $M_\perp$-dependence of the single-pion spectrum is very strongly affected by resonance decays[47]. Since resonance decays contribute more to pions than to kaons they may also affect the $M_\perp$-scaling arguments. The work by the Marburg group[38] on resonance decay effects on HBT in the context of hydrodynamical simulations indicates, within the standard Cartesian framework and without accounting for the cross-term, a possible additional $M_\perp$-dependence of the transverse radius. However, a systematic analysis of resonance contributions to HBT as a function of various characteristic source parameters is only now becoming available [48,49].

Formalism

Not much is known directly from experiment about the amount of resonance production in heavy-ion collisions. For pp collisions at similar energies a thermal model, where resonances are produced with thermal abundances in mutual chemical equilibrium, appears to work surprisingly well[50,47,51]. We therefore assume the same to hold for heavy-ion collisions. We also assume for simplicity that all hadrons decouple at the same point. Thus all resonances are assumed to have the same emission function (80-82), adjusted only for the particle rest mass m_i, its spin J_i, and its chemical potential μ_i. In chemical equilibrium μ_i is given in terms of the two independent chemical potentials μ_b and μ_s which account for conservation of baryon number and strangeness:

$$\mu_i = b_i \mu_b + s_i \mu_s \,, \tag{103}$$

where b_i, s_i are the baryon number and strangeness for resonance species i. μ_s is determined in terms of T and μ_b by the condition of strangeness neutrality of the collision region [52] which cannot be violated on the time scale of strong interactions. For illustration we will below consider the case $\mu_b = \mu_s = 0$.

I will now describe very shortly the formal steps for calculating the resonance contributions to the correlation function; for more details I refer the reader to Ref.[48]. The total source distribution of pions (and similarly for other stable-particle species, although for pions the resonance contributions are most important) can be written as

$$S_\pi(x,p) = S_\pi^{\mathrm{dir}}(x,p) + \sum_{r \neq \pi} S_{r \to \pi}(x,p) \,, \tag{104}$$

where the first term on the r.h.s. is the contribution from the directly emitted pions (Eq. (80) with $J = 0, m = m_\pi$) and the sum contains all contributions from resonance feed-down:

$$
\begin{aligned}
S_{r \to \pi}(x,p) &= M_r \int_{s_-}^{s_+} ds\, g(s) \int \frac{d^3 P}{E_p}\, \delta\left(p \cdot P - M_r E^*\right) \\
&\times \int d^4 X \int_0^\infty d\tau\, \Gamma\, e^{-\Gamma \tau}\, \delta^{(4)}\left(x - \left(X - \frac{P}{M_r}\tau\right)\right) S_r^{\mathrm{dir}}(X,P)\,. \tag{105}
\end{aligned}
$$

Here capital letters indicate coordinates associated with the parent resonance r, lower case letters are associated with the decay pion. s is the the invariant mass of the other, unobserved decay products; in an n-body decay, it can vary between $s_- = (\sum_{i=2}^n m_i)^2$ and $s_+ = (M_r - m_\pi)^2$. $g(s)$ is the decay phase space for the $(n-1)$ unobserved particles; for the isotropic 2-body decay of an unpolarized resonance it is given by

$$g(s) = \frac{b_{r \to \pi}}{4\pi p^*}\, \delta(s - m_2^2)\,, \tag{106}$$

167

(where $b_{r\to\pi}$ is the branching ratio for the decay channel), and for isotropic 3-body decays by[53]

$$g(s) = \frac{M_r b_{r\to\pi}}{2\pi\, s} \frac{\sqrt{[s-(m_2+m_3)^2][s-(m_2-m_3)^2]}}{Q(M_r,m_\pi,m_2,m_3)}, \qquad (107)$$

$$Q(M_r,m_\pi,m_2,m_3) = \int_{s_-}^{s_+} \frac{ds'}{s'}\sqrt{(M_r+m_\pi)^2-s'}\sqrt{s_+-s'}$$
$$\times \sqrt{s_--s'}\sqrt{(m_2-m_3)^2-s'}.$$

p^*, E^* are the momentum and energy of the decay pion in the resonance rest frame,

$$E^* = \sqrt{m_\pi^2 + p^{*2}}, \qquad p^* = \frac{\sqrt{[(M_r+m_\pi)^2-s][(M_r-m_\pi)^2-s]}}{2\,M_r}, \qquad (108)$$

and functions of s only. The τ-integration in (105) extends over the exponential decay probability of the resonance with total decay width Γ. The 4-dimensional δ-function of the space-time coordinate X ensures that for the pion to appear at point x from a resonance decaying at time τ, the parent resonance with momentum P must have been emitted from the source at point $X-(P/M_r)\tau$.

The integration over the resonance momentum P is restricted by the energy-momentum constraint $\delta(p\cdot P - M_r E^*)$. In the coordinate system where the momentum p of the decay pion is given by

$$p^\mu = (m_\perp \cosh y, p_\perp, 0, m_\perp \sinh y), \qquad (109)$$

(see (89)), the resonance momentum P is parametrized as

$$P^\mu = (M_\perp \cosh Y, P_\perp \cos\Phi, P_\perp \sin\Phi, M_\perp \sinh Y). \qquad (110)$$

For $p_\perp \neq 0$ the δ-function can be used to fix the azimuthal angle Φ of the resonance momentum P to

$$\Phi_\pm = \pm\tilde\Phi \quad \text{with} \quad \cos\tilde\Phi = \frac{E\,E_P - p_L P_L - E^* M}{p_\perp P_\perp} = \frac{m_\perp M_\perp \cosh(Y-y) - E^* M}{p_\perp P_\perp}. \qquad (111)$$

Let us denote by $P^\pm$ the two values of P obtained by inserting the two solutions (111) into (110). After doing the Φ- and X-integrations in (105) one thus obtains

$$S_{r\to\pi}(x,p) = M_r \int_{s_-}^{s_+} ds\, g(s) \int_{Y_-}^{Y_+} dY \int_{M_{\perp,-}^2}^{M_{\perp,+}^2} dM_\perp^2 \int_0^\infty d\tau\, \Gamma e^{-\Gamma\tau}$$

$$\times \frac{\frac{1}{2}\sum_\pm S_r^{\rm dir}\left(x - \frac{P^\pm}{M_r}\tau, P^\pm\right)}{\sqrt{p_\perp^2(M_\perp^2 - M_r^2) - [E^* M_r - m_\perp M_\perp \cosh(Y-y)]^2}}, \qquad (112)$$

with the kinematic limits

$$M_{\perp,\pm} = \overline{M}_\perp \pm \Delta M_\perp \qquad (113)$$
$$\equiv \frac{E^* M_r m_\perp \cosh(Y-y)}{m_\perp^2 \cosh^2(Y-y) - p_\perp^2} \pm \frac{M_r p_\perp \sqrt{E^{*2} + p_\perp^2 - m_\perp^2 \cosh^2(Y-y)}}{m_\perp^2 \cosh^2(Y-y) - p_\perp^2}$$

$$Y_\pm = y \pm \Delta Y \equiv y \pm \ln\left(\frac{p^*}{m_\perp} + \sqrt{1 + \frac{p^{*2}}{m_\perp^2}}\right) \qquad (114)$$

resulting from the zeroes of the square root in the denominator (which, incidentally, can also be written as $p_\perp P_\perp |\sin\tilde\Phi|$). – For the limiting case $p_\perp = 0$, the constraint $p\cdot P = M_r E^*$ cannot be used to do the Φ-integration. One then uses it to do the $M_\perp$-integral:

$$
\begin{aligned}
S_{r\to\pi}(x;y,p_\perp = 0) &= M_r \int_{s_-}^{s_+} ds\, g(s) \int_0^\pi d\Phi \int_{Y_-}^{Y_+} dY\, \frac{M_r E^*}{m^2 \cosh^2(Y - y)} \\
&\quad \times \int d\tau\, \Gamma e^{-\Gamma\tau} S_r^{\mathrm{dir}}\left(x - \frac{P}{M_r}\tau, P\right)\Bigg|_{M_\perp = \frac{M_r E^*}{m_\pi \cosh(Y - y)}}.
\end{aligned}
\tag{115}
$$

For the more generic case $p_\perp \neq 0$ a few further manipulations are useful in practice[48]: Rewriting the square root in (112) as

$$
\frac{1}{\sqrt{m_\perp^2 \cosh^2(Y - y) - p_\perp^2}} \frac{1}{\sqrt{(\Delta M_\perp)^2 - (M_\perp - \overline{M}_\perp)^2}}
\tag{116}
$$

and introducing new integration variables $v \in [-1, 1]$, $\zeta \in [-\pi, \pi]$ via

$$
\begin{aligned}
M_\perp &= \overline{M}_\perp + \Delta M_\perp \cos\zeta, \tag{117} \\
Y &= y + v\,\Delta Y, \tag{118}
\end{aligned}
$$

Eq. (112) can be further transformed into

$$
S_{R\to\pi}(x, p) = \sum_\pm \int_{\mathbf{R}} \int_0^\infty d\tau\, \Gamma e^{-\Gamma\tau} S_R^{\mathrm{dir}}\left(x - \frac{P^\pm}{M}\tau, P^\pm\right),
\tag{119}
$$

with the following shorthand for the integration over the resonance momenta:

$$
\int_{\mathbf{R}} \equiv M_r \int_{s_-}^{s_+} ds\, g(s) \int_{-1}^1 \frac{\Delta Y\, dv}{\sqrt{m_\perp^2 \cosh^2(v\,\Delta Y) - p_\perp^2}} \int_0^\pi d\zeta\, \left(\overline{M}_\perp + \Delta M_\perp \cos\zeta\right).
\tag{120}
$$

For the calculation of the correlation function we need the Fourier transform of the emission function

$$
\tilde{S}_{r\to\pi}(q, p) = \int d^4x\, e^{iq\cdot x} S_{r\to\pi}(x, p) = \sum_\pm \int_{\mathbf{R}} \frac{1}{1 - i\frac{q\cdot P^\pm}{M_r \Gamma}} \tilde{S}_r^{\mathrm{dir}}(q, P^\pm),
\tag{121}
$$

and must evaluate

$$
C(\boldsymbol{q}, \boldsymbol{K}) = 1 + \frac{|\tilde{S}_\pi^{\mathrm{dir}}(q, K)|^2 + 2\sum_{r\neq\pi} \mathrm{Re}\,[\tilde{S}_\pi^{\mathrm{dir}}(q, K)\tilde{S}_{r\to\pi}(q, K)] + |\sum_{r\neq\pi} \tilde{S}_{r\to\pi}(q, K)|^2}{|\tilde{S}_\pi(0, K)|^2},
\tag{122}
$$

where the denominator includes all contributions. Numerically, this is a rather involved expression. If the resonance contributions are small compared to the direct term one can use the Grassberger approximation[54] in which the last term in the numerator is neglected. For heavy-ion collisions this is not good enough since about 50% of all pions come from resonance decays. Instead, we can try to exploit the connection from Lecture 2 between the half-widths of the correlation function and the space-time variances which are now given by

$$
\langle \tilde{x}_\mu \tilde{x}_\nu \rangle(\boldsymbol{K}) = \frac{\sum_r \int d^4x\, \tilde{x}_\mu \tilde{x}_\nu S_{r\to\pi}(x, K)}{\sum_r \int d^4x\, S_{r\to\pi}(x, K)}.
\tag{123}
$$

Here the sum now runs over all contributions, including the direct pions. It is instructive to rewrite the average over the emission function in the following form:

$$\langle x_\nu \rangle(\boldsymbol{K}) = \sum_r f_r(\boldsymbol{K}) \, \langle x_\nu \rangle_r(\boldsymbol{K}) \,,$$

$$\langle x_\mu x_\nu \rangle(\boldsymbol{K}) = \sum_r f_r(\boldsymbol{K}) \, \langle x_\mu x_\nu \rangle_r(\boldsymbol{K}) \,. \tag{124}$$

Here we introduced the single-particle fractions[38]

$$f_r(\boldsymbol{K}) = \frac{\int d^4x \, S_{r\to\pi}(x,K)}{\sum_r \int d^4x \, S_{r\to\pi}(x,K)} = \frac{dN_\pi^r/d^3K}{dN_\pi^{\text{tot}}} \,, \quad \sum_r f_r(\boldsymbol{K}) = 1 \,, \tag{125}$$

which give the fraction of single pions with momentum $\boldsymbol{K}$ resulting from decay channel r, and the average $\langle \ldots \rangle_r$ with the effective pion emission function arising from this particular channel:

$$\langle \ldots \rangle_r(\boldsymbol{K}) = \frac{\int d^4x \ldots S_{r\to\pi}(x,K)}{\int d^4x \, S_{r\to\pi}(x,K)} \,. \tag{126}$$

The variances (123) can then be rewritten as

$$\langle \tilde{x}_\mu \tilde{x}_\nu \rangle = \sum_r f_r \, \langle \tilde{x}_\mu \tilde{x}_\nu \rangle_r + \sum_{r,r'} f_r (\delta_{r,r'} - f_{r'}) \langle x_\mu \rangle_r \langle x_\nu \rangle_{r'} \,. \tag{127}$$

The first term has an easy intuitive interpretation: each resonance decay channel r contributes an effective emission function $S_{r\to\pi}$. The full variance is calculated by weighting the variance (homogeneity length) of the emission function from a particular decay channel with the fraction f_r with which this channel contributes to the single-particle spectrum. The second term in (127) is due to the fact that in general the effective emission functions from the various decay channels have different saddle points; it somewhat spoils the intuitive interpretation of (123).

Influence on HBT radii and non-Gaussian features

It turns out that, contrary to the situations discussed before, in the case of long-lived resonances the expressions (123) are not very useful for a quantitative understanding of the correlator, although certain qualitative features can still be extracted relatively easily. The reason for this is best explained by considering the simple example of only one longlived resonance in a 1-dimensional space. Let us model the emission function for the direct pions by a Gaussian in coordinate space with width R_{dir} and (somewhat unrealistically) the emission function of the pions from the decaying resonance by a second Gaussian with much larger radius R_{halo} (assuming that the resonance travels on average a distance of order R_{halo} before it decays), with weights ε and $(1-\varepsilon)$, respectively:

$$S_\pi(x,K) = S_\pi^{\text{dir}}(x,K) + S_{r\to\pi}(x,K) = (1-\varepsilon) \, e^{-x^2/(2R_{\text{dir}}^2)} + \varepsilon \, e^{-x^2/(2R_{\text{halo}}^2)} \,. \tag{128}$$

Then the correlator is given by

$$C(q,K) - 1 = (1-\varepsilon)^2 \, e^{-R_{\text{dir}}^2 q^2} + \varepsilon^2 \, e^{-R_{\text{halo}}^2 q^2} + 2\varepsilon(1-\varepsilon) \, e^{-(R_{\text{dir}}^2 + R_{\text{halo}}^2)q^2/2} \,. \tag{129}$$

If ε is small, but R_{halo} is large, then the correlator is a superposition of a large, broad Gaussian with width $1/R_{\text{dir}}$ and weight $(1-\varepsilon)^2$, a second, narrower Gaussian with width $\sqrt{2/(R_{\text{dir}}^2 + R_{\text{halo}}^2)}$ and smaller weight $2\varepsilon(1-\varepsilon)$, and a third, extremely narrow Gaussian with width $1/R_{\text{halo}}$ and tiny weight ε^2. Obviously, the rough structure of the

170

correlator will be determined by the large and broad direct contribution; the two other contributions will, however, modify its functional form:

(i) If the resonance is shortlived such that $R_{halo} \gtrsim R_{dir}$ its effect on the correlator will be minor; its shape will remain roughly Gaussian, with a width somewhere between $1/R_{dir}$ and $1/R_{halo}$, depending on the weight ε of the resonance contribution.

(ii) If the resonance lifetime and thus R_{halo} are extremely large, the second and third term in (129) will be very narrow and, due to the finite two-track resolution of every experiment, may escape detection; then the correlator looks again Gaussian with a width $1/R_{dir}$, but at $q = 0$ it will not approach the value 2, but $1 + (1 - \varepsilon)^2 < 2$. The correlation appears to be incomplete, with an "chaoticity parameter" $\lambda = (1 - f_r)^2 = (1 - \varepsilon)^2$.

(iii) If the resonance lifetime is in between such that $R_{halo} \gg R_{dir}$ but $1/R_{halo}$ being still large enough to be experimentally resolved, all three Gaussians contribute, and the full correlator deviates strongly from a single Gaussian.

In cases (ii) and (iii) the space-time variances give misleading or outright wrong results for the width of the correlation function. As noted in connection with Eq. (53), they reproduce the curvature of the correlator at $q = 0$ which for our toy model is

$$\frac{1}{2} \left. \frac{\partial^2 C(q)}{\partial q^2} \right|_{q=0} = (1 - \varepsilon) R_{dir}^2 + \varepsilon R_{halo}^2 . \tag{130}$$

In case (ii), for not too small values of ε, this is dominated by the second term although the resonance contribution is not even visible in the correlator! On a quantitative level, the situation is not very much better for case (iii).

For the case (ii) of very long-lived resonances there is, of course, an easy way to save the usefulness of the space-time variances: if one simply leaves them out from the sum over resonances in (127), but only includes them via an "chaoticity parameter"

$$\lambda(\boldsymbol{K}) = \left(1 - \sum_{r=\text{longlived}} f_r(\boldsymbol{K}) \right)^2 , \tag{131}$$

the roughly Gaussian contributions to the correlator from the direct pions and short-lived resonances are still correctly reproduced. The real head ache comes from resonances with an intermediate lifetime which lead to a large halo but can still be experimentally resolved. They cause appreciable deviations from a Gaussian behaviour for the correlator and cannot be reliably treated by the method of space-time variances.

In nature there is only one such resonance: the ω-meson, with a lifetime of approximately 20 fm/c. All other resonances either live so shortly (typically 1 fm/c) that they hardly modify the correlator, or so long that their contribution to the correlator cannot be resolved such that they only affect λ. At low $K_\perp$, however, up to 10% of all pions come from ω-decays ($f_\omega(\boldsymbol{K} = 0) \approx 0.1$), and their non-Gaussian effects on the correlator can be clearly seen. On account of the ω a full numerical evaluation of the correlator (122) or a treatment with more powerful analytical methods which can deal with the non-Gaussian features of the correlator (q-variances, see Ref.[48]) become indispensible.

K-dependence of correlator including resonance decays

I now show some numerical results for the correlation functions resulting from the emission function (80), but now including the resonance contributions. The complete spectrum of relevant resonances is included, and in the decays the 2- and 3-body decay kinematics is fully taken into account. The HBT radii are extracted from a Gaussian fit to the numerically-calculated correlation function. A detailed technical discussion is given in Ref.[48].

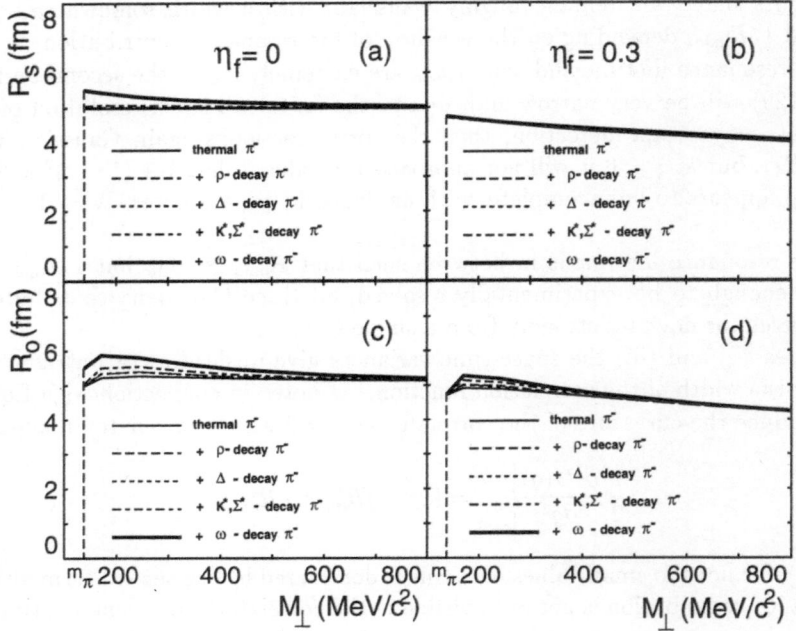

Figure 7. The influence of resonance decays on the $M_\perp$-dependence of R_s (a,b) and R_o (c,d) for $Y_{cm} = 0$ pion pairs. a,c: no transverse flow; b,d: transverse flow rapidity $\eta_f = 0.3$. The Gaussian transverse radius is here $R = 5$ fm, and $T = 150$ MeV. The HBT radii are extracted from unidirectional fits to the correlator in the respective direction of $\boldsymbol{q}$. (Figure taken from Ref. [48].)

Fig. 7 shows results for the standard Cartesian parameters R_s and R_o from 1-dimensional fits to the numerically computed correlator in the respective (q_s or q_o) directions (setting the other components of $\boldsymbol{q}$ to zero). One sees that the effects of the short-lived resonances with lifetimes of order 1 fm/c on R_s are essentially negligible, both at vanishing and at non-zero transverse flow. Only the ω with its intermediate lifetime of 20 fm/c affects R_s, but only for vanishing transverse flow. There it induces a weak $M_\perp$-dependence at small $M_\perp$ even in the absence of transverse flow; at $M_\perp > 500$ MeV the contribution of the ω dies out, and R_s again becomes $M_\perp$-independent (which would not be the case if it were affected by flow). At $\eta_f = 0.3$ and 0.6[48] not even the ω generates any additional $M_\perp$-dependence! $-R_o$ shows some effects from the additional lifetime of the resonances, in particular from the long-lived ω. Resonances with much longer lifetimes than the ω (in particular all weak decays) cannot be resolved experimentally in the correlator and have no effect on the HBT radii.

The weak effect of resonances on $R_s = R_\perp$ seems surprising: due to their non-zero lifetime they should be able to propagate outside the original source before decay and form a pion "halo"[38,55]. This effect is, however, much weaker than naively expected: most of the resonances are not very fast, and the halo thickness is thus only a fraction of the resonance lifetime. At finite transverse flow an additional effect comes into play: it turns out that then the effective size of the emission function for directly emitted resonances is *smaller* than that for direct pions[48]! At $\eta_f=0.3$ and 0.6 this even slightly overcompensates the halo effect, and altogether the resonances change neither the size nor the $M_\perp$-dependence of R_s.

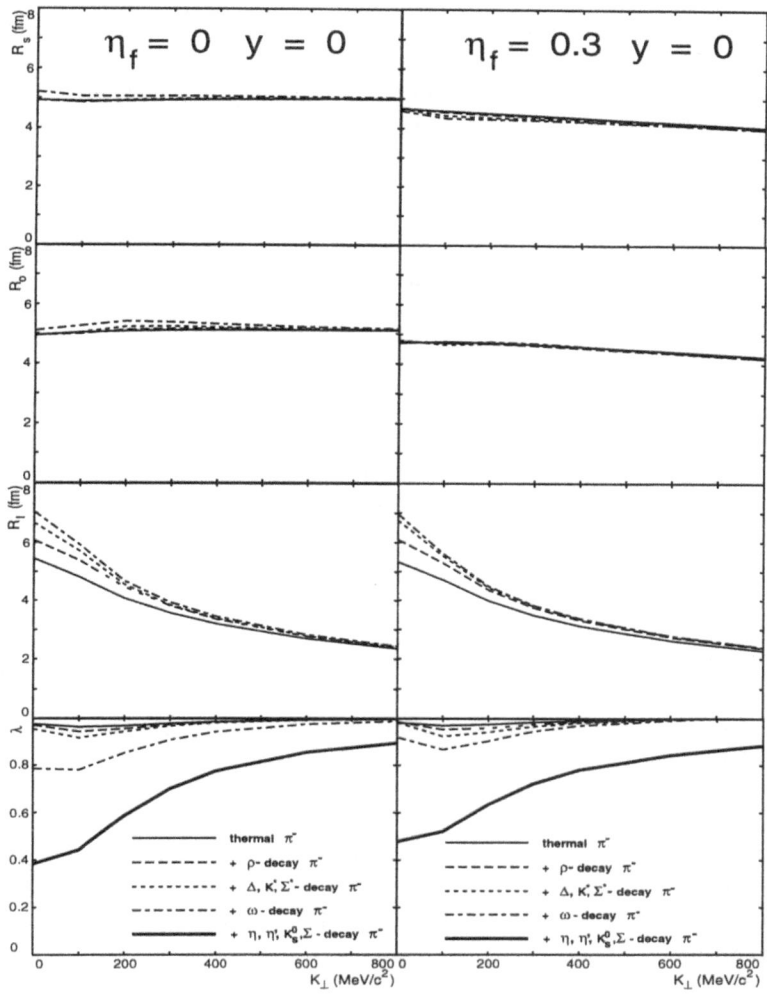

Figure 8. Same as Fig. 7, but now with parameters extracted from a complete 5-dimensional fit to the correlator (see text). Top row: R_s; second row: R_o; third row: R_l; bottom row: effective chaoticity parameter λ. Left column: no transverse flow; right column: transverse expansion with $\eta_f = 0.3$. The cross term R_{ol} vanishes for $Y = 0$ pairs. (Figure taken from Ref. [48].)

In Fig. 8 also the longitudinal radius R_l is shown. Here even the shortlived resonances are seen to make an effect. It can be essentially traced to a lifetime effect: Since the pions from the short-lived resonances appear typically 1 fm/c later than the direct pions, the source has in the mean time expanded longitudinally to a situation with a smaller longitudinal velocity gradient (the latter goes like $1/\tau$). Thus the resulting pion source has a larger longitudinal length of homogeneity and features a larger value R_l.

The "chaoticity parameter" λ

As noted above, the ω-decays make the correlator non-Gaussian. Slight non-Gaussian features exist even without resonance decays in the longitudinal direction, induced by the strong longitudinal flow, and for large transverse flow η_f become also visible in the transverse direction. A Gaussian fit to such a slightly non-Gaussian correlator in general does not extrapolate to the correct value at $q = 0$, but introduces an effective "chaoticity parameter" λ. In 1-dimensional fits along the three cartesian directions of q the different degrees of non-Gaussicity lead to different values of λ. This situation is aggravated by the resonance contributions, in particular from the ω which affects the correlator differently in each direction. Still a different value of λ is found in a 5-dimensional fit to the correlator, using the 4 standard Cartesian HBT radii and λ as fit parameters, because now λ has to "compromise" between the values found in the unidirectional fits. From the last row in Figs. 8 one sees that even without the contribution from the ω and the longlived resonances the effective λ is below 1 by up to 10%; inclusion of the ω reduces it further to about 80-85% at $K = 0$. Thus not only the very longlived resonances affect λ as anticipated above, but so do to some extend the medium- and shortlived resonances and even flow.

Comparing Figs. 7 and Fig. 8 one sees that this compromise in λ between the 3 different unidirectional fit values and the one resulting in the 5-dimensional fit also changes the fitted HBT radii. On the same (small) level as already observed for the contribution from ω decays, it even affects the $M_\perp$-dependence of $R_\perp$. Both effects, the non-Gaussian features of the correlator and the effective "chaoticity parameter" which differs from $\lambda = 1$, vanish at large K because the resonance decay pions are concentrated at low K [47].

FINAL REMARKS

In these three lectures I have presented to you two-particle intensity interferometry for relativistic heavy-ion collisions both as an intellectually stimulating problem and as a powerful practical method. Its application to nuclear collisions has turned out to be much more difficult than expected from the astrophysical analogue (two-photon intensity interferometry of stars). But at the same time, due to the dynamical nature of the problem, the physics of heavy-ion collisions is very much richer, and I have tried to show you that the HBT method is up to the task of clarifying a lot of this physics in a rather direct manner.

The key to our understanding of the HBT method and how to apply it to dynamical situations are the model-independent expressions derived in the second lecture, which express the HBT width parameters in terms of second order space-time variances of the emission function. They provide the basis of a detailed physical interpretation of the measured HBT radii. They show that generally the HBT radius parameters do not measure the full geometric extension of the source, but regions of homogeneity inside the effective emission function for particles with certain fixed momenta. For expanding systems these are usually smaller than the naive geometric source size and decreasing functions of the pair momentum. For systems with finite lifetime the HBT parameters usually mix the spatial and temporal structure of the source, and their unfolding requires model studies.

With the new YKP parametrization a method has been found which, for systems with dominant longitudinal expansion, cleanly factorises the longitudinal and transverse spatial from the temporal homogeneity length. The effective source lifetime is directly fitted by the parameter R_0; it is generically a function of the pair momentum and largest for pairs which are slow in the CMS. Another fit parameter, the YK velocity, measures directly the longitudinal velocity of the emitting fluid element, and its dependence on the pair rapidity allows for a direct determination of the longitudinal expansion of the source. Without transverse expansion, the YKP radius parameters show exact $M_\perp$-scaling. The breaking of this scaling and the $M_\perp$-dependence of the transverse radius parameter $R_\perp$ allow for a determination of the transverse expansion velocity of the source. Resonance decays were shown to mostly affect the lifetime parameter and, as a consequence, the longitudinal homogeneity length. They leave the $M_\perp$-dependence of $R_\perp$ nearly unchanged and thus do not endanger the extraction of the transverse flow via HBT.

With this new and detailed understanding of the method, I believe that HBT interferometry has a begun a new and vigorous life as a powerful tool for reconstructing the geometric and dynamic space-time characteristics of the collision zone from the measured momentum spectra.

Acknowledgements: I wish to express my thanks to my collaborators on this project, S. Chapman, J.R. Nix, B. Tomášik, U.A. Wiedemann, and Wu Yuanfang, who each contributed valuable pieces to the results presented in these lectures. Working with them has always been a pleasure. I would also like to acknowledge stimulating discussions with H. Appelshäuser, T. Csörgő, D. Ferenc, M. Gaździcki, B. Jacak, and P. Seyboth. Last but not least I would like to thank the organizers of this summer school for the invitation to come here and the students for their enthusiasm and for a pleasant and entertaining week in Dronten with many interesting conversations. This work was supported by grants from BMBF, DFG, and GSI.

REFERENCES

1. D. Boal, C.K. Gelbke, and B. Jennings, Rev. Mod. Phys. **62** (1990) 553.
2. *Quark Matter '96*, edited by P. Braun-Munzinger et al., Nucl. Phys. A **610** (1996), in press.
3. M. Gyulassy, S.K. Kauffmann, and L.W. Wilson, Phys. Rev. C**20** (1979) 2267.
4. A. Fetter and J.D. Walecka, *Quantum Theory of Many-Particle Systems*, McGraw-Hill, New York (1971).
5. See I.V. Andreev, M. Plümer, and R.M. Weiner, Int. J. Mod. Phys. A**8** (1993) 4577, and references therein.
6. D.K. Srivastava and J.I. Kapusta, Phys. Lett. B**307** (1993) 1.
7. G. Baym and P. Braun-Munzinger, Los Alamos eprint archive nucl-th/9606055.
8. K. Kolehmainen and M. Gyulassy, Phys. Lett. B**180** (1986) 203.
9. S. Chapman and U. Heinz, Phys. Lett. B**340** (1994) 250.
10. E. Shuryak, Phys. Rev. B**44** (1973) 387; Sov. J. Nucl. Phys. **18** (1974) 667.
11. S. Pratt, Phys. Rev. Lett. **53** (1984) 1219; and Phys. Rev. D**33** (1986) 1314.
12. S. Chapman, P. Scotto, and U. Heinz, Heavy Ion Physics **1** (1995) 1.
13. J. Aichelin, Phys. Rep. **202** (1991) 233.
14. S. Padula, M. Gyulassy, and S. Gavin, Nucl. Phys. B**329** (1990) 357.
15. R. Hanbury Brown and R.Q. Twiss, Nature **178** (1956) 1046.
16. S. Chapman, J.R. Nix, and U. Heinz, Phys. Rev. C**52** (1995) 2694.
17. U.A. Wiedemann, P. Scotto and U. Heinz, Phys. Rev. C**53** (1996) 918.
18. S. Chapman, P. Scotto, and U. Heinz, Phys. Rev. Lett. **74** (1995) 4400.
19. S.V. Akkelin and Y.M. Sinyukov, Phys. Lett. B**356** (1995) 525.

20. T. Csörgő and B. Lörstad, Phys. Rev. C**54** (1996) 1396; and Nucl. Phys. A**590** (1995) 465c.

21. U. Heinz, B. Tomášik, U.A. Wiedemann, and Y.-F. Wu, Phys. Lett. B382 (1996) 181.

22. Y.-F. Wu, U. Heinz, B. Tomášik, and U.A. Wiedemann, Los Alamos eprint archive nucl-th/9607044, submitted to Phys. Rev. C.

23. A.N. Makhlin and Yu.M. Sinyukov, Z. Phys. C**39** (1988) 69.

24. G. Bertsch, Nucl. Phys. A**498** (1989) 173.

25. S. Pratt, T. Csörgö, and J. Zimányi, Phys. Rev. C**42** (1990) 2646.

26. M. Herrmann and G.F. Bertsch, Phys. Rev. C**51** (1995) 328.

27. T. Csögő and B. Lörstad, Heavy Ion Physics **4** (1996) 221.

28. B. Tomášik and U. Heinz, Regensburg preprint TPR-96-16, in preparation.

29. T. Csörgő and S. Pratt, in *Proceedings of the Workshop on Relativistic Heavy Ion Physics at Present and Future Accelerators*, Budapest, 1991, edited by T. Csörgő et al. (MTA KFKI Press, Budapest, 1991), p. 75.

30. D.H. Rischke and M. Gyulassy, Los Alamos eprint archive nucl-th/9606039, Nucl. Phys. A (1996), in press.

31. M.A. Lisa et al., Phys. Rev. Lett. **71** (1993) 2863.

32. M.A. Lisa et al., Phys. Rev. C**49** (1994) 2788.

33. NA35 Coll., T. Alber et al., Z. Phys. C**66** (1995) 77;
    NA35 Coll., T. Alber et al., Phys. Rev. Lett. **74** (1995) 1303.

34. NA44 Coll., H. Beker et al., Phys. Rev. Lett. **74** (1995) 3340.

35. K. Kadija for the NA49 Coll., in *Quark Matter '96*, (Heidelberg, 20.-24.5.1996), edited by P. Braun-Munzinger et al., Nucl. Phys. A, in press.

36. F. Yano and S. Koonin, Phys. Lett. B**78** (1978) 556.

37. M.I. Podgoretskiĭ , Sov. J. Nucl. Phys. **37** (1983) 272.

38. B.R. Schlei et. al., Phys. Lett. **B293** (1992) 275; J. Bolz et. al., Phys. Lett. **B300** (1993) 404; and Phys. Rev. D**47** (1993) 3860.

39. F. Cooper and G. Frye, Phys. Rev. D**10** (1974) 186.

40. A.N. Makhlin and Y.M. Sinyukov, Sov. J. Nucl. Phys. **46** (1987) 345.

41. S. Chapman and U. Heinz, unpublished notes (1994).

42. J. Aichelin, Los Alamos eprint archive nucl-th/9609006, Nucl. Phys. A, in press.

43. J.D. Bjorken, Phys. Rev. D**27** (1983) 140.

44. U. Heinz, B. Tomášik, U.A. Wiedemann, and Y.-F. Wu, Heavy Ion Physics **4** (1996) 249.

45. S. Chapman an J.R. Nix, Phys. Rev. C**54** (1996) 866.

46. NA49 Coll., T. Alber et al., Nucl. Phys. A**590** (1995) 453c; T. Alber, PhD thesis, MPI für Physik, München (1995), unpublished.

47. J. Sollfrank, P. Koch, and U. Heinz, Phys. Lett. B**252** (1990) 256; and Z. Phys. C**52** (1991) 593; E. Schnedermann, J. Sollfrank, and U. Heinz, Phys. Rev. C**48** (1993) 2462.

48. U.A. Wiedemann and U. Heinz, Regensburg preprint TPR-96-14, in preparation; U.A. Wiedemann, Invited lecture at the Triangle Meeting, School and Workshop on Heavy Ion Collisions, 1.-5. Sept. 1996, Bratislava, to appear in the Proceedings in Acta Phys. Slov.

49. U. Heinz, Los Alamos eprint archive nucl-th/9608002, in Ref. ².

50. R. Hagedorn, Rivista Nuovo Cimento, vol. 6, serie 3, no. 10 (1983) 1.

51. F. Becattini, Z. Phys. C**69** (1996) 485, and private communication.

52. J. Letessier, A. Tounsi, U. Heinz, J. Sollfrank, J. Rafelski, Phys. Rev. D**51** (1995) 3408.

53. R. Hagedorn, *Relativistic Kinematics*, W.A. Benjamin, New York (1963).

54. P. Grassberger, Nucl. Phys. B**120** (1977) 231.

55. T. Csörgő, B. Lörstad, and J. Zimányi, Z. Phys. C**71** (1996) 491.

CHIRAL SYMMETRY AND ITS RESTORATION IN RELATIVISTIC HEAVY-ION COLLISIONS

Volker Koch

Nuclear Science Division
Lawrence Berkeley National Laboratory
University of California
Berkeley, CA, 94720, U.S.A.

INTRODUCTION

Chiral symmetry is a symmetry of QCD in the limit of vanishing quark masses. We know, however, that the current-quark masses are finite. But compared with hadronic scales the masses of the two lightest quarks, up and down, are very small, so that chiral symmetry may be considered an approximate symmetry of the strong interactions.

Long before QCD was known to be the theory of strong interactions, phenomenological indications for the existence of chiral symmetry came from the study of the nuclear beta decay. There one finds, that the weak coupling constants for the vector and axial vector hadronic currents, C_V and C_A, did not (in case of C_V) or only slightly (25% in case of C_A) differ from those for the leptonic counterparts. Consequently strong interaction 'radiative' corrections to the weak vector and axial vector 'charge' are absent. The same is true for the more familiar case of the electric charge, and there we know that it is its conservation, which protects it from radiative corrections. Analogously, we expect the weak vector and axial vector charge, or more generally, currents, to be conserved due to some symmetry of the strong interaction. In case of the vector current, the underlying symmetry is the well known isospin symmetry of the strong interactions and thus the hadronic vector current is identified with the isospin current. The identification of the axial current, on the other hand is not so straightforward. This is due to another, very important and interesting feature of the strong interaction, namely that the symmetry associated with the conserved axial vector current is 'spontaneously broken'. By that, one means that while the Hamiltonian possesses the symmetry, its ground state does not. An important consequence of the spontaneous breakdown of a symmetry is the existence of a massless mode, the so called Goldstone-boson. In our case, the Goldstone boson is the pion. If chiral symmetry were a perfect symmetry of QCD, the pion should be massless. Since chiral symmetry is only approximate, we expect the pion to have a finite but small (compared to all other hadrons) mass. This is indeed the case!

The fact that the pion is a Goldstone boson is of great practical importance. Low-energy/temperature hadronic processes are dominated by pions and thus all observables can be expressed as an expansion in pion masses and momenta. This is the basic idea of chiral perturbation theory, which is very successful in describing threshold pion physics.

At high temperatures and/or densities one expects to 'restore' chiral symmetry. By that one means, that, unlike the ground state, the state at high temperature/density posses the same symmetry as the Hamiltonian (the symmetry of the Hamiltonian, of course, will not be changed). As a consequence of this so called 'chiral restoration' we expect the absence of any Goldstone modes and thus the pions, if still present, should become as massive as all other hadrons[a]. To create a system of restored chiral symmetry in the laboratory is one of the major goals of the ultra-relativistic heavy-ion experiments.

These lectures are intended to serve as an introduction into the ideas of chiral symmetry. Emphasis will be put on the ideas and concepts rather than formalism. Consequently, most arguments presented will be heuristic and/or based on simple effective models. References will be provided for those seeking more rigorous derivations.

In the first section we briefly review the role of symmetries in quantum-field theory. Then we will introduce the chiral-symmetry transformations and derive some results, such as the Goldberger-Treiman relation. In the second section we will present the linear sigma-model as the most simple effective chiral model. Using this rather intuitive model we will discuss explicit chiral symmetry breaking. As an application we will consider pion-nucleon scattering. The third section will be devoted to the so called non-linear sigma-model, which then serves as a basis for the introduction into chiral perturbation theory. In the last section we will give some examples for chiral symmetry in the physics of hot and dense matter.

If not stated otherwise, natural units, i.e. $\hbar = c = 1$ and the conventions of Bjorken and Drell [1] for metric, gamma-matrices etc are used.

SYMMETRIES

General Considerations

In this section we briefly review role of symmetries in a field theory. For a detailed and thorough exposition of this subject as well the basic introduction into quantum-field theory we refer the reader to the standard textbooks such as the one by Bjorken and Drell [1] or Ramond [2].

A field theory is usually written down in the Lagrangian formulation. The Lagrange-function is given by the spatial integral over the Lagrangian density, $\mathcal{L}$, or Lagrangian, as we shall call it from now on, which depends on the fields $\Phi_i(x)$ and its derivatives.

$$ L = \int d^3x \, \mathcal{L}(\Phi_i(x,t), \partial_\mu \Phi_i(x,t), t). \tag{1} $$

Here the index i denotes different fields such as e.g. the different charge states of the

---

[a]If, of course, chiral restoration and deconfinement take place at the same temperature, as current lattice gauge calculations suggest, the concept of hadrons in the restored phase may become meaningless.

pion. The action S is the time integral over the Lagrange function

$$S = \int_{t_1}^{t_2} dt\, L = \int d^4x\, \mathcal{L}(\Phi_i(x,t), \partial_\mu \Phi_i(x,t), t)\,. \tag{2}$$

Requiring that the variation of the action, δS, vanishes leads to the equations of motion for the fields, i.e. to the dynamics of the theory. They are obtained from the Lagrangian by

$$\frac{\partial \mathcal{L}}{\partial \Phi_i} - \partial_\mu \left(\frac{\partial \mathcal{L}}{\partial(\partial_\mu \Phi_i)} \right) = 0\,. \tag{3}$$

We should point out, that Lorentz invariance implies that the action S and, thus, the Lagrangian $\mathcal{L}$ transform like Lorentz-scalars.

One of the big advantages of the Lagrangian formulation is that symmetries of the Lagrangian lead to conserved quantities (currents). This we already know from classical mechanics. For example, if the Lagrange function is independent of space and time, momentum and energy are conserved, respectively.

Let us assume that L is symmetric under a transformation of the fields

$$\Phi_i \longrightarrow \Phi_i + \delta \Phi_i \tag{4}$$

meaning

$$\mathcal{L}(\Phi_i + \delta \Phi_i) = \mathcal{L}(\Phi_i)\,, \tag{5}$$

then one can show (for details see e.g. ref. [1]) that the current

$$J_\mu = \frac{\partial \mathcal{L}}{\partial(\partial_\mu \Phi_i)} \delta \Phi_i \tag{6}$$

is conserved, i.e. $\partial^\mu J_\mu = 0$.

As an example, let us discuss the case of a unitary transformation on the fields, such as e.g. an isospin rotation among pions. For obvious reasons unitary transformations are the most common ones, and the chiral symmetry transformations also belong to this class.

$$\Phi_i \longrightarrow \Phi_i - i\Theta^a T^a_{ij} \Phi_j\,, \tag{7}$$

where Θ^a corresponds to the rotation angle and T^a_{ij} is a matrix, usually called the generator of the transformation (isospin matrix in case of isospin rotations). The index a indicates that there might be several generators associated with the symmetry transformation (in case of isospin rotations, we have three isospin matrices). Equation (7) corresponds to the expansion for small angles of the general transformation

$$\vec{\Phi} \longrightarrow e^{-i\Theta^a \hat{T}^a} \vec{\Phi}\,, \tag{8}$$

where the vector on $\vec{\Phi}$ indicates the several components of the field Φ such as π^+, π^- and π^0. From eq. (6) and eq. (7) we find the following expression for the conserved currents

$$J^a_\mu = -i \frac{\partial \mathcal{L}}{\partial(\partial_\mu \Phi_j)} T^a_{jk} \Phi_k\,, \tag{9}$$

181

where we have divided by the angle Θ^a. This current is often referred to as a Noether current, after E. Noether who first showed its existence[b]. Of course, a conserved current leads to a conserved charge

$$Q = \int d^3x J_0(x); \quad \frac{d}{dt}Q = 0. \tag{10}$$

Example: Massless fermions

As an example for the Noether current, let us consider the Lagrangian of two flavors of massless fermions. Since we will only discuss transformations on the fermions, the results will be directly applicable to massless QCD.

The Lagrangian is given by

$$\mathcal{L} = i\bar{\psi}_j \not{\partial} \psi_j, \tag{11}$$

where the index 'j' refers to the two different flavors, let's say 'up' and 'down', and $\not{\partial}$ is the usual shorthand notation for $\partial_\mu \gamma^\mu$.

(i) Consider the following transformation

$$\Lambda_V : \; \psi \longrightarrow e^{-i\frac{\vec{\tau}}{2}\vec{\Theta}} \psi \simeq (1 - i\frac{\vec{\tau}}{2}\vec{\Theta})\psi, \tag{12}$$

where $\vec{\tau}$ refers to the Pauli (iso)spin- matrices, and where we have switched to a iso-spinor notation for the fermions, $\psi = (u, d)$. The conjugate field, $\bar{\psi}$ transforms under Λ_V as follows

$$\bar{\psi} \longrightarrow e^{+i\frac{\vec{\tau}}{2}\vec{\Theta}}\bar{\psi} \simeq (1 + i\frac{\vec{\tau}}{2}\vec{\Theta})\bar{\psi} \tag{13}$$

and, hence, the Lagrangian is invariant under Λ_V

$$\begin{aligned} i\bar{\psi}\not{\partial}\psi &\longrightarrow i\bar{\psi}\not{\partial}\psi - i\vec{\Theta}\left(\bar{\psi}i\not{\partial}\frac{\vec{\tau}}{2}\psi - \bar{\psi}\frac{\vec{\tau}}{2}i\not{\partial}\psi\right) \\ &= i\bar{\psi}\not{\partial}\psi. \end{aligned} \tag{14}$$

Following eq. (9) the associated conserved current is

$$V_\mu^a = \bar{\psi}\,\gamma_\mu \frac{\tau^a}{2}\,\psi \tag{15}$$

and is often referred to as the 'vector-current'.

(ii) Next consider the transformation

$$\Lambda_A : \qquad \psi \longrightarrow e^{-i\gamma_5 \frac{\vec{\tau}}{2}\vec{\Theta}}\psi = (1 - i\gamma_5 \frac{\vec{\tau}}{2}\vec{\Theta})\psi, \tag{16}$$

$$\Rightarrow \quad \bar{\psi} \longrightarrow e^{-i\gamma_5 \frac{\vec{\tau}}{2}\vec{\Theta}}\bar{\psi} \simeq (1 - i\gamma_5 \frac{\vec{\tau}}{2}\vec{\Theta})\bar{\psi}, \tag{17}$$

where we have made use of the anti-commutation relations of the gamma matrices, specifically, $\gamma_0\gamma_5 = -\gamma_5\gamma_0$. The Lagrangian transforms under Λ_A as follows

$$\begin{aligned} i\bar{\psi}\not{\partial}\psi &\longrightarrow i\bar{\psi}\not{\partial}\psi - i\vec{\Theta}\left(\bar{\psi}i\partial_\mu\gamma^\mu\gamma_5\frac{\vec{\tau}}{2}\psi + \bar{\psi}\gamma_5\frac{\vec{\tau}}{2}i\partial_\mu\gamma^\mu\psi\right) \\ &= i\bar{\psi}\not{\partial}\psi, \end{aligned} \tag{18}$$

$$\tag{19}$$

---

[b]Note, that some of the Noether currents are not conserved on the quantum-level. In other words, not every symmetry of the classical field theory has a quantum analog. If this is not the case one speaks of anomalies. For a discussion of anomalies, see e.g. [3].

where the second term vanishes because γ_5 anti-commutes with γ_μ. Thus the Lagrangian is also invariant under Λ_A with the conserved 'axial vector' current

$$A_\mu^a = \bar{\psi}\gamma_\mu\gamma_5\frac{\tau}{2}\psi. \tag{20}$$

In summary, the Lagrangian of massless fermions, and, hence, massless QCD, is invariant under both transformations, Λ_V and Λ_A[c] symmetry is what is meant by chiral symmetry[d]. The chiral symmetry is often referred to by its group structure as the $SU(2)_V \times SU(2)_A$ symmetry.

Now let us see, what happens if we introduce a mass term,

$$\delta\mathcal{L} = -m\left(\bar{\psi}\psi\right). \tag{21}$$

¿From the above, $\delta\mathcal{L}$ is obviously invariant under the vector transformations Λ_V but *not* under Λ_A

$$\Lambda_A: \quad m\left(\bar{\psi}\psi\right) \longrightarrow m\bar{\psi}\psi - 2im\vec{\Theta}\left(\bar{\psi}\frac{\vec{\tau}}{2}\gamma_5\psi\right). \tag{22}$$

Thus, Λ_A is not a good symmetry, if the fermions (quarks) have a finite mass. But as long as the masses are small compared to the relevant scale of the theory one may treat Λ_A as an approximate symmetry, in the sense, that predictions based under the assumption of the symmetry should be reasonably close to the actual results[e].

In case of QCD we know that the masses of the light quarks are about $5 - 10\,\text{MeV}$ whereas the relevant energy scale given by $\Lambda_{QCD} \simeq 200\,\text{MeV}$ is considerably larger. We, therefore, expect that Λ_A should be an approximate symmetry and that the axial current should be approximately (partially) conserved. This slight symmetry breaking due to the quark masses is the basis of the so called Partial Conserved Axial Current hypothesis (PCAC). Furthermore, as long as the symmetry breaking is small, one would also expect, that its effect can be described in a perturbative approach. This is carried out in a systematic fashion in the framework of chiral perturbation theory.

Chiral Symmetry and PCAC

Chiral transformation of mesons

In order to develop a better feeling for the meaning of the symmetry transformations Λ_V and Λ_A, let us find out how pions and rho-mesons transform under these operations. To this end, let us consider combinations of quark fields, which carry the quantum numbers of the mesons under consideration. This should give us the correct transformation properties:

pion-like state: $\vec{\pi} \equiv i\bar{\psi}\vec{\tau}\gamma_5\psi$;	sigma-like state: $\sigma \equiv \bar{\psi}\psi$
rho-like state: $\vec{\rho}_\mu \equiv \bar{\psi}\vec{\tau}\gamma_\mu\psi$;	a_1-like state: $\vec{a}_{1\mu} \equiv \bar{\psi}\vec{\tau}\gamma_\mu\gamma_5\psi$

---

[c]Note, that the above Lagrangian is also invariant under the operations $\psi \to exp(-i\Theta)\psi$ and $\psi \to exp(-i\gamma_5\Theta)\psi$. The first operation is related to the conservation of the baryon number while the second symmetry is broken on the quantum level. This is referred to as the U(1) axial anomaly, which is a real breaking of the symmetry in contrast to the spontaneous breaking discussed below (see e.g. 3).

[d]Often, people talk about 'chiral' symmetry but actually only refer to the axial transformation Λ_A. This is due to its special role is plays, since it is spontaneously broken in the ground state.

[e]A wheel which is slightly bent and thus not perfectly invariant under rotations, can for most practical purposes still be considered as being round, as long as the bending is small compared to the radius of the wheel.

183

Here, the vector again indicates the iso-vector nature of the mesons such as pion and rho etc. i.e. these particles transform like a vector under *isospin* - rotations. In addition particles, which transform like vectors under a Lorentz transformation, have an additional Lorentz index μ. These are the vector mesons ρ and a_1, which carry a total spin of one.

(i) vector transformations Λ_V, see eqs. (12,13):

$$
\begin{aligned}
\pi_i : \quad i\bar{\psi}\tau_i\gamma_5\psi \quad &\longrightarrow \quad i\bar{\psi}\tau_i\gamma_5\psi + \Theta_j\left(\bar{\psi}\tau_i\gamma_5\frac{\tau_j}{2}\psi - \bar{\psi}\frac{\tau_j}{2}\tau_i\gamma_5\psi\right) \\
&= \quad i\bar{\psi}\tau_i\gamma_5\psi + i\Theta_j\epsilon_{ijk}\,\bar{\psi}\gamma_5\tau_k\psi,
\end{aligned}
\tag{23}
$$

where we have used the commutation relation between the τ matrices $[\tau_i, \tau_j] = 2i\epsilon_{ijk}\tau_k$. In terms of pions this can be written as

$$
\vec{\pi} \longrightarrow \vec{\pi} + \vec{\Theta} \times \vec{\pi},
\tag{24}
$$

which is nothing else but an isospin rotation, namely the isospin direction of the pion is rotated by Θ. The same result one obtains for the ρ-meson

$$
\vec{\rho_\mu} \longrightarrow \vec{\rho_\mu} + \vec{\Theta} \times \vec{\rho_\mu}.
\tag{25}
$$

Consequently, the vector-transformation Λ_V can be identified with the isospin rotations and the conserved vector current with the isospin current, which we know to be conserved in strong interactions.

(i) axial transformations Λ_A, see eqs. (16,17):

$$
\begin{aligned}
\pi_i : \quad i\bar{\psi}\tau_i\gamma_5\psi \quad &\longrightarrow \quad i\bar{\psi}\tau_i\gamma_5\psi + \Theta_j\left(\bar{\psi}\tau_i\gamma_5\gamma_5\frac{\tau_j}{2}\psi + \bar{\psi}\gamma_5\frac{\tau_j}{2}\tau_i\gamma_5\psi\right) \\
&= \quad i\bar{\psi}\tau_i\gamma_5\psi + \Theta_i\bar{\psi}\psi,
\end{aligned}
\tag{26}
$$

where we have made use of the anti-commutation relation of the τ matrices $\{\tau_i, \tau_j\} = 2\delta_{ij}$ and of $\gamma_5\gamma_5 = 1$. In terms of the mesons this reads:

$$
\vec{\pi} \longrightarrow \vec{\pi} + \vec{\Theta}\sigma
\tag{27}
$$

and similarly for the σ-meson

$$
\sigma \longrightarrow \sigma - \vec{\Theta}\vec{\pi}.
\tag{28}
$$

The pion and the sigma-meson are obviously rotated into each other under the axial transformations Λ_A. Similarly the rho rotates into the a_1

$$
\vec{\rho_\mu} \longrightarrow \vec{\rho_\mu} + \vec{\Theta} \times \vec{a_{1_\mu}}.
\tag{29}
$$

Above we just have convinced ourselves that Λ_A is a symmetry of the QCD Hamiltonian. Naively, this would imply, that states which can be rotated into each other by this symmetry operation should have the same Eigenvalues, i.e the same masses. This, however, is clearly not the case, since $m_\rho = 770\,\text{MeV}$ and $m_{a_1} = 1260\,\text{MeV}$. We certainly do not expect that the slight symmetry breaking due to the finite current-quark masses is responsible for this splitting. This should lead to mass differences which are small compared to the masses themselves. In case of the ρ and a_1, however, the mass difference is of the same order as the mass of the ρ. The resolution to this problem will be the spontaneous breakdown of the axial symmetry. Before we discuss what is meant by that, let us first convince ourselves, that the axial vector is conserved to a good approximation, so that the axial symmetry must be present somehow.

Pion decay and PCAC

Let us first consider the weak decay of the pion. In the simple Fermi theory the weak interaction Hamiltonian is of the current-current type, where both currents are a sum of axial and vector currents, as we have defined them above (see e.g. [4]). Because of parity, the weak decay of the pion is controlled by the matrix element of the axial current between the vacuum and the pion $< 0|A_\mu|\pi >$. This matrix element must be proportional to the pion momentum, because this is the only vector around

$$< 0|A_\mu^a(x)|\pi^b(q) >= if_\pi q_\mu \delta^{ab} e^{-iq \cdot x} \tag{30}$$

and the proportionality constant $f_\pi = 93 MeV$ is determined from experiment[f]. Here, the indices a and b refer to isospin whereas μ again indicates the Lorentz vector character of the axial current.

Let us now take the divergence of eq. (30)

$$< 0|\partial^\mu A_\mu^a(x)|\pi^b(q) >= -f_\pi q^2 \delta^{ab} e^{-iq \cdot x} = -f_\pi m_\pi^2 \delta^{ab} e^{-iq \cdot x} . \tag{31}$$

To the extent, that the pion mass is small compared to hadronic scales, the axial current is approximately conserved. Or in other words, the smallness of the pion mass is directly related to the partial conservation of the axial current, i.e. to the fact that the axial transformation is an approximate symmetry of QCD. In the literature the above relation (31) is often referred to as the PCAC relation. The above relations (30,31) also suggest, that the axial current carried by a pion is

$$A_{\mu, pion}^a = f_\pi \partial_\mu \Phi^a(x) \tag{32}$$

or that the divergence of the axial vector current can be identified with the pion field (up to a constant). Here $\Phi^a(x)$ is the pion field. Sometimes this relation between pion field and axial current is also referred to as the PCAC relation.

Goldberger-Treiman relation

There is more evidence for the conservation of the axial current. Let us consider the axial current of a nucleon. This is simply given by (see eq. (20))

$$A_{\mu, nucleon}^a = g_a \bar{\psi}_N \gamma_\mu \gamma_5 \frac{\tau^a}{2} \psi_N , \tag{33}$$

where $\psi_N = (proton, neutron)$ is now an iso-spinor representing proton and neutron. The factor $g_a = 1.25$, is due to the fact, that the axial current of the nucleon is renormalized by 25%, as seen in the weak beta decay of the neutron. Since the nucleon has a large mass M_N, we do not expect that its axial current is conserved, and indeed by using the free Dirac equation for the nulceon one can show that

$$\partial^\mu A_{\mu, nucleon}^a = i g_a M_N \bar{\psi}_N \gamma_5 \tau^a \psi_N \neq 0 , \tag{34}$$

which vanishes only in case of vanishing nucleon mass. We know, however, that the nucleon interacts strongly with the pion. Therefore, let us assume that the total axial current is the sum of the nucleon and the pion contribution. Using the PCAC-relation (32) and equ. (33) we have

$$A_\mu^a = g_a \bar{\psi}_N \gamma_\mu \gamma_5 \frac{\tau^a}{2} \psi_N + f_\pi \partial_\mu \Phi^a . \tag{35}$$

---

[f] The are several definitions of f_π around, depending on whether factors of 2, $\sqrt{2}$ are present in eq. (30).

If we require, that the total current is conserved, $\partial^\mu A_\mu = 0$, we obtain

$$\partial^\mu \partial_\mu \Phi^A = -g_a\, i \frac{M_N}{f_\pi} \bar\psi_N \gamma_5 \tau^a \psi_N\,, \tag{36}$$

where we have used (34). This is nothing else but a Klein Gordon equation for a massless boson (pion) coupled to the nucleon. Hence, requiring the conservation of the total axial current immediately leads us to the prediction that the pion should be massless. This is exactly what we also concluded from the weak pion decay. If we now allow for a finite pion mass, which is equivalent to requiring that the divergence of the axial current is consistent with the PCAC result (31), then we arrive at the Klein Gordon equation for a pion coupled to the nucleon

$$\left(\partial^\mu \partial_\mu + m_\pi^2\right) \Phi = -g_a\, i \frac{M_N}{f_\pi} \bar\psi_N \gamma_5 \tau \psi_N\,, \tag{37}$$

where the pion-nucleon coupling constant is given by

$$g_{\pi NN} = g_a \frac{M_N}{f_\pi} \simeq 12.6\,. \tag{38}$$

This is to be compared with the value for the pion-nucleon coupling as extracted e.g. from pion-nucleon scattering experiments

$$g_{\pi NN}^{exp} = 13.4\,, \tag{39}$$

which is in remarkable close agreement, considering the fact that eq. (38) relates the strong-interaction pion-nucleon coupling $g_{\pi NN}$ with quantities extracted from the weak interaction, namely g_a and f_π. Of course, the reason why this works is that there is some symmetry, namely chiral symmetry, at play, which allows to connect seemingly different pieces of physics. Equation (38) is usually called the Goldberger-Treiman relation.

Spontaneous breakdown of chiral symmetry

There appears to be some contradiction: On the one hand the meson-mass spectrum does not reflect the axial vector symmetry. On the other hand, the weak pion decay seems to be consistent with a (partially) conserved axial vector current. Also the success of the Goldberger-Treiman relation indicates that the axial vector current is conserved and, hence, that the axial transformation Λ_A is a symmetry of the strong interactions.

The solution to this puzzle is that the axial vector symmetry is *spontaneously* broken. What does one mean by that? One speaks of a spontaneously broken symmetry, if a symmetry of the Hamiltonian is not realized in the ground state.

This is best illustrated in a classical mechanics analog. In fig. 1 we have two rotationally invariant potentials ('interactions'). In (a) the ground state is right in the middle, and the potential plus ground state are still invariant under rotations. In (b), on the other hand, the ground state is at a finite distance away from the center. The point at the center is a local maximum of the potential and thus unstable. If we put a little ball in the middle, it will roll down somewhere and find its ground state some place in the valley which represents the true minimum of the potential. By picking one point in this valley (i.e picking the ground state), the rotational symmetry is obviously broken. Potential plus ground state are not symmetric anymore. The symmetry has been broken *spontaneously* by choosing a certain direction to be the ground state.

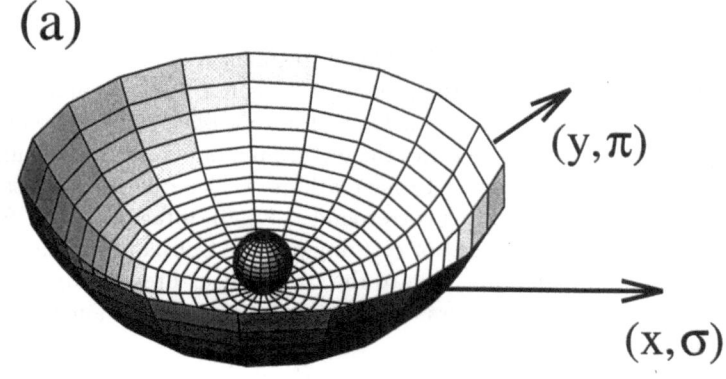

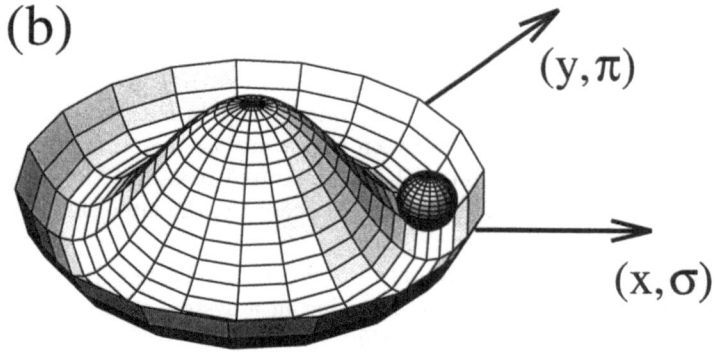

Figure 1. Effective potentials. (a) No spontaneous breaking of symmetry. (b) Spontaneous breaking of symmetry.

However, effects of the symmetry are still present. Moving the ball around in the valley (rotational excitations) does not cost any energy, whereas radial excitations do cost energy.

Let us now use this mechanical analogy in order to understand what the spontaneous breakdown of the axial vector symmetry of the strong interaction means. Assume, that the effective QCD-hamiltonian at zero temperature has a form similar to that depicted in fig. 1b, where the (x,y)-coordinates are replaced by $(\sigma, \vec{\pi})$-fields. The spacial rotations are then the mechanical analog of the axial vector rotation Λ_A, which rotates $\vec{\pi}$ into σ (see eq. (27)). Since the ground state is not at the center but at some finite distance away from it, one of the fields will have a finite expectation value. This can only be the σ-field, because it carries the quantum numbers of the vacuum. In the quark language, this means we expect to have a finite scalar quark condensate $< \bar{q}q > \neq 0$. In this picture, pionic excitation correspond to small 'rotations' away from the ground state along the valley, which do not cost any energy. Consequently the mass of the pion should be zero. In other words, due to the spontaneous breakdown of chiral symmetry, we predict a vanishing pion mass. Excitations in the σ-direction correspond to radial excitations and therefore are massive.

This scenario is in perfect agreement with what we have found above. The spontaneous breakdown of the axial vector symmetry leads to different masses of the pion and sigma. However, since the interaction itself is still symmetric, pions become massless, which is exactly what we find from the PCAC relation, provided that the axial current is perfectly conserved. Thus the mesonic-mass spectrum as well as the PCAC– and the Goldberger-Treiman relation are consistent with a spontaneous breakdown of the axial vector symmetry Λ_A. The pion appears as a massless mode (Goldstone boson) as a result of the symmetry of the interaction.

Incidentally, the assumption of a spontaneously broken axial vector symmetry also explains the mass difference between the ρ- and a_1 meson and one predicts that $m_{a_1} = \sqrt{2}m_\rho$ in good agreement with the measured masses. The derivation of this result, however, is too involved to be presented here and the interested reader is referred to the literature [5, 6].

One expects that at high temperature/densities the finite expectation value of the scalar quark condensate melts away, resulting in a system where chiral symmetry is not spontaneously broken anymore. In this, as it is often called, chirally restored phase pion/sigma as well as rho/a_1, if they exist[g], should be degenerate and the pion looses its identity as a Goldstone boson, i.e. it will become massive. The effective interaction in this phase would then have a shape similar to fig 1(a). It is one of the major goals of the ultrarelativistic heavy-ion program to create and identify a macroscopic sample of this phase in the laboratory.

In the following section we will construct a chiral invariant Lagrangian, the so called 'Linear sigma-model', in order to see how the concept of spontaneous breakdown of chiral symmetry is realized in the framework of a simple model. We will also discuss how to incoorporate the effect of the finite quark masses leading to the explicit breaking of chiral symmetry.

---

[g]If deconfinement and chiral restoration occur at the same temperature, it may become meaningless to talk about mesons above the critical temperature.

LINEAR SIGMA-MODEL

Chiral limit

In this section we will construct a simple chirally-invariant model involving pions and nucleons, the so called linear sigma-model. This model was first introduced by Gell-Mann and Levy in 1960 [7], long before QCD was known to be the theory of the strong interaction. In order to construct such a model, we have to write down a Lagrangian which is a Lorentz-scalar and which is invariant under the vector- and axial vector transformations, Λ_V and Λ_A.

In the previous section, we have shown, that the pion transforms under Λ_V and Λ_A as (27),

$$\Lambda_V : \pi_i \longrightarrow \pi_i + \epsilon_{ijk}\Theta_j\pi_k\,, \qquad \Lambda_A : \pi_i \longrightarrow \pi_i + \Theta_i\sigma\,. \tag{40}$$

Similarly one can also show, that the σ-field transforms like

$$\Lambda_V : \sigma \longrightarrow \sigma\,, \qquad \Lambda_A : \sigma \longrightarrow \sigma - \Theta_i\pi_i\,. \tag{41}$$

Since Λ_V is simply an isospin rotation, the squares of the fields are invariant under this transformation

$$\Lambda_V : \quad \pi^2 \longrightarrow \pi^2\,, \qquad \sigma^2 \longrightarrow \sigma^2\,, \tag{42}$$

whereas under Λ_A they transform like

$$\Lambda_A : \quad \vec{\pi}^2 \longrightarrow \vec{\pi}^2 + 2\sigma\Theta_i\pi_i\,, \qquad \sigma^2 \longrightarrow \sigma^2 - 2\sigma\Theta_i\pi_i\,. \tag{43}$$

However, the combination $(\vec{\pi}^2 + \sigma^2)$ is invariant under both transformations, Λ_V *and* Λ_A

$$(\vec{\pi}^2 + \sigma^2) \overset{\Lambda_V,\Lambda_A}{\longrightarrow} (\vec{\pi}^2 + \sigma^2)\,. \tag{44}$$

Since this combination is also a Lorentz-scalar, we can build a chirally invariant Lagrangian around this structure:

- Pion-nucleon interaction:
  The standard pion nucleon interaction involves a pseudo-scalar combination of the nucleon field multiplied by the pion field:

$$g_\pi \left(i\bar{\psi}\gamma_5\vec{\tau}\psi\right)\vec{\pi}\,, \tag{45}$$

  where from now on we denote the pion-nucleon coupling constant simply by g_π. Under the chiral transformations this transforms exactly like π^2, because the term involving the nucleon has the same quantum numbers as the pion. Chiral invariance requires that there must be another term, which transforms like σ^2, in order to have the invariant structure (44). The simplest choice is a term of the form,

$$g_\pi \left(\bar{\psi}\psi\right)\sigma\,, \tag{46}$$

  so that the interaction term between nucleons and the mesons is

$$\delta\mathcal{L} = -g_\pi \left[(\bar{\psi}\gamma_5\vec{\tau}\psi)\,\vec{\pi} + (\bar{\psi}\psi)\sigma\right]\,. \tag{47}$$

- Nucleon-mass term:
We know that an explicit nucleon-mass term breaks chiral invariance (see eq. 22). The nucleon mass is also too large to be simply a result of the small explicit chiral symmetry breaking as reflected in the PCAC relation (31). The simplest[h] way to give the nucleon a mass without breaking chiral symmetry, is to exploit the coupling of the nucleon to the σ-field (46), which has the structure of a nucleon-mass term. This, however, requires that the σ-field as a *finite vacuum expectation value*,

$$< \sigma > = \sigma_0 = f_\pi \, , \qquad (48)$$

where choice of $\sigma_0 = f_\pi$ is dictated by the Goldberger-Treiman relation (38) in the limit of $g_a = 1$. A finite vacuum expectation value for the σ-field immediately implies, that chiral symmetry will be spontaneously broken, as discussed in the last section. In order for our model to generate such an expectation value, we have to introduce a potential for the sigma-field, which has its minimum at $\sigma = f_\pi$. This brings us to the next ingredient of our model.

- Pion - sigma potential:
The potential, which generates the vacuum expectation value of the σ-field has to be a function of the invariant structure (44) in order to be chirally invariant. The simplest choice is:

$$V = V(\pi^2 + \sigma^2) = \frac{\lambda}{4} \left((\pi^2 + \sigma^2) - f_\pi^2 \right)^2 . \qquad (49)$$

This potential, which is plotted in fig. (2) (see also fig. (1) for a three-dimensional view) indeed has its minimum at $\sigma = f_\pi$ for $\pi = 0$. Due to its shape, it is often referred to as the 'Mexican hat potential'.

- Kinetic-energy terms:
Finally we have to add kinetic-energy terms for the nucleons and the mesons which have the form $i\bar{\psi}\partial\!\!\!/\psi$ and $\frac{1}{2}(\partial_\mu\pi\partial^\mu\pi + \partial_\mu\sigma\partial^\mu\sigma)$, respectively. Both are chirally invariant. The first term is just the Lagrangian of free mass less fermions, which we have shown to be invariant. The second term again has the invariant structure (44).

Putting everything together, the Lagrangian of the linear sigma-model reads (remember that the potential V enters with a minus-sign into the Lagrangian):

$$\begin{aligned}
\mathcal{L}_{L.S.} = \ & i\bar{\psi}\partial\!\!\!/\psi - g_\pi \left(i\bar{\psi}\gamma_5\vec{\tau}\psi\,\vec{\pi} + \bar{\psi}\psi\,\sigma \right) \\
& - \frac{\lambda}{4} \left((\pi^2 + \sigma^2) - f_\pi^2 \right)^2 + \frac{1}{2}\partial_\mu\pi\partial^\mu\pi + \frac{1}{2}\partial_\mu\sigma\partial^\mu\sigma .
\end{aligned} \qquad (50)$$

What are the properties of this model? Let us start with the ground state. As already mentioned, in the ground state the σ-field has a finite expectation value, whereas the pion has none, because of parity. Furthermore, the nucleon obtains its mass from its interaction with the sigma-field. But what are the masses of the σ and π-mesons? There are no explicit mass terms for the σ- and π-fields in the Lagrangian (50), but, as with the nucleon, there could be some coupling to the expectation value of the σ-field,

---

[h] Actually one can allow for an explicit nucleon-mass term if one also includes the chiral partner of the nucleon, which is believed to be the $N^*(1535)$. This is an interesting alternative approach which is discussed in detail in ref. [8]

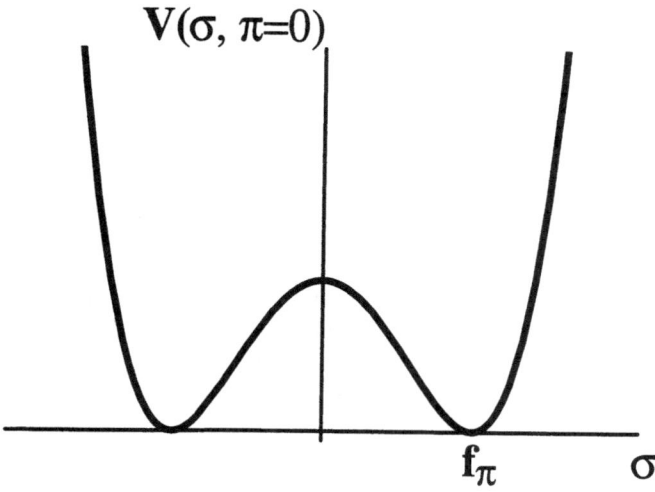

Figure 2. Potential of linear sigma-model

which gives rise to mass terms. From the structure of the potential (see figs.(1) and (2)) as well as from our discussion of the spontaneous breakdown of chiral symmetry, we expect the pion to be massless and the σ-meson to become massive. In order to verify that, let us expand the potential (49) for small fluctuations around the ground state,

$$\sigma = \sigma_0 + (\delta\sigma); \quad \pi = (\delta\pi). \tag{51}$$

Actually, it is these fluctuations $((\delta\sigma),(\delta\pi))$, which are to be be identified with the observed particles (σ- and π-meson). Since a bosonic-mass term is quadratic in the fields, let us expand the potential up to quadratic order in the fluctuations $(\delta\sigma)(\delta\pi)$. Expanding around a minimum, the linear order vanishes, and we have:

$$V(\sigma,\pi) = \lambda f_\pi^2 (\delta\sigma)^2 + \mathcal{O}(\delta^3), \tag{52}$$

where we have used that $\sigma_0 = f_\pi$. Comparing with the Lagrangian of a free boson we identify the mass of the sigma to be (remember that $L = T - V$)

$$m_\sigma^2 = 2\lambda f_\pi^2 \neq 0. \tag{53}$$

We find no mass term for the pion in agreement with our expectation, that the pion should be the massless Goldstone boson of the spontaneously broken chiral symmetry.

In summary, the properties of the ground state of the linear sigma-model are:

$$
\begin{aligned}
<\sigma> &= \sigma_0 = f_\pi, & (54)\\
<\pi> &= 0, & (55)\\
M_N &= g_\pi \sigma_0 = g_\pi f_\pi, & (56)\\
m_\sigma^2 &= 2\lambda f_\pi^2 \neq 0, & (57)\\
m_\pi &= 0. & (58)
\end{aligned}
$$

191

Before we conclude this section, let us calculate the conserved axial current and check if the PCAC-relation is satisfied in our model. The infinitesimal axial transformations of the nucleon, pion and sigma-fields are given by (see (16), (40) and (41))

$$\psi \longrightarrow \psi - i\gamma_5 \frac{\tau^a}{2}\Theta^a\psi\,, \tag{59}$$

$$\pi^i \longrightarrow \pi^i + \Theta^a\delta^{i,a}\sigma\,, \tag{60}$$

$$\sigma \longrightarrow \sigma - \Theta^a\pi^a\,. \tag{61}$$

Comparing with the general form (7) for unitary transformations, we find that the generator of the axial transformation T^a act on the fields in the following way

$$T^a\psi \;=\; \gamma_5 \frac{\tau^a}{2}\psi\,, \tag{62}$$

$$T^a\pi^j \;=\; i\sigma\delta^{a,j}\,, \tag{63}$$

$$T^a\sigma \;=\; -i\pi^a\,. \tag{64}$$

Using the expression for the conserved current (9) the conserved axial current is given by

$$A^a_\mu = \bar{\psi}\gamma_\mu\gamma_5\frac{\tau^a}{2}\psi - \pi^a\partial_\mu\sigma + \sigma\partial_\mu\pi^a\,. \tag{65}$$

In order to check the PCAC-relation, we again expand the fields around the ground state (see eq. (51))

$$A^a_\mu = \bar{\psi}\gamma_\mu\gamma_5\frac{\tau^a}{2}\psi - (\delta\pi^a)\partial_\mu(\delta\sigma) + (\delta\sigma)\partial_\mu(\delta\pi^a) + f_\pi\partial_\mu(\delta\pi^a)\,, \tag{66}$$

where we have used that $\sigma_0 = f_\pi$. Since the PCAC-relation involves the matrix element $< 0|A^a_\mu|\pi^j >$ only the last term of (66) contributes. The other terms would require either nucleons or sigma-mesons in the final or initial state. Thus, as far as the PCAC relation is concerned, the axial current reduces to ($(\delta\pi) = \pi$)

$$A^a_\mu(x)_{PCAC} = f_\pi\partial_\mu\pi(x)\,, \tag{67}$$

in agreement with the PCAC-result in eq. (32).

Explicit breaking of chiral symmetry

So far we have assumed that the axial vector symmetry is a perfect symmetry of the strong interactions. From our discussion in section we know, however, that the small but finite current-quark masses of the up and down quark break the axial vector symmetry explicitly. This explicit breaking of the symmetry should not be confused with the spontaneous breakdown we have discussed before. In case of a spontaneous breaking of a symmetry the Hamiltonian is still symmetric, whereas in case of an explicit breaking, already the Hamiltonian is not symmetric.

One may wonder if the whole concept of spontaneous symmetry breaking makes any sense if already the Hamiltonian is not symmetric. The answer to that, again, depends on the scales involved. If the explicit symmetry breaking is small, i.e. if the quark masses are small compared to to relevant energy scale of QCD, as we believe they are, then it will be sensible to apply the notion of a spontaneously broken symmetry.

To illustrate that, let us again utilize our little mechanics analogy, which we have developed in the previous section. An explicit symmetry breaking would imply that both potentials of fig. (1) are not invariant under rotation. This could for instance be achieved by slightly tilting them towards, say, the x-direction. As a result, also the ground state of potential (a) is away from the center $(x, y = 0)$. But the dislocation is small compared to that due to the spontaneous breaking. Furthermore, as long as the potentials are tilted only slightly, rotational excitation (pions) in potential (b) are still considerably softer than the radial ones (sigma-mesons). So in this sense, we expect the effect due to the spontaneous breakdown of chiral symmetry to dominate the dynamics, as long as the explicit breaking is small. In the linear sigma-model, the mass scale generated by the spontaneous breakdown is the nucleon mass, whereas that generated by the explicit breakdown is the mass of the pion, as we shall see. Thus, indeed the explicit breaking is small, and our picture, developed under the assumption of perfect axial vector symmetry, will survive the introduction of the explicit breaking to a very good approximation.

After these remarks let us now introduce a symmetry-breaking term into the linear sigma-model. In QCD, we know, that the symmetry is explicitly broken by a quark-mass term

$$\delta \mathcal{L}_{X\chi SB} = -m\bar{q}q \, , \tag{68}$$

where the subscript $X\chi SB$ stands for explicit chiral symmetry breaking. If we identify, as we have done before, the scalar quark-field combination $\bar{q}q$ with the σ-field, this would suggest the following symmetry-breaking term in the sigma-model

$$\delta \mathcal{L}_{SB} = \epsilon \sigma \, , \tag{69}$$

where ϵ is the symmetry breaking parameter. This term clearly is not invariant under the axial transformation Λ_A but preserves the vector symmetry Λ_V. Including this term, the potential V (49) now has the form

$$V(\sigma, \pi) = \frac{\lambda}{4} \left((\pi^2 + \sigma^2) - v_0^2 \right)^2 - \epsilon \sigma \, , \tag{70}$$

where we now have replaced f_π of eq. (49) by a general parameter v_0, which in limit of $\epsilon \to 0$ will go to f_π. The effect of the symmetry-breaking term is to tilt the potential slightly towards the positive σ direction, and thus to break the symmetry (see fig. (3)).

What are the consequences of this additional term? First of all, the minimum has shifted slightly. If we require that the value of the new minimum is still f_π in order to preserve the Goldberger-Treiman relation, we find for the parameter v_0 to leading order in ϵ

$$v_0 = f_\pi - \frac{\epsilon}{2\lambda f_\pi^2} \, . \tag{71}$$

Also the mass of the sigma is slightly changed

$$m_\sigma^2 = \left. \frac{\partial^2 V}{\partial \sigma^2} \right|_{\sigma_0} = 2\lambda f_\pi^2 + \frac{\epsilon}{f_\pi} \, . \tag{72}$$

But most importantly, the pion now acquires a finite mass

$$m_\pi^2 = \left. \frac{\partial^2 V}{\partial \pi^2} \right|_{\sigma_0} = \frac{\epsilon}{f_\pi} \neq 0 \, , \tag{73}$$

which fixes the parameter ϵ

$$\epsilon = f_\pi m_\pi^2 \, . \tag{74}$$

193

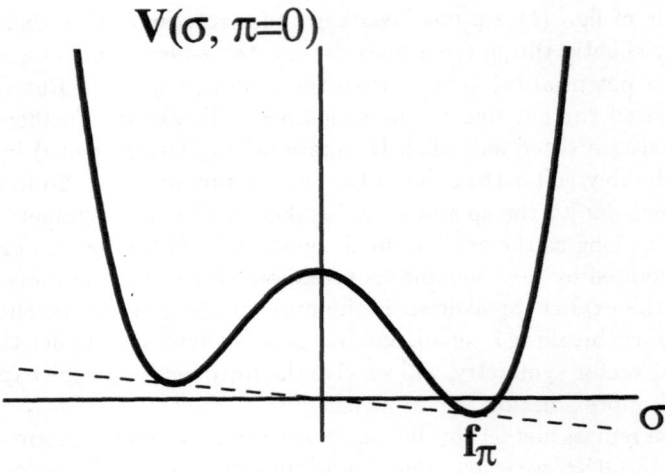

Figure 3. Potential of linear sigma-model with explicit symmetry breaking

Thus, the square of the pion mass is directly proportional to the symmetry breaking parameter ϵ as we would have expected it from our previous discussion.

Due to our choice of $\sigma_0 = f_\pi$, the nucleon mass is not changed, which, however, does not mean that there is no contribution to the nucleon mass from the explicit symmetry breaking. If we split the nucleon mass into a contribution from the symmetric part of the potential ($\sim v_0$) and one from the symmetry-breaking term ($\sim \epsilon$),

$$M_N = g_\pi \sigma_0 = g_\pi \left(v_0 + \frac{\epsilon}{2\lambda f_\pi^2} \right), \tag{75}$$

we find that the contribution from the symmetry breaking, which is often referred to as the pion-nucleon sigma-term[i], is given by

$$\Sigma_{\pi N} = \delta M_N^{X \chi SB} = g_\pi \frac{\epsilon}{2\lambda f_\pi^2} \simeq g_\pi f_\pi \frac{m_\pi^2}{m_\sigma^2}. \tag{76}$$

The pion-nucleon sigma-term can be measured in pion-nucleon scattering experiments and its is currently believed to be [10] $\Sigma_{\pi N}(0) = 35 \pm 5 \text{MeV}$.

Since chiral symmetry is now explicitly broken, the axial vector current is not conserved anymore. The functional form of the axial current is the same, however, as in the symmetric case, eq. (65), because the symmetry-breaking term (69) does not involve any derivatives (see eq. (9)). Its divergence is related to the variation of the symmetry-breaking term in the Lagrangian

$$\partial^\mu A_\mu^a = \epsilon \, \delta(\sigma) = -f_\pi m_\pi^2 \pi^a, \tag{77}$$

---

[i]This definition of the pion-nucleon sigma term should be taken with some care. For a rigorous definition see e.g. [10, 9]. In the framework of the sigma-model, this definition, however, is correct to leading order in ϵ.

which leads directly to the PCAC relation (31). Here $\delta(\sigma)$ denotes the variation of the σ-field with respect to the axial vector transformation Λ_A, not the fluctuation around the ground state. As in equ. (9) the angle Θ^a has been divided out.

The main effect of the explicit chiral symmetry breaking was to give the pion a mass. But we can utilize the symmetry breaking further to derive[j] some rather useful relations between expectation values of the scalar quark operator $\bar{q}q$ and measurable quantities like f_π, m_π, and $\Sigma_{\pi N}$.

When we introduced the symmetry-breaking term into our model, we had required that it has the same transformation properties under the chiral transformations as the QCD-symmetry-breaking term. The overall strength of the symmetry breaking, ϵ we then adjusted to reproduce the ground state properties, namely the pion mass. Therefore, it seems reasonable to expect, that that the vacuum expectation value of the symmetry-breaking terms in QCD (68) and in the effective model (69) are the same.

$$< 0|\,\epsilon\sigma\,|0> \;=\; < 0|-m\bar{q}q|0> \,. \tag{78}$$

If we insert for $\epsilon = m_\pi^2 f_\pi$ and use $< 0|\sigma|0 >= f_\pi$ we arrive at the so called Gell-Mann - Oakes - Renner (GOR) relation [11, 9]

$$m_\pi^2 f_\pi^2 = -\frac{m_u + m_d}{2} < 0|\bar{u}u + \bar{d}d|0 > \,, \tag{79}$$

where we have written out explicitly the average quark mass, m, and the quark operator $\bar{q}q$. The GOR relation is extremely useful, since it relates the quark condensate with f_π and/or the pion mass with the current-quark mass.

Similarly, but less convincingly, one can argue, that the contribution to the nucleon mass due to chiral symmetry breaking, $\Sigma_{\pi N}$, is the expectation value of the symmetry breaking Hamiltonian $\delta H_{X\chi SB} = -\delta\mathcal{L}_{X\chi SB}$ between nucleon states. This leads to the exact expression of the pion-nucleon sigma-term in terms of QCD variables [10, 12]

$$\Sigma_{\pi N} = \frac{m_u + m_d}{2} < N|\bar{u}u + \bar{d}d|N > \,. \tag{80}$$

This relation will turn out to be very helpful in order to estimate the change of the chiral condensate in nuclear matter at finite density.

S-wave pion-nucleon scattering

In order to see how chiral symmetry affects the dynamics, let us, as an example, study pion-nucleon scattering in the sigma-model. Let us begin by introducing some notation.

The invariant scattering amplitude $T(q, q')$ is commonly decomposed into a scalar and a vector part[k] (see fig. (4) for the notation of momenta)

$$T(q, q') = A(s, t) + \frac{1}{2}\gamma^\mu (q_\mu + q'_\mu)B(s, t)\,, \tag{81}$$

where (s, t) are the usual Mandelstam variables, and q and q' denote the incoming and outgoing pion - four-momenta. The relativistic scattering amplitude is related to the more familiar scattering amplitude in the center of mass frame, $\mathcal{F}(\vec{q}, \vec{q}')$ by

$$\chi^+ \mathcal{F} \chi' = \frac{M_N}{4\pi\sqrt{s}}\bar{u}(p, \sigma)T u(p', \sigma')\,. \tag{82}$$

---

[j]These 'derivations' are merely heuristic, but I feel they nicely demonstrate the physics which is going on. For a rigorous derivation see e.g. [9].

[k]For details see e.g. the appendix of [13].

195

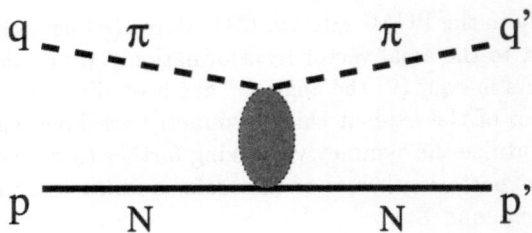

Figure 4. Pion-nucleon scattering amplitude.

Here χ are Pauli-spinors for the nucleon representing spin and isospin and $u(p,\sigma)$ stands for a relativistic spinors for a nucleon of momentum p, and spin σ.

The scattering amplitude can be decomposed into isospin-even and -odd components

$$T_{ab} = T^+ \delta_{ab} + \frac{1}{2}[\tau_a, \tau_b]T^-, \tag{83}$$

where the indices a, b refer to the isospin[l].

Now let us calculate the pion-nucleon scattering amplitude in the sigma-model. At tree level the diagrams shown in fig. (5) contribute to the amplitude. The first two processes represent the simple absorption and re-emission of the pion by the nucleon. Provided, that there is a coupling between pion and nucleon, one would have written down these diagrams immediately, without any knowledge of chiral symmetry. The third diagram (c), which involves the exchange of a sigma-meson, is a direct result of chiral symmetry, and, as well shall see, is crucial in order to give the correct value for the amplitude.

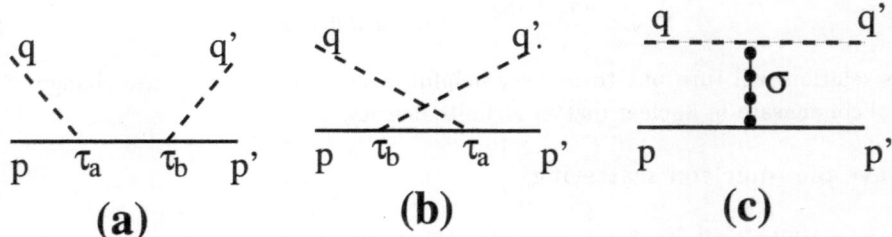

Figure 5. Diagrams contributing to the pion-nucleon scattering amplitude T_{ab}.

In the following, we will restrict ourselves to the forward scattering amplitudes, i.e. $q = q'$ and $p = p'$. Using standard Feynman-rules (see e.g. [1]), the above diagrams can be evaluated in a straightforward fashion. For diagram (a) we obtain

$$\bar{u}(p)T_{ab}^{(a)}u(p)$$
$$= g_\pi^2 \bar{u}(p)\,\tau_a\gamma_5\,\frac{(p+q)^\mu\gamma_\mu + m}{(p+q)^2 - m^2}\,\tau_b\gamma_5\,u(p)$$
$$= \bar{u}(p)\left[(\delta_{ab} + \frac{1}{2}[\tau_a, \tau_b])(-g_\pi^2\frac{q^\mu\gamma_\mu}{s - m^2})\right]u(p), \tag{84}$$

---

[l]Notice, that the isospin-odd amplitude is the *negative* of what in the literature is commonly called the iso-vector amplitude whereas the isospin-even amplitude is identical to the so called isoscalar one (see [13]).

where we have used that $\gamma_5 \gamma_\mu = -\gamma_\mu \gamma_5$, $\gamma_5^2 = 1$, $\tau_a \tau_b = \delta_{ab} + \frac{1}{2}[\tau_a, \tau_b]$, and the Dirac equation $(p_\mu \gamma^\mu - m)u(p) = 0$. Obviously, diagram (a) contributes only to the vector piece of the amplitude, B, and the isospin-even and -odd amplitudes are the same

$$B^+_{(a)} = B^-_{(a)} = -\frac{g_\pi^2}{s - M_N^2}. \tag{85}$$

The contribution of the crossed or u-channel (diagram (b)) one obtains by replacing

$$s \rightarrow u, \tag{86}$$
$$(\tau_a \tau_b) \rightarrow (\tau_b \tau_a), \tag{87}$$
$$q \rightarrow -q, \tag{88}$$

with the result

$$B^+_{(b)} = \frac{g_\pi^2}{u - M_N^2} = -B^-_{(b)}. \tag{89}$$

Here isospin-even and -odd amplitudes have the opposite sign.

It is instructive to calculate the scattering amplitude resulting from the first two diagrams only. If we didn't know about chiral symmetry, and, hence, the existence of the σ-exchange diagram, this is what we would naively obtain. At threshold ($\vec{q} = 0$), the combined amplitudes are

$$\nu B^+_{(a)+(b)} = -\frac{g_\pi^2}{M_N}\left(\frac{1}{1 - \frac{m_\pi^2}{4M_N^2}}\right), \tag{90}$$

$$\nu B^-_{(a)+(b)} = g_\pi^2 \frac{m_\pi}{2M_N^2}\left(\frac{1}{1 - \frac{m_\pi^2}{4M_N^2}}\right). \tag{91}$$

Using equations (81) and (82), the resulting s-wave isospin-even and isospin-odd scattering length a_0 would be

$$a_0^+((a)+(b)) = -\frac{g_\pi}{4\pi f_\pi(1 + \frac{m_\pi}{M_N})}(1 + \mathcal{O}(\frac{m_\pi^2}{M_N^2})) \simeq -1.4\, m_\pi^{-1}, \tag{92}$$

$$a_0^-((a)+(b)) = \frac{m_\pi}{8\pi f_\pi^2(1 + \frac{m_\pi}{M_N})}(1 + \mathcal{O}(\frac{m_\pi^2}{M_N^2})) \simeq 0.078\, m_\pi^{-1}, \tag{93}$$

where we have made of the Goldberger-Treiman relation $g_\pi f_\pi = M_N$. This is to be compared with the experimental values of [13]

$$a_0^+(exp) = -0.010(3)\, m_\pi^{-1}, \qquad a_0^-(exp) = 0.091(2)\, m_\pi^{-1}. \tag{94}$$

While we find reasonable agreement for the isospin-odd amplitude, the isospin even amplitude is off by two orders of magnitude! A different choice of the pion-nucleon coupling g_π would not fix the problem, but just shift it from one amplitude to the other. Before we evaluate the remaining diagram (c), let us point out that in the chiral limit, i.e. $m_\pi = 0$, the isospin-odd amplitude vanishes.

In order to evaluate the σ-exchange diagram, we need to extract the pion-sigma coupling from our Lagrangian. This is done by expanding the potential V (49) up to third order in the field fluctuations $((\delta\pi)$ and $(\delta\sigma))$. The terms proportional to $(\delta\pi)^2(\delta\sigma)$ then give the desired coupling.

$$\delta\mathcal{L}_{\pi\pi\sigma} = -\lambda f_\pi(\delta\pi)^2(\delta\sigma). \tag{95}$$

197

The resulting amplitude is then given by

$$\bar{u}(p)T_{ab}^{(c)}u(p) = -g_\pi \frac{2\lambda f_\pi}{t - m_\sigma^2} \delta_{ab}.$$ (96)

It only contributes to the scalar part of the amplitude, A, and only in the isospin-even channel. Using $2\lambda f_\pi^2 = m_\sigma^2 - m_\pi^2$ (see eqs. (72, 73)) we find

$$A_{(c)}^+ = -\frac{g_\pi}{f_\pi} \frac{m_\sigma^2 - m_\pi^2}{m_\sigma^2 - t} = \frac{g_\pi}{f_\pi} \left(1 - \frac{t - m_\pi^2}{t - m_\sigma^2}\right).$$ (97)

To leading order, the contribution to the s-wave scattering lengths of diagram (c) is

$$a_0^+((c)) = \frac{g_\pi}{4\pi f_\pi(1 + \frac{m_\pi}{M_N})}(1 + \mathcal{O}(\frac{m_\pi^2}{m_\sigma^2})),$$ (98)

$$a_0^-((c)) = 0.$$ (99)

Thus to leading order the contribution of the σ-exchange diagram (c) *exactly* cancels that of the nucleon-pole diagrams (a) and (b) and the total isospin-even scattering length vanishes,

$$a_0^+ = 0 + \mathcal{O}(\frac{m_\pi^2}{M_N^2}, \frac{m_\pi^2}{m_\sigma^2}),$$ (100)

in much better agreement with experiment. The cancellation between the large individual contributions to the isospin-even amplitude is a direct consequence of chiral symmetry, which required the σ-exchange diagram. In the chiral limit, this cancellation is perfect, i.e. the isospin-even scattering amplitude vanishes identically, because the corrections $\sim m_\pi$ are zero in this case.

Furthermore, since the third diagram (c) does not contribute to the isospin-odd amplitude, the good agreement found above still holds. In other words, with the 'help' of chiral symmetry both amplitudes are reproduced well.

NON-LINEAR SIGMA-MODEL

One of the disturbing features of the linear sigma-model is the existence of the σ-field, because it cannot really be identified with any existing particle. Furthermore, at low energies and temperatures one would expect that excitations in the σ-direction should be much smaller than pionic ones, which in the chiral limit are massless (see fig. (1)). This is supported by our results for the pion-nucleon scattering, where in the final result the mass of the sigma-meson only showed up in next to leading order corrections, which vanish in the chiral limit.

Let us, therefore, remove the σ-meson as a dynamical field by sending its mass to infinity. Formally this can be achieved by assuming an infinitely large coupling λ in the linear sigma-model. As a consequence the mexican-hat potential gets infinitely steep in the sigma-direction (see figure below). This confines the dynamics to the circle, defined by the minimum of the potential.

$$\sigma^2 + \pi^2 = f_\pi^2.$$ (101)

This additional condition removes one degree of freedom, which close to the ground state, where $< \sigma >= f_\pi$ is the sigma-field, and we are left with pionic excitations only.

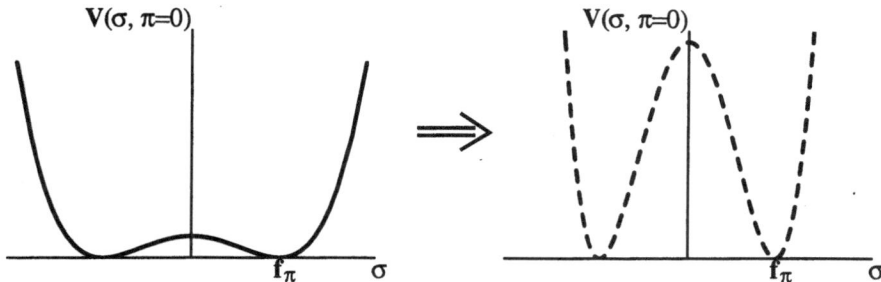

Because of the above constraint (101), the dynamics is now restricted to rotation on the so called chiral circle (actually it is a sphere). Therefore, the fields can be expressed in terms of angles $\vec{\Phi}$,

$$
\begin{aligned}
\sigma(x) &= f_\pi \cos(\frac{\Phi(x)}{f_\pi}) = f_\pi + \mathcal{O}(\Phi^2)\,, \\
\vec{\pi}(x) &= f_\pi \hat{\Phi} \sin(\frac{\Phi(x)}{f_\pi}) = \vec{\Phi}(x) + \mathcal{O}(\Phi^3)\,,
\end{aligned}
\tag{102}
$$

which to leading order can be identified with the pion field. Here $\Phi = \sqrt{\vec{\Phi}\vec{\Phi}}$ and $\hat{\Phi} = \frac{\vec{\Phi}}{\Phi}$. Clearly, this ansatz fulfills the constraint (101). Equivalently, one can chose a complex notation for the fields, as it is commonly done in the literature

$$
U(x) = e^{i\frac{\vec{\tau}\vec{\Phi}(x)}{f_\pi}} = \cos(\frac{\Phi(x)}{f_\pi}) + i\vec{\tau}\hat{\Phi} \sin(\frac{\Phi(x)}{f_\pi}) = \frac{1}{f_\pi}(\sigma + i\vec{\tau}\vec{\pi})\,,
\tag{103}
$$

where U represents a unitary (2×2) matrix. The constraint (101) is then equivalent to

$$
\frac{1}{2}tr(U^+U) = \frac{1}{f_\pi}(\sigma^2 + \pi^2) = 1\,.
\tag{104}
$$

Since chiral symmetry corresponds to a symmetry with respect to rotation around the chiral circle, all structures of the form

$$
tr(U^+U),\ tr(\partial_\mu U^+ \partial^\mu U),\ldots
\tag{105}
$$

are invariant. Already at this point it becomes obvious that we eventually will need some scheme, which tells us which structures to include and which ones not. This will lead us to the ideas of chiral perturbation theory in the following section.

Let us continue by rewriting the Lagrangian of the linear sigma-model (50) in terms of the new variables U or Φ. After a little algebra we find that the kinetic-energy term of the mesons is given by

$$
\frac{1}{2}\partial_\mu\sigma\partial^\mu\sigma = \frac{f_\pi}{4}tr(\partial_\mu U^+ \partial^\mu U)\,.
\tag{106}
$$

Next, we realize that nucleon-meson coupling term can be written as

$$
\begin{aligned}
-g_\pi\left(\bar{\psi}\psi\,\sigma + \bar{\psi}\gamma_5\vec{\tau}\psi\,\vec{\pi}\right) &= -g_\pi\bar{\psi}\left[f_\pi\left(\cos(\frac{\Phi}{f_\pi}) + i\gamma_5\vec{\tau}\hat{\Phi}\sin(\frac{\Phi}{f_\pi})\right)\right]\psi \\
&= -g_\pi\bar{\psi}\left(f_\pi e^{i\gamma_5\frac{\vec{\tau}\vec{\Phi}(x)}{f_\pi}}\right)\psi \\
&= -g_\pi f_\pi\bar{\psi}\Lambda\Lambda\psi\,,
\end{aligned}
\tag{107}
$$

199

where we have defined

$$\Lambda \equiv e^{i\gamma_5 \frac{\vec{\tau}\vec{\Phi}(x)}{2f_\pi}} \, . \qquad (108)$$

If we now redefine the nucleon fields

$$\psi_W = \Lambda\psi \, , \qquad (109)$$

$$\Rightarrow \bar{\psi}_W = \psi^+ \Lambda^+ \gamma^0 \overset{\{\gamma_0,\gamma_5\}=0}{=} \psi^+ \gamma^0 \Lambda = \bar{\psi}\Lambda \, , \qquad (110)$$

the interaction term (107) can be simply written as

$$-g_\pi f_\pi \bar{\psi}\Lambda\Lambda\psi = -g_\pi f_\pi \bar{\psi}_W \psi_W = -M_N \bar{\psi}_W \psi_W \, , \qquad (111)$$

where we have used the Goldberger-Treiman relation (38). In terms of the new fields, ψ_W, the entire interaction term as been reduced to the nucleon-mass term. If we want to identify the nucleons with the redefined fields ψ_W we also have to rewrite the nucleon kinetic-energy term in terms of those fields.

$$\bar{\psi}i\partial\!\!\!/\psi = \bar{\psi}_W \Lambda^+ i\partial\!\!\!/\Lambda^+ \psi_W \, . \qquad (112)$$

Since Λ is space-dependent through the fields $\Phi(x)$, the derivative also acts on Λ, giving rise to additional terms. After some straightforward algebra, one finds

$$\bar{\psi}_W \Lambda^+ i\partial\!\!\!/\Lambda^+ \psi_W = \bar{\psi}_W \left(i\partial\!\!\!/ + \gamma^\mu V_\mu + \gamma^\mu \gamma_5 A_\mu \right) \psi_W \, , \qquad (113)$$

with

$$V_\mu = \frac{1}{2} \left[\xi^+ \partial_\mu \xi + \xi \partial_\mu \xi^+ \right] \, , \qquad (114)$$

$$A_\mu = \frac{i}{2} \left[\xi^+ \partial_\mu \xi - \xi \partial_\mu \xi^+ \right] \, , \qquad (115)$$

$$\xi = e^{i\frac{\vec{\tau}\vec{\Phi}(x)}{2f_\pi}} \Rightarrow U = \xi\xi \, . \qquad (116)$$

We do not need to transform the potential of the linear sigma-model, $V(\pi, \sigma)$, since it vanishes on the chiral circle due to the constraint condition (101). Putting everything together, the Lagrangian of the non-linear sigma-model, which is often referred to as the Weinberg-Lagrangian, reads in the above variables

$$\mathcal{L}_W = \bar{\psi} \left(i\partial\!\!\!/ + \gamma^\mu V_\mu + \gamma^\mu \gamma_5 A_\mu - M_N \right) \psi + \frac{f_\pi}{4} tr(\partial_\mu U^+ \partial^\mu U) \, , \qquad (117)$$

where we have dropped the subscript from the nucleon fields. Clearly, this Lagrangian depends non-linearly on the fields $\vec{\Phi}$. It is instructive to expand the Lagrangian for small fluctuations Φ/f_π around the ground state. This gives

$$\begin{aligned}
\mathcal{L}_W \simeq\ & \bar{\psi}(i\partial\!\!\!/ - M_N)\psi + \frac{1}{2}(\partial_\mu \vec{\Phi})^2 \\
& + \frac{1}{2f_\pi} (\bar{\psi}\gamma_\mu \gamma_5 \vec{\tau}\psi)\partial^\mu \vec{\Phi} - \frac{1}{4f_\pi^2} (\bar{\psi}\gamma_\mu \vec{\tau}\psi) \cdot \left(\vec{\Phi} \times (\partial^\mu \vec{\Phi}) \right) \, ,
\end{aligned} \qquad (118)$$

where $\vec{\Phi}$ is now to be identified with the pion field. Comparing with the linear sigma-model, the σ-field has disappeared and the coupling between nucleons and pions has been changed to a pseudo-vector-one, involving the derivatives (momenta) of the pion-field. In addition, an explicit isovector coupling-term has emerged. From this Lagrangian it is immediately clear that the s-wave pion-nucleon scattering amplitude

200

vanishes in the chiral limit, because all couplings involve the pion four-momentum, which at threshold is zero in case of massless pions. Thus, the important cancellation between the nucleon pole-diagrams and the σ-exchange diagram, which we found in the linear sigma-model, has been moved into the derivative coupling of the pion through the above transformations.

On the level of the expanded Lagrangian (118), the explicit breaking of chiral symmetry is introduced by an explicit pion-mass term. Consequently corrections to the scattering lengths due to the nucleon pole diagrams should be of the order of m_π^2, since two derivative couplings are involved. However, the coupling $\delta\mathcal{L} = -\frac{1}{4f_\pi^2}(\bar\psi\gamma_\mu\vec\tau\psi)\cdot$ $\left(\vec\Phi\times(\partial^\mu\vec\Phi)\right)$, which contributes to first order to the isospin-odd amplitude, should give rise to a term $\sim \frac{m_\pi}{f^2}$ in agreement with our previous findings (93). Not too surprisingly one finds, that the above Lagrangian gives exactly the same results for the scattering-length as the linear sigma-model, except, that corrections $\sim \frac{1}{m_\sigma^2}$ are absent, because we have assumed that the mass of the σ-meson is infinite. However, the full Lagrangian (117) would give rise to many more terms, if we expand to higher orders in the fields Φ, which then would lead to loops etc. How to control these corrections in a systematic fashion will be the subject of the following section, where we discuss the ideas of chiral perturbation theory.

BASIC IDEAS OF CHIRAL PERTURBATION THEORY

In the previous sections we were concerned with the most simple chiral Lagrangian in order to see how chiral symmetry enters into the dynamics. As we have already pointed out, many more chirally-invariant terms can be included into the Lagrangian and thus we need some scheme which tells us what to include and what not. This scheme is provided by chiral perturbation theory.

Roughly speaking, the essential idea of chiral perturbation theory is to realize that at low energies the dynamics should be controlled by the lightest particles, the pions, and the symmetries of QCD, chiral symmetry. Therefore, S-matrix elements, i.e. scattering amplitudes, should be expandable in a Taylor series of the pion-momenta and masses, which is also consistent with chiral symmetry. This scheme will be valid until one encounters a resonance, such as the ρ-meson, which corresponds to a singularity of the S-matrix. Practically speaking, above the resonance a Breit-Wigner distribution cannot be expanded in a Taylor series.

It is not too surprising that such a scheme works. Imagine, we did not know anything about QED. We still could go ahead and parameterize the, say, electron-proton scattering amplitude in powers of the momentum transfer t. In this case the Taylor coefficients would be related to the total charge, the charge radius etc. With this information we could write down an effective proton-electron Lagrangian, where the couplings are fixed by the above Taylor-coefficients, namely the charge and the charge radius. This effective theory will, of course, reproduce the results of QED up to the order, which has been fixed by experiment. It is in this sense, the effective Lagrangian, obtained in chiral perturbation theory, should be understood: namely as a method of writing S-matrix elements to a given order in pion-momentum/mass. And to the order considered, the the effective Lagrangian obtained with chiral-perturbation theory should be equivalent with QCD [14, 15].

It should be stressed, that chiral perturbation theory is not a perturbation theory in the usual sense, i.e., it is not a perturbation theory in the QCD-coupling constant. In this respect, it is actually a non-perturbative method, since it takes already infinitely many orders of the QCD coupling constant in order to generate a pion. Instead, as already pointed out, chiral perturbation theory is an expansion of the S-matrix elements in terms of pion-momenta/masses.

From the above arguments one could get the impression that chiral perturbation theory has no predictive power, since it represents simply a power expansion of measured scattering amplitudes. Although this may true in some cases, one could easily imagine that one fixes the effective Lagrangian from some experiments and then is able to calculate other observables. For example, imagine that the effective pion-nucleon interaction has been fixed from pion nucleon-scattering experiments. This interaction can then be used to calculate e.g. the photoproduction of pions.

To be specific, let us discuss the case of pure pionic interaction, i.e. without any nucleons. As pointed out in the previous section, chiral invariance requires that the effective Lagrangian has to be built from structures involving U^+U (104) such as

$$tr(\partial_\mu U^+ \partial^\mu U), \quad tr(\partial_\mu U^+ \partial^\mu U)tr(\partial_\mu U^+ \partial^\mu U), \quad tr[(\partial_\mu U^+ \partial^\mu U)^2], \dots \qquad (119)$$

Furthermore, each $U = e^{i\frac{\vec{\tau}\vec{\Phi}(x)}{f_\pi}}$ contains any power of the pion-field Φ, which may give rise to loops etc. To specify which of the above terms should be included into the effective Lagrangian and how much each term should be expanded in terms of the pion field, one has to count the powers of pion momenta contributing to the desired process (scattering amplitude).

Consider a given Feynman-diagram contributing to the scattering amplitude. It will have a certain number L of loops, a certain number V_i of vertices of type i involving d_i derivatives of the pion field an a certain number of internal lines I_p. The power D of the pion momentum q, this diagram will have at the end, can be determined as follows:

- each loop involves an integral over the internal momenta $\int d^4q \sim q^4$,

- each internal pion line corresponds to a pion propagator, and thus contributes as $\frac{1}{q^2}$,

- each vertex V_i involving d_i derivatives of the pion field, contributes like q^{d_i} .

Consequently, the total power of q, q^D is given by

$$D = 4L - 2I_P + \sum_i V_i d_i \,. \qquad (120)$$

This can be simplified by using the general relation between the numbers of loops, internal lines and vertices of a given diagram

$$L = I_p - \sum_i V_i + 1 \,, \qquad (121)$$

to give

$$D = 2 + 2L + \sum_i V_i(d_i - 2) \,. \qquad (122)$$

With this formula we can determine to which order of the Taylor expansion of the scattering amplitude a given diagram contributes.

In order to see how this counting rule leads to an effective Lagrangian of a given order, we best study the simple example of pion-pion scattering. Since $U^+U = 1$ does not contribute to the dynamics, the simplest contribution to the effective Lagrangian is given by

$$\mathcal{L}_2 = \frac{f_\pi}{4} tr(\partial_\mu U^+ \partial^\mu U), \qquad (123)$$

where the subscript denotes the number of derivatives involved. Since we are discussing pion-pion scattering, we have to expand at least up to fourth order in the pion fields,

$$\mathcal{L}_2 = \frac{1}{2}(\partial_\mu \Phi)^2 + \frac{1}{6f_\pi^2}\left[(\Phi\partial_\mu\Phi)^2 - \Phi^2(\partial_\mu\Phi\partial^\mu\Phi)\right] + \mathcal{O}(\Phi^6), \qquad (124)$$

where the second term contributes to the pion-pion scattering amplitude. Although this term has two contributions, for the purposes of power counting, the second term may be considered as one vertex function, because both contributions have the same number of derivatives. Thus, to lowest order we have just one diagram, which is shown in fig. (6). It has no loops, $L = 0$, and the vertex function carries two derivatives of the pion field. Using the above counting rule (122), the order of this diagram is $D = 4$.

We can easily convince ourselves that there are no more terms contributing to this order. Including terms into the Lagrangian with four derivatives of the pions field such as e.g. $tr[(\partial_\mu U^+ \partial^\mu U)^2]$ immediately leads to $D \leq 6$. Also expanding the above Lagrangian (123) up to sixth order in the pion field leads to $D \leq 6$, because two of the pion fields have to be combined into a loop, since we are only considering a process with four external pions.

Obviously, the order of the effective Lagrangian depends on the process under consideration. Whereas a term involving six pion fields contributes to the order $D \leq 6$ to pion-pion scattering, it would contribute to order $D = 4$ to a process with three initial and three final pions. Of course, having realized, that we are actually parameterizing s-matrix elements, this is not such a surprise.

Figure 6. Leading order diagram for π-π scattering.

As already mentioned, to order $D = 6$ we have contributions from different sources. First of all, from higher-derivative terms in the Lagrangian and secondly, from the expansion to higher order in the pion fields, giving rise to loops. The beauty of chiral perturbation theory is that the effects of loops can be systematically be absorbed into renormalized couplings and masses. For details see e.g [3].

By now, the astute reader will have asked himself: How do I know that a momentum is small or, in other words, what is the expansion scale? There are several answers on the market. Georgi [16] argues, based on renormalization arguments, that the scale should be $4\pi f_\pi \sim 1\,\mathrm{GeV}$, whereas others suggest [3, 17] that the mass of the lowest lying resonance should give the scale, since this is the energy, where the entire game ceases to work. This seems to be a reasonable argument and, assuming that there is no σ-meson of mass $\sim 500\,\mathrm{MeV}$, the mass of the ρ-meson should provide a reasonable benchmark.

So far we have worked in the chiral limit, i.e. assuming that the pion mass vanishes. The explicit breaking of chiral symmetry is introduced by terms of the form $\sim tr(U^+ +$

Table 1. Pion-pion scattering length

	Experiment	Lowest Order	First Two Orders
$a_0^0 m_\pi$	$0.26 \ \pm 0.05$	0.16	0.20
$a_0^2 m_\pi$	-0.028 ± 0.012	-0.045	-0.041
$a_1^1 m_\pi^3$	0.038 ± 0.002	0.030	0.036

U) and and the simplest symmetry breaking is

$$\delta \mathcal{L}_{X\chi SB} = \frac{f_\pi^2 m_\pi^2}{4} tr(U^+ + U) \simeq 4 - \frac{1}{2} m_\pi^2 \Phi^2 + \mathcal{O}(\Phi^4), \qquad (125)$$

which to leading order in the pion-fields corresponds to a pion-mass term (the constant term does not contribute to the dynamics). Again, one can have many symmetry-breaking terms involving the above structure, such as

$$tr(U^+ + U), \quad tr(\partial_\mu U^+ \partial^\mu U) tr(U^+ + U) \dots \qquad (126)$$

so that an ordering scheme is necessary. Therefore, in the realistic case of explicit chiral symmetry breaking, the scattering amplitudes are not only expanded in terms of the pion momenta, but also in terms of the pion masses. The counting-rule is the same as given above (122), where d_i now gives the number of derivatives *and* pion masses of a given vertex of type i. The total effective Lagrangian for pion-pion scattering to order $D = 4$ is then given by

$$\mathcal{L}_2^{(4)} = \frac{1}{2}(\partial_\mu \Phi)^2 - \frac{1}{2} m_\pi^2 \Phi^2 + \frac{1}{6f_\pi^2} \left[(\vec{\Phi} \cdot \partial_\mu \vec{\Phi})^2 - \Phi^2 (\partial_\mu \vec{\Phi} \cdot \partial^\mu \vec{\Phi}) \right] + \frac{m_\pi^2}{24 f_\pi^2} (\vec{\Phi} \cdot \vec{\Phi})^2 . \quad (127)$$

In principle the 'adjustable' parameters of this Lagrangian are the pion-mass and the pion-decay constant, which have to be fixed to the experimental values.

The resulting pion-pion scattering length and volumes are then given by [18]

$$a_0^0 = \frac{7 m_\pi}{32 \pi f_\pi^2}, \quad a_0^2 = -\frac{m_\pi}{16 \pi f_\pi^2}, \quad a_1^1 = \frac{1}{24 \pi f_\pi^2 m_\pi} , \qquad (128)$$

where the subscript denotes the angular momentum and the superscript the isospin of the amplitude. As shown in table (1) [3], the leading order results agree reasonably well with experiment and are improved by the next to leading order corrections. Apparently we do not find perfect agreement with experiment even for the s-wave scattering lengths, although already to leading order we haven taken into account terms quadratic in the momenta, so that higher orders in the pion momentum will not improve the situation. However, remember that we not only expand in terms of the pion momenta, but also, as a result of the explicit symmetry breaking, in terms of the pion mass, which in principle can contribute to any order to the s-wave scattering length.

As already pointed out in the beginning of this section, chiral perturbation theory or, more precisely, the expansion in momenta breaks down once we get close to a resonance. This one easily understands by looking at the Breit-Wigner formula for the scattering amplitude involving a resonance.

$$f(E) \sim \frac{\Gamma/2}{E_R - E - i\Gamma/2} . \qquad (129)$$

For energies, which are small compared to the resonance energy, $E \ll E_R$ this amplitude may be expanded in terms of a power series and the concept of chiral perturbation theory works well

$$f(E) \sim \frac{\Gamma/2}{E_R}\left(1 + \frac{E + i\Gamma/2}{E_R} + \ldots\right); \quad E \ll E_R. \tag{130}$$

However, once we get close to the resonance energy, we need to expand to higher and higher order until at $E \geq E_R$ the power-series in E seizes to converge. To be specific, we expect that in the the isovector p-wave channel, which is dominated by the ρ-meson resonance, the chiral perturbation expansion should fail for energies $E \sim m_\rho$.

Finally, let us include the nucleons into the chiral counting. Naively, one would think, that this should destroy the entire concept, because the nucleon has a large mass, which is of the order of the expansion scale. However, since at low energies the scattering amplitude may also be calculated in a non-relativistic framework, we do not expect the nucleon mass to enter directly, but, to leading order, only via the kinetic energy $\sim \frac{p^2}{2M_N}$, which is small compared to that of the pion at the same momentum. Therefore, chiral perturbation theory should also work with nucleons present (for details see. [19]). The above argument can be formalized by realizing that the nucleon only enters into the amplitudes through the nucleon propagator (see e.g. the results for pion-nucleon scattering). At low momenta, the nucleon propagator contributing to diagram (a) of fig. (5) can be written as

$$\frac{\gamma_\mu(p^\mu + q^\mu) + M_N}{(p+q)^2 - M_N^2} \simeq \frac{\gamma_0 M_N + M_N}{2M_N q} = \frac{\Lambda}{q}(1 + \mathcal{O}(\frac{q}{M_N})), \tag{131}$$

where

$$\Lambda = \frac{\gamma_0 M_N + M_N}{2M_N} = \begin{pmatrix} 1 & 0 \\ 0 & 0 \end{pmatrix}. \tag{132}$$

projects on positive-energy states. Hence, to leading order each nucleon propagator contributes like $\frac{1}{q}$ to the power of pion momentum of the scattering amplitude. This leads to the following counting rule, which now also includes the nucleons [19]

$$D = 2 + 2L - \frac{1}{2}E_N + \sum_i V_i(d_i + \frac{1}{2}n_i - 2) \tag{133}$$

Here the notation is as in eq. (122) and E_N denotes the number of external nucleon lines and n_i the number of nucleon fields of vertex i, which is typically $n_i = 2$.

For the simple nucleon-pole diagram using pseudovector coupling we thus would have: $L = 0$, $E_N = 2$, $d = 1$, $n = 2$ such that, $d + \frac{1}{2}n - 2 = 0$ and, $D = 1$.

On top of the expansion in terms of pion-momenta and pion masses, from eq. (131) we, therefore, also have an expansion in the velocity of the nucleons $v \sim \frac{q}{M_N}$. This is carried out in a systematic fashion in the so called Heavy Baryon Chiral Perturbation Theory, as introduced by Jenkins and Manohar [20]. This approach essentially corresponds to a systematic non-relativistic expansion for the nucleon wave-function, on the basis that the nucleon (baryon) is heavy compared to the momenta involved. We should mention, that the effect of the nucleon can also be included in a fully covariant fashion as discussed by Gasser et al. [21].

Including the nucleon gives rise to additional structures which explicitly break the chiral symmetry, such as

$$\delta\mathcal{L} = a\,tr(U^+ + U)\bar{\psi}\psi \simeq a(1 - \frac{\phi^2}{2f_\pi^2})\bar{\psi}\psi. \tag{134}$$

To leading order, this is just a contribution to the nucleon mass, which allows us to identify the coefficient a with the sigma-term $\Sigma_{\pi N}$

$$\delta \mathcal{L} = -\Sigma_{\pi N} \, tr(U^+ + U) \bar{\psi} \psi \simeq -\Sigma_{\pi N} \, \bar{\psi} \psi + \frac{\Sigma_{\pi N}}{2 f_\pi^2} \, \bar{\psi} \psi \, \phi^2 \,. \tag{135}$$

The next to leading term in the above expression is an *attractive* interaction between pion and nucleon, which contributes to the order $D = 2$ to the amplitude. This term by itself is quite large and would lead to a wrong prediction for the s-wave pion-nucleon amplitude. However, there are additional terms contributing to the same order, which in the heavy-fermion expansion comes from the nucleon-pole diagrams. The coefficients of these terms then need to be chosen such, that the resulting scattering length acquire the small value observed in experiment [22].

APPLICATIONS

In this last section, we want to discuss a few applications of chiral symmetry relevant for the physics of dense and hot matter. First we will discuss the temperature and density dependence of the quark condensate. We will conclude with some general remarks on the properties of vector mesons in matter.

Change of the quark-condensate in hot and dense matter

Temperature dependence

One of the applications of chiral perturbation theory relevant to the physics of hot and dense matter is the calculation of the temperature dependence of the quark condensate. Here we just want derive the leading order result. A detailed discussion, which includes also higher-order corrections can be found in ref. [23]. The basic idea is to realize that the operator of the quark-condensate, $\bar{q}q$, enters into the QCD-Lagrangian via the quark-mass term. Thus, we may write the QCD-Hamiltonian as

$$H = H_0 + m_q \bar{q}q \,. \tag{136}$$

The quark condensate at finite temperature is then given by the following statistical sum

$$< \bar{q}q >_T = \frac{\sum_i < i | \bar{q}q \, e^{-H/T} | i >}{\sum_i < i | e^{-H/T} | i >} \,. \tag{137}$$

Since $\partial H / \partial m_q = \bar{q}q$, this can be written as

$$< \bar{q}q >_T = T \frac{\partial}{\partial m_q} \ln Z(m_q) \,, \tag{138}$$

where the partition function Z is given by $Z = \sum_i < i | e^{-H/T} | i >$.

In chiral perturbation theory we do not calculate the partition function of QCD, but rather that of the effective Lagrangian. To make contact with the above relations, we utilize the Gell-Mann Oakes Renner relation (79). To leading order in the pion mass the derivative with respect to the quark mass, therefore, can be written as

$$\frac{\partial}{\partial m_q} = -\frac{< 0 | \bar{q}q | 0 >}{f_\pi^2} \frac{\partial}{\partial m_\pi^2} \,. \tag{139}$$

Next to leading order contributions arise, among others, from the quark-mass dependence of the vacuum condensate.

To leading order the partition function is simply given by that of a non-interacting pion gas

$$\ln Z = \ln Z_0 + \ln Z_{\pi-\text{gas}} = \ln Z_0 + \frac{3}{(2\pi)^3} \int d^3p \, \ln(1 - \exp(-E/T)), \qquad (140)$$

where Z_0 stands for the vacuum contribution, which we, of course, cannot calculate in chiral perturbation theory, since we are only concerned with fluctuations around that vacuum. Thus the temperature dependence of the quark condensate in the chiral limit is given by

$$\begin{aligned} <\bar{q}q>_T &= <0|\bar{q}q|0> - T\frac{<0|\bar{q}q|0>}{f_\pi^2}\frac{\partial}{\partial m_\pi^2}Z_{\pi-\text{gas}}\bigg|_{m_\pi \to 0} \\ &= <0|\bar{q}q|0> (1 - \frac{T^2}{8f_\pi^2}). \end{aligned} \qquad (141)$$

Thus to leading order, the quark condensate drops like $\sim T^2$, i.e. at low temperatures the change in the condensate is small.

Corrections include the effect of pion interactions, which in the chiral limit are proportional to the pion momentum and thus contribute to higher orders in the temperature. Including contributions up to three loops, one finds see e.g. [23]

$$\frac{<\bar{q}q>_T}{<\bar{q}q>_0} = 1 - c_1\left(\frac{T^2}{8f_\pi^2}\right) - c_2\left(\frac{T^2}{8f_\pi^2}\right)^2 - c_3\left(\frac{T^2}{8f_\pi^2}\right)^3 \ln(\frac{\Lambda_q}{T}) + \mathcal{O}(T^8). \qquad (142)$$

For N_f flavors of massless quarks the coefficients are given in the chiral limit by

$$c_1 = \frac{2}{3}\frac{N_f^2 - 1}{N_f} \qquad c_2 = \frac{2}{9}\frac{N_f^2 - 1}{N_f^2} \qquad c_3 = \frac{8}{27}(N_f^2 + 1)N_f. \qquad (143)$$

The scale Λ_q can be fixed from pion scattering data to be $\Lambda_q = 470 \pm 110 \, \text{MeV}$. In fig. (7) we show the temperature dependence of the quark-condensate as predicted by the above formula. Currently, lattice gauge calculations predict a critical temperature $T_c \simeq 150 \, \text{MeV}$, above which the quark condensate has disappeared. At this temperature chiral-perturbation theory predicts only a drop of about 50 %, which gets even smaller once pion masses are included [23]. However, we do not expect chiral perturbation to work well close to the critical temperature. The strength of this approach is at low temperatures. The prediction, that to leading order the condensate drops quadratic in the temperature is a direct consequence of chiral symmetry and can be used to check chiral models as well as any other conjectures involving the change of the quark-condensate, such as e.g. the change of hadron masses.

Density dependence

For low densities, the density dependence of the quark condensate can also be determined in a model independent way[m]. We expect that to leading order in density the change in the quark condensate is simply given by the amount of quark condensate in a nucleon multiplied by the nuclear density,

$$<\bar{q}q>_\rho = <\bar{q}q>_0 + <N|\bar{q}q|N> \rho + \text{higher orders in } \rho. \qquad (144)$$

---

[m] Again, we give a heuristic argument. A rigorous derivation based on the Hellmann-Feynman theorem can be found e.g. in [24, 25].

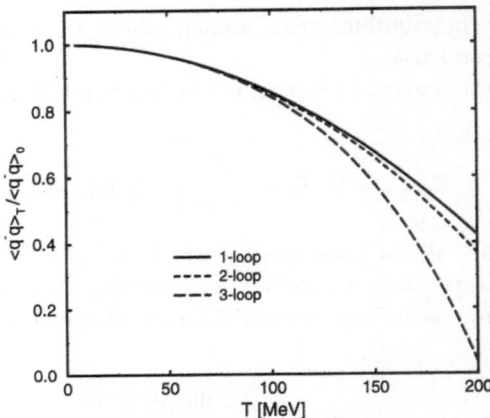

Figure 7. Temperature dependence of the quark condensate from chiral perturbation theory (chiral limit).

All we need to know is the matrix element of $\bar{q}q$ between nucleon states. This matrix element, however, enters into the pion-nucleon sigma-term, eq. (80)

$$< N|\bar{q}q|N >= \frac{\Sigma_{\pi N}}{m_q} = -\Sigma_{\pi N} \frac{< \bar{q}q >_0}{m_\pi^2 f_\pi^2} , \qquad (145)$$

where we also have made use of the GOR-relation (79), namely $m_q = -\frac{m_\pi^2 f_\pi^2}{<\bar{q}q>_0}$. Thus we predict, that the quark condensate drops *linearly* with density, as compared to the quadratic temperature dependence found above

$$< \bar{q}q >_\rho =< \bar{q}q >_0 \left(1 - \frac{\Sigma_{\pi N}}{m_\pi^2 f_\pi^2}\rho + \ldots\right). \qquad (146)$$

Corrections to higher order in density arise, among others, from nuclear binding effects. These have been estimated [26, 27] to be at most of the order of 15 % for densities up to twice nuclear-matter density. Assuming a value for the sigma-term of $\Sigma_{\pi N} \simeq 45\,\text{MeV}$ we find that the condensate has dropped by about 35 % at nuclear-matter density

$$< \bar{q}q >_\rho =< \bar{q}q >_0 \left(1 - 0.35\frac{\rho}{\rho_0}\right). \qquad (147)$$

Thus finite density is very efficient in reducing the quark condensate and we should expect that any in medium modification due to a dropping quark condensate should already be observable at nuclear-matter density. The above findings also suggest, that chiral restoration, i.e. the vanishing of the quark-condensate, is best achieved in heavy-ion collisions at bombarding energies, which still lead to full stopping of the nuclei.

Masses of vector mesons

Finally, let us briefly discuss what chiral symmetry tells us about the masses of vector mesons in the medium. Vector mesons, such as the ρ-meson, are of particular interest, because they decay into dileptons. Therefore, possible changes of their masses in medium are accessible to experiment.

Using current algebra and PCAC, Dey et al. [28] could show, that at finite temperature the mass of the rho-meson does not change to order T^2. Instead to order T^2 the vector-correlation function gets an admixture from the axial vector correlation function,

$$C_V(T) = (1 - \epsilon) C_V(T = 0) + \epsilon C_A(T = 0), \qquad (148)$$

with $\epsilon = \frac{T^2}{6f_\pi^2}$. The imaginary part of this vector-correlation function is directly related to the dilepton-production cross section. As depicted in fig. (8), the above result, therefore, predicts that to leading order in the temperature the dilepton invariant-mass spectrum develops a peak at the mass of the a_1-meson in addition to that at the mass of the ρ. At the same time, the contribution at the ρ-peak is reduced in comparison to the free case. Furthermore, the position of the peaks is not changed to this order in temperature. This general result is also confirmed by calculations in chiral models, which have been extended to include vector mesons [29, 30]. Notice, that the above finding also rules out that the mass of the ρ-meson scales linearly with the quark condensate, because previously we found that the quark condensate already drops to order T^2, whereas the mass of the ρ does not change to this order.

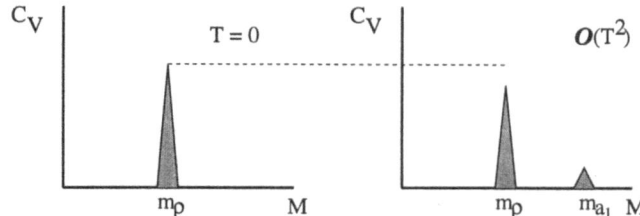

Figure 8. Vector-spectral functions at $T = 0$ and to leading order in temperature as given by equ. (148).

Corrections to higher order in the temperature, however, are not controlled by chiral symmetry alone and, therefore, one finds model dependencies. Pisarski [30] for instance predicts in the framework of a linear sigma-model with vector mesons, that to order T^4 the mass of the ρ decreases and that of the a_1 increases. Song [29, 31], on the other, uses a non-linear σ-model and finds a slight increase of the ρ-mass as well as a dropping of the a_1-mass. At the critical temperature, both again agree qualitatively in that the masses of a_1 and ρ become degenerate at a value which is roughly given by the average of the vacuum masses $\simeq 1\,GeV$. This agreement, again, is a result of chiral symmetry.

At and above the critical temperature, where chiral symmetry is not anymore spontaneously broken, chiral symmetry demands that the vector and axial vector correlation functions are the same. One way to realize that is by the having the same masses for the vector (ρ) and axial vector (a_1). However, this is not the only possibility! As was nicely discussed in a paper by Kapusta and Shuryak [32], there are at least three qualitatively different possibilities, which are sketched in fig. (9).

1. The masses of ρ and a_1 are the same. In this case, clearly the vector and axial vector correlation functions are the same. Note, however, that we cannot make any statement about the value of the common mass. It may be zero, as suggested by some people, it may be somewhere in between the vacuum masses, as the chiral models seem to predict and it my be even much larger than the mass of the a_1.

209

2. We may have a complete mixing of the spectral functions. Thus, both the vector and axial vector spectral functions have peaks of equal strength at both the mass of the ρ and the mass of the a_1, leading to two peaks of equal strength in the dilepton spectrum (modulo Boltzmann-factors of course). One example would be given by the low temperature result (148) with the mixing parameter $\epsilon(T_c) = \frac{1}{2}$. Using the low-temperature result for $\epsilon = \frac{T^2}{6f_\pi^2}$ would give a critical temperature of $T_c = \sqrt{3}f_\pi \simeq 164\,MeV$, which is surprisingly close to the value given by recent lattice calculations.

3. Both spectral functions could be smeared over the entire mass range. Due to thermal broadening of the mesons and the onset of deconfinement, the structure of the spectral function may be washed out and it becomes meaningless to talk about mesonic states.

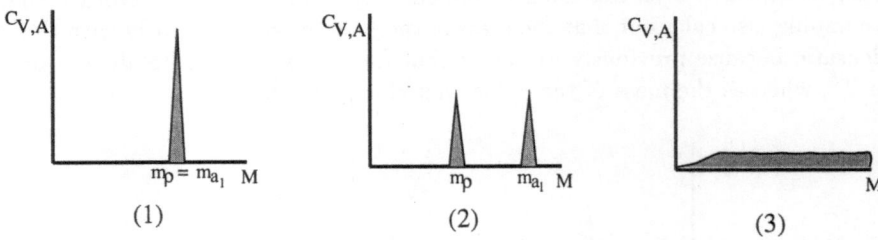

(1)   (2)   (3)

Figure 9. Several possibilities for the vector and axial vector spectral functions in the chirally restored phase.

To summarize, the only unique prediction derived from chiral symmetry (current algebra) about the temperature dependence of the ρ-mass, is that it does not change to order T^2, i.e. at low temperatures. Furthermore, at and above the critical temperature, chiral symmetry requires that the vector and axial vector spectral functions are identical, which, however, does not necessarily imply, that both exhibit just one peak, located at the same position. Corrections of order T^4 cannot be obtained from chiral symmetry alone.

Finally let us point out that the above findings do not rule out scenarios which relate the mass of the ρ with the temperature dependence of the bag-constant or gluon condensate, such as proposed by Pisarski [33] and Brown and Rho [34]. These ideas, however, involve concepts which go beyond chiral symmetry, such as the melting of the gluon condensate. Consequently in these scenarios, a certain behavior of the mass of the ρ-meson can only indirectly be brought in connection with chiral restoration.

Acknowledgments: I would like to thank C. Song for useful discussions concerning the effects of chiral symmetry on the vector mesons. The organizers and the students of the Dronten summer school deserve all the credit for its success. They generated a great atmosphere and I hope the students took home as much from my lectures as I did from their questions and comments. Last but not least, I want to thank the chef of the cafeteria, Mr. Rinus Lansberger, for kindly sharing his QCD-Hamiltonian, which made the presentation of the lectures so much easier. This work was supported by the Director, Office of Energy Research, Office of High-Energy and Nuclear-Physics Division of the Department of Energy, under Contract No. DE-AC03-76SF00098.

REFERENCES

1. J.D. Bjorken and S.D. Drell, *Relativistic Quantum Fields*, Mc Graw Hill, New York, 1965.
2. P. Ramond, *Field Theory*, Addison-Wesley, New York, 1989.
3. J.F. Donoghue, E. Golowich, and B.R. Holstein, *Dynamics of the Standard Model*, Cambridge University Press, Cambridge, U.K., 1992.
4. E.D. Commins and P.H. Bucksbaum, *Weak interactions of leptons and quarks*, Cambridge University Press, Cambridge, U.K., 1983.
5. S. Weinberg, *Phys. Rev. Lett.* **18** (1967) 507.
6. J.J. Sakurai, *Currents and mesons*, University of Chicago Press, Chicago, 1969.
7. M. Gell-Mann and M. Levy, *Nuovo Cimento* **16** (1960) 53.
8. C. DeTar and T. Kunihiro, *Phys. Rev.* **D39** (1989) 2805.
9. V. de Alfaro, V. Fubini, S. Furlan, and G. Rossetti, *Currents in hadron physics*, North-Holland, Amsterdam, 1973.
10. H. Höhler, *Pion-nucleon scattering (ed. H. Schopper), Landolt-Börnstein*, volume 9, Springer, Berlin, 1983.
11. M. Gell-Mann, R.J. Oakes, and B. Renner, *Phys. Rev.* **175** (1968) 2195.
12. J. Gasser and H. Leutwyler, *Phys. Rep.* **87** (1982) 77.
13. T. Ericson and W. Weise, *Pions and Nuclei*, Clarendon Press, Oxford, 1988.
14. S. Weinberg, *Physica* **96 A** (1979) 327.
15. H. Leutwyler, Principles of chiral perturbation theory, In *Proceedings of Hadron 94 Workshop, Gramado, Brazil 1994, hep-ph/9406283*; World Scientific, 1995.
16. H. Georgi, *Weak interactions and modern particle theory*, Benjamin/Cummings, Menlo Park, Ca., 1984.
17. H. Leutwyler, Foundations and scope of chiral pertubation theory, In *Proceedings Workshop on 'Chiral Dynamics: Theory and Experiment', MIT 1994 (Lecture Notes in Physics 452), hep/ph9409423*, A.M. Bernstein and B.R. Holstein, editors; Springer, 1995.
18. S. Weinberg, *Phys. Rev. Lett.* **17** (1966) 616.
19. S. Weinberg, *Nucl. Phys.* **B363** (1991) 3.
20. E. Jenkins and A. Manohar, *Nucl. Phys.* **B368** (1992) 190.
21. J. Gasser, M.E. Sanio, and A. Svarc, *Nucl. Phys.* **B307** (1988) 779.
22. V. Thorsson and A. Wirzba, *Nucl. Phys.* **A589** (1995) 633.
23. P. Gerber and H. Leutwyler, *Nucl. Phys.* **B321** (1989) 387.
24. E.G. Drukarev and E.M. Levin, *Nucl. Phys.* **A511** (1990) 679.
25. T.D. Cohen, R.J. Furnstahl, and D.K. Griegel, *Phys. Rev.* **C45** (1992) 1881.
26. G.Q. Li and C. M . Ko, *Phys. Lett.* **B388** (1994) 118.
27. R. Brockmann and W. Weise, The chiral condensate in matter, In *Proceedings of the Interanational Workshop on Hadrons in Nuclear Matter, Hirschegg 1995*, H. Feldmeier and W. Nörenberg, editor, volume XXIII, page 12, 1995.
28. M. Dey, V.L. Eletzky, and B.L. Ioffe, *Phys. Lett.* **B252** (1990) 620.
29. C. Song, *Phys. Rev.* **D48** (1993) 1375.
30. R.D. Pisarski, *Phys. Rev.* **D52** (1995) 3373.
31. C. Song, *Phys. Rev.* **D53** (1993) 1375.
32. J.I. Kapusta and E.V. Shuryak, *Phys. Rev.* **D49** (1994) 4694.
33. R.D. Pisarski, *Phys. Lett.* **B110** (1982) 155.
34. G.E. Brown and M. Rho, *Phys. Rev. Lett.* **66** (1991) 2720.

UTRA-RELATIVISTIC HEAVY ION PHYSICS FROM AGS TO LHC

Jurgen Schukraft

CERN, Div. PPE,
CH-1211 Geneva 23
Switzerland

ABSTRACT

Within the short time span of less than 20 years, the physics of ultra-relativistic heavy-ion collisions will have evolved from light ion reactions at a laboratory energy of a few GeV/nucleon, first explored at the Brookhaven AGS starting in 1986, to using heavy projectiles at a center-of-mass energy of several TeV/nucleon at the CERN LHC starting around 2004. The main topics of this new and rapidly evolving field, its current status and some results from the ongoing fixed-target program, as well as a preview of the experimental program planned at the LHC can be found in the following publications:

J. Schukraft, 'Utra-relativistic heavy-ion collisions: searching for the Quark-Gluon Plasma', Int. Nuclear Physics Conference, Wiesbaden, Germany (1992), Nucl. Phys. A553 (1993) 31c.

J. Schukraft, 'Utra-relativistic heavy-ion collisions: searching for the Quark-Gluon Plasma', Nuclear Physics News Vol. 2, No. 2 (1992) 14.

H.R. Schmidt and J. Schukraft, 'The physics of ultra-relativistic heavy-ion collisions', J. Phys. G: Nucl. Part. Phys. 19 (1993) 1705.

J. Schukraft, 'Heavy Ion Physics at the future colliders RHIC and LHC', Fifth Int. Conf. on Nucleus-Nucleus Collisions, Taormina, Italy (1994), Nucl. Phys. A583 (1995) 673c.

J. Schukraft, 'Little bang at big colliders: Heavy ion physics at RHIC and LHC', in Proc. International Conference on Nuclear Physics at the Turn of the Millennium: Structure of Vacuum and Elementary Matter, Wilderness/George, South Africa (1996), to be publ. by World Scientific Publ. Co., Singapore, 1996.

J. Schukraft, 'The ALICE heavy-ion experiment at the CERN LHC', Quark-matter '93 conference, Borlange, Sweden (1993), Nucl. Phys. A566 (1994) 311c.

J. Schukraft, 'The ALICE heavy ion experiment', Nuclear Physics News Vol. 6, No. 1 (1996) 16.

QUARK WAVE FUNCTIONS WITHIN NUCLEONS AND MESONS, AND EXCLUSIVE PROCESSES IN QCD

Carl E. Carlson

Physics Department
College of William and Mary
Williamsburg, VA 23187, U.S.A.

INTRODUCTION

Most of the lectures in this summer school are devoted to nuclei—their wave functions, correlations within them, how to measure such correlations, etc. We can do similar things at the quark level within a single nucleon or meson. These lectures will be devoted to just that, particularly to the quarkic wave functions of hadrons.

There will be a lot of talk about perturbative QCD (pQCD) and about exclusive reactions in these lectures. Perturbation theory we understand and think it applies if the momentum transfer is large. The calculation of an exclusive reaction depends upon pQCD and also upon the wave function of the hadrons. And here lies our opportunity. Our goal is to use pQCD and exclusive reactions to learn the wave functions of the hadrons, and thus get a step closer to understanding binding within QCD.

The topics to be covered in these lectures include:

- The quarkic wave functions of hadrons, and calculations of amplitudes and form factors for exclusive reactions.

- Sudakov form factors: what they are and how they help maintain the reliability of pQCD reactions.

- QCD sum rules: where the (theoretical) wave functions come from.

Correlations and Clustering Phenomena in Subatomic Physics
Edited by Harakeh *et al.*, Plenum Press, New York, 1997

THE WAVE FUNCTIONS

A hadron—for example, a proton—wave function consists of many Fock components, as in

$$|p\rangle = \sum_{N=3}^{\infty} \psi_N(x_i, k_{iT}, \lambda_i) | q(x_1, k_{1T}, \lambda_1), q(x_2, k_{2T}, \lambda_2) \ldots\rangle$$
$$= \psi_{qqq} |qqq\rangle + \psi_{qqqg} |qqqg\rangle + \cdots \tag{1}$$

The Fock component with the fewest number of constituents is the "valence" component. The notation for the momentum uses the light-front notation and formalism[1, 2], namely

$$k^{\pm} = k^0 \pm k^3$$
$$k_T = k_{\perp} = (k^x, k^y) . \tag{2}$$

All constituents are on-shell in this formalism,

$$k^2 = k^+ k^- - k_T^2 = m^2, \tag{3}$$

and the mass m of the quarks will be generally taken to be zero. If the momentum of the proton is p, we will often though not always take $p_T = 0$, whence

$$p = [p^+, p^-, p_T] = [p^+, m_N^2/p^+, 0]. \tag{4}$$

The momenta of the proton's constituents will often be expressed using a light-cone momentum fraction x_i whose definition can be read from

$$k_i = [k_i^+, k_i^-, k_{iT}] = [x_i p^+, \frac{m^2 + k_{iT}^2}{x_i p^+}, k_{iT}]. \tag{5}$$

The light-front formalism shares a feature with non-relativistic perturbation theory in that the four momentum is not conserved at every vertex. Three momentum components are conserved, the fourth is not. In the case of light-front perturbation theory the three conserved components are k^+ and k_T (k_T has two components); k^- is not conserved at each vertex though it must be conserved overall. The combination of k^+ conservation and of having all quarks on-shell will give us a simple picture of an exclusive reaction, as will be seen shortly.

We wish to learn the individual Fock wave functions of the hadron, starting as a practical matter with the valence quark wave function. Inclusive reactions, for example $e + p \rightarrow e +$ anything, is overly crude for this purpose because it allows Fock components with any number of constituents to participate in the reaction, and the cross section depends on integrals over a sum $\sum_N |\psi_N|^2$.

Exclusive reactions, reactions where all particles are observed as in $e + p \rightarrow e + p$, at high-momentum transfers depend only on the valence Fock component of the wave function. Higher components are suppressed by powers of M^2/Q^2 where M is some mass scale or inverse-size scale and Q is some momentum-transfer scale. Since this statement is important, we will justify it in two ways.

One way to see the suppression of higher Fock components is in a physical picture of how a high-momentum-transfer process works[3]. Let us for simplicity consider the reaction $e + \pi \rightarrow e + \pi$ and think in the Breit or brick-wall frame. In the Breit frame the energy transfer is zero, the particle reverses its direction of motion as if it bounced back from a brick wall. The virtual photon enters the reaction with momentum $\vec{q}$, the incoming pion has momentum $-\vec{q}/2$, and the outgoing pion has momentum $\vec{q}/2$. The entering hadron—the pion—may be thought of as a beam of constituents and moving in the same direction. To turn the hadron around *and keep it intact*, all constituents must interact and be turned. The virtual γ hits one quark, which must then share the momentum it has acquired with all the other quarks. See Fig.1 with its accompanying text. In particular note that the hadron is lorentz contracted into a disk-like volume. The hadron and its quarks enter from the left in this picture. If one quark is hit, it reverses direction and moves backward at the speed of light. An interaction must transfer momentum to reverse the direction of next quark *before* the first quark has moved the thickness of the hadron, or else it will be too late to reform the hadron. Thus the transverse separation between the struck quark and the one momentum is being transferred to must be less than the thickness of the disk, which is about MR/Q (where $Q = |\vec{q}|$). The probability of finding the second quark close enough is about

$$\text{probability} \approx \frac{\pi(MR/Q)^2}{\text{full area}}$$

$$\approx M^2/Q^2. \tag{6}$$

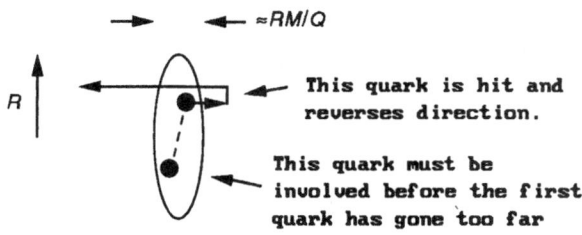

Figure 1. Physical picture of scattering at high-momentum transfer.

Since the form factor F_π is just the probability of the struck hadron remaining intact, we have

$$F_\pi \approx \frac{M^2}{Q^2} \qquad \text{for a } q\bar{q} \text{ state,} \tag{7}$$

and this generalizes to

$$F_\pi \approx \left(\frac{M^2}{Q^2}\right)^{N-1} \qquad \text{for an } N-\text{constituent state.} \tag{8}$$

Thus the minimum Fock state, the valence state, dominates at high Q.

We will check this scaling behavior against data, after we proceed part way into the "real calculation"[4], which now begins.

We will choose a frame where $q^+ = 0$, or

$$q = [0, q^-, q_T]. \tag{9}$$

In this frame, we can have diagrams like the first one in Fig. 2, but not like the second one, where a photon breaks into a $q\bar{q}$ pair (the second of Fig. 2) or a $q\bar{q}$ pair turns into a photon (not shown). This is because the quarks and antiquarks are on-shell so that

$$k^+ = k^0 + k^3 > 0, \tag{10}$$

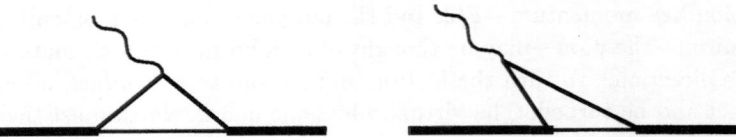

Figure 2. Light cone frame possibilities and impossibilities. The heavy solid lines are pions and the normal solid lines are quarks or antiquarks.

and because the "+" components of momentum are conserved.

In this frame,

$$Q^2 \equiv -q^2 = q_T^2. \tag{11}$$

We will take the initial hadron moving along the z-axis, and shall neglect all masses. The amplitude for $\gamma_V + \pi \to \pi$ is illustrated in Fig. 3. It is a convolution of a pion wave function in, a "hard-scattering amplitude" T for $\gamma_V + q\bar{q} \to q\bar{q}$, and a pion wave function out. The initial pion and struck quark momenta are

$$p = [p^+, 0, 0]$$
$$k_1 = [x_1 p^+, \frac{k_T^2}{x_1 p^+}, k_T], \tag{12}$$

and the initial wave function is denoted $\psi(x, k_T)$.

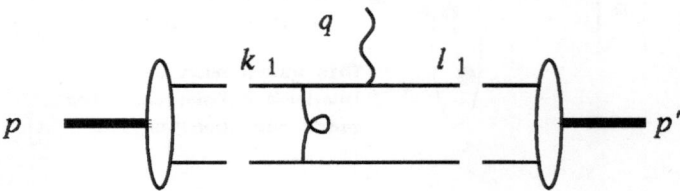

Figure 3. Model for the pion form factor. The looped line represents a gluon.

The outgoing-pion momentum is

$$p' = p + q = [\, p^+, \frac{q_T^2}{p^+}, q_T] \,. \tag{13}$$

The outgoing $q\bar{q}$ will have momentum fractions $y_{1,2}$, and *if* they have parallel momenta, they must share q_T proportionately as in

$$\ell_1 = [\, y_1 p^+, \frac{(y_1 q_T)^2}{y_1 p^+}, y_1 q_T] = y_1 p',$$
$$\ell_2 = [\, y_2 p^+, \frac{(y_2 q_T)^2}{y_2 p^+}, y_2 q_T] = y_2 p' \,. \tag{14}$$

In general, the outgoing quarks will have a transverse-momentum relative to each other and to the outgoing pion. We will call this ℓ_T and now

$$\ell_1 = [\, y_1 p^+, \frac{(y_1 q_T + \ell_T)^2}{y_1 p^+}, y_1 q_T + \ell_T],$$
$$\ell_2 = [\, y_2 p^+, \frac{(y_2 q_T - \ell_T)^2}{y_2 p^+}, y_2 q_T - \ell_T], \tag{15}$$

218

with the outgoing wave function $\psi(y, \ell_T)$.

The form factor is defined from the matrix element of the electro-magnetic current,

$$\langle \pi(p') | J_\mu^{\text{e.m.}} | \pi(p) \rangle = F_\pi(Q^2)(p + p')_\mu \, , \tag{16}$$

where $q = p' - p$. We then get

$$F_\pi = \int dx \, d^2 k_T \, dy \, d^2 \ell_T \, \psi^*(y, \ell_T) T(x, k_T, y, \ell_T, Q) \psi(x, k_T). \tag{17}$$

(Some factors of 2 and π have not been indicated.) Within T, the transverse momenta of the quarks are negligible for $Q = |\vec{q}_T| \to \infty$. For example, we can work out that the four-momentum squared of the exchanged gluon is

$$q_G^2 = x_2 y_2 q^2 + O(\langle k_T \rangle q_T) \tag{18}$$

where $\langle k_T \rangle$ is the root-mean-square (RMS) transverse momentum of the valence quarks. We drop the last term in the above equation and others like it in the evaluation of T which amounts to assuming that T is independent. This allows us to write

$$F_\pi = \int dx \, dy \left(\int d^2 \ell_T \, \psi^*(y, \ell_T) \right) T(x, 0, y, 0, Q) \left(\int d^2 k_T \, \psi(x, k_T) \right), \tag{19}$$

suggesting the definition of the "distribution amplitude"[1],

$$\phi_\pi(x) = \int d^2 k_T \, \psi(x, k_T) \, , \tag{20}$$

giving finally,

$$F_\pi = \int dx \, dy \, \phi_\pi^*(y) T_H(x, y, Q) \phi_\pi(x) \tag{21}$$

where T_H is, except for removing a factor $(p + p')_\mu$, the matrix element of the electro-magnetic current between two parallel moving $q\bar{q}$-states.

Let us derive the counting rules again. The Q dependence comes from T, and may be ferreted out using a few rules. One lowest order diagram contributing to T is shown in figure 3. The rules are[5]

- Each fermion line gives a factor Q. This comes from the $\bar{u} \ldots u$, where u is a Dirac spinor

$$u(p) = \sqrt{E + m} \begin{pmatrix} \chi \\ \frac{\vec{\sigma} \cdot \vec{p}}{E+m} \chi \end{pmatrix} = O(\sqrt{Q}), \tag{22}$$

where χ is (the transpose of) $(1, 0)$ or $(0, 1)$ or a linear combination thereof, and the size quoted is for $m \to 0$ and in the Breit frame. Thus

$$\bar{u} \ldots u \propto Q \quad \text{(or is zero)} \quad \text{as } m \to 0. \tag{23}$$

- Each internal fermion propagator gives an additional $1/Q$.

- Each internal gluon propagator gives $1/Q^2$.

- From the definition, the matrix element, which corresponds to the diagram we are evaluating, is of order $F_\pi(Q^2) \cdot Q$.

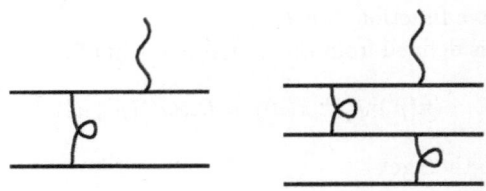

Figure 4. Figures for scaling calculation.

Thus overall, the form factor scales like

$$F_\pi \propto \frac{1}{Q} Q^2 \frac{1}{Q} \frac{1}{Q^2} = \frac{1}{Q^2}. \tag{24}$$

The generalization to an N-constituent state, illustrated as the second diagram in Fig. 4, is

$$F_N \propto \frac{1}{Q} Q^N \frac{1}{Q^{N-1}} \left(\frac{1}{Q^2} \right)^{N-1} = \left(\frac{1}{Q^2} \right)^{N-1}. \tag{25}$$

The success of the scaling predictions may be judged from a compilation of form-factor data for the nucleon for elastic and transitions, plotted by Stoler[6] and shown in Fig. 5. The form factors should fall like Q^{-4}, so if divided by the dipole form factor, they should be flat at high Q. Three out of four work well; the $N \to \Delta$ transition form factor is an exceptional case which has an explanation within pQCD[7, 8] that will be given later.

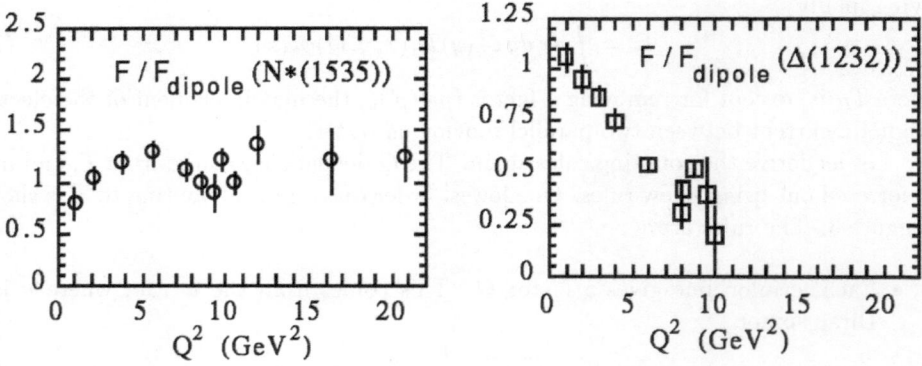

Figure 5. Scaling of real form factors. The form factors F are defined analogously to F_1 for elastic scattering and should scale like $1/Q^4$ according to pQCD. F_{dipole} is the dipole form factor. The S_{11} works well, as do the not shown nucleon elastic case and the bump at 1688 MeV. The Δ fails: but see later in the text.

More accurately, in addition to the power behavior there is at least some additional Q^2 dependence from the QCD coupling parameter. For the pion this means

$$Q^2 F_\pi \propto const. \times \alpha_s(Q^2) \tag{26}$$

and for $\gamma_V + p \to$ baryon, the elastic or transition form factor satisfies,

$$Q^4 F \propto const. \times \alpha_s^2(Q^2). \tag{27}$$

The extra logarithmic fall off from the coupling parameter can to some extent also been seen in the data.

While we are discussing scaling of form factors, it is also convenient to mention the scaling rules for general processes like[1]

$$\gamma_V + p \to \pi^+ + n$$
$$\gamma_V + d \to p + n. \tag{28}$$

Amplitudes for these processes have two independent variables, say t and s. Fix the ratio t/s, and then just consider scaling in s. Momenta will be of scale $\sqrt{s}$ or $\sqrt{t}$, and we can use the rules derived earlier if these momenta are large.

Consider $\gamma_V + p \to \pi^+ + n$. A lowest order diagram involving the lowest Fock states is shown in Fig. 6. There are

- 4 quark lines (8 external quarks),

- 3 internal quark propagators, and

- 3 gluon propagators.

Thus if Q is a suitable momentum scale, the matrix element is

$$\mathcal{M} \propto Q^4 \frac{1}{Q^3} \left(\frac{1}{Q^2} \right)^3 = \frac{1}{Q^5} \propto s^{-5/2}. \tag{29}$$

The differential cross section then goes like

$$\frac{d\sigma}{dt} = \frac{\mathcal{M}^2}{16\pi s^2} \propto \frac{1}{s^7} \tag{30}$$

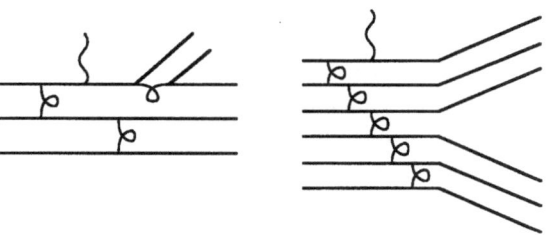

Figure 6. Scaling of exclusive reactions.

The general result is that

$$\frac{d\sigma}{dt} \propto s^{2-(\text{number of external fields})}, \tag{31}$$

and this applied to deuteron photodistintegration, also illustrated in Fig. 6, leads to

$$\frac{d\sigma}{dt} \propto s^{-11}. \tag{32}$$

Data for the $\gamma p \to \pi^+ n$ example are shown[9] in Fig. 7. Pion quasi-elastic photo-production seems to exhibit QCD scaling behavior when we are above the resonance region, and deuteron photodisintegration does the same if E_γ is above 1 GeV.

Scaling cannot prove QCD. But scaling depends on having short distance interactions, so in the regime where scaling is working—if the observed scaling is not mere accident and if QCD is the right theory—then perturbation-theory calculations in QCD should be valid.

We can now proceed to finish the pion form factor calculation.

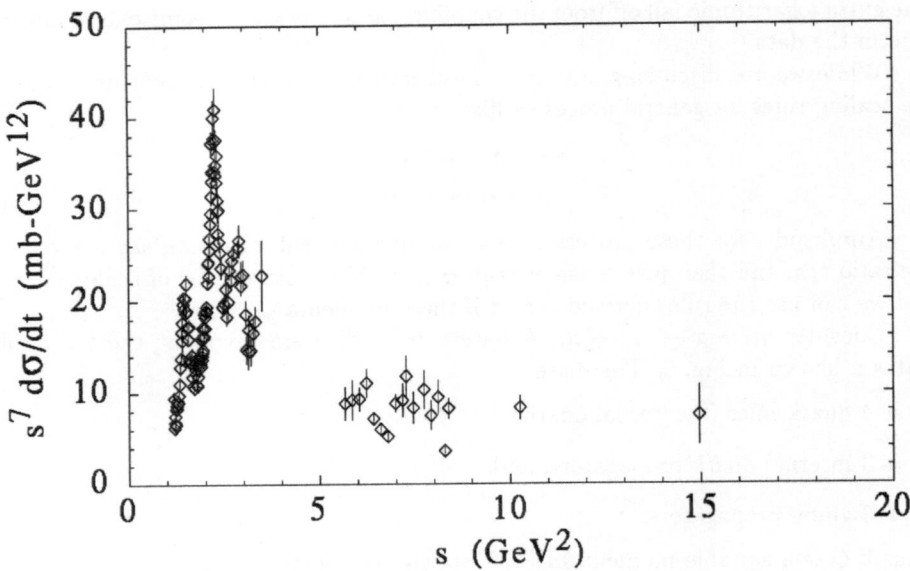

Figure 7. Data for scaling of exclusive reactions: $\gamma + p \to \pi^+ + n$.

A HARD SCATTERING CALCULATION

We began the calculation of the pion form factor in the last section, but then contented ourselves with extracting the scaling behavior for the form factor and other quantities, and checking the calculated scaling against data. We will now finish the calculation of the hard-scattering kernel, T_H, for F_π.

The kernel T_H is by itself Lorentz invariant and we can calculate it using Feynman perturbation theory even if we use light-front variables and ideas for the rest of the calculation. The basic diagram is again in Fig. 4. The incoming quark and antiquark have colors i and j and for T_H have parallel momenta k_1 and k_2. When this diagram is embedded into a calculation with a pion, the quark ends are knit together with a projection

$$\text{Pion} \longrightarrow \frac{\delta_{ij}}{\sqrt{3}} \frac{\gamma_5 \not{p}}{\sqrt{2}} \quad \left[\text{with an } \int_0^1 dx \, \phi_\pi(x) \ldots \text{ somewhere}\right]. \tag{33}$$

The $\delta_{ij}/\sqrt{3}$ represents the color wave function, which is often written in terms of the colors red, green, and blue as

$$\frac{1}{\sqrt{3}} \left(R\bar{R} + G\bar{G} + B\bar{B} \right). \tag{34}$$

The pion is a spin-zero particle made from spin-1/2 particles, and this gives the projection

$$\frac{1}{\sqrt{2}} \left(u_\uparrow(k_2)\bar{v}_\downarrow(k_1) - u_\downarrow(k_2)\bar{v}_\uparrow(k_1) \right) \longrightarrow \frac{1}{\sqrt{2}} \gamma_5 \not{p}. \tag{35}$$

The reader may want to prove this as an exercise; if so, note that the "$-$" in the above equation depends on one's conventions for the v-spinors, and that some factors of $\sqrt{x_1 x_2}$ may be hidden by the lack of an equal sign. We can now show (take this as another exercise) that

$$T_H = \frac{64\pi}{3} \frac{\alpha_s}{Q^2} \frac{1}{x_2 y_2}, \tag{36}$$

222

where we have used the fact that the charges of the quarks in a π^+ add to one, and the fact that the distribution amplitude must be invariant under $x_1 \leftrightarrow x_2$ or $y_1 \leftrightarrow y_2$. The latter follows informally from what one is used to from S-state positronium wave functions, or formally from the charge-conjugation properties of the pion. Thus

$$F_\pi(Q^2) = \frac{64\pi}{3} \frac{\alpha_s}{Q^2} I_\pi^2, \tag{37}$$

where

$$I_\pi = \int_0^1 \frac{dx_1}{x_2} \phi_\pi(x). \tag{38}$$

At this point I will interrupt to say something about higher-order corrections, one of which is illustrated in Fig. 8. This will be just a short note on a long but good story[3, 10].

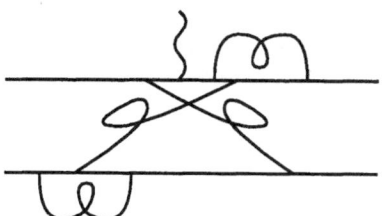

Figure 8. One higher-order diagram

There are lots of higher-order corrections, and they generally have a form

$$\left(\alpha_s(Q^2) \ln \frac{Q^2}{\mu^2} \right)^n \times \text{ lowest order}, \tag{39}$$

where n is the order beyond lowest and where μ is some renormalization scale. Since

$$\alpha_s(Q^2) \propto \frac{1}{\ln \frac{Q^2}{\Lambda^2}}, \tag{40}$$

it does look like the terms in the higher-order corrections form a convergent series. However, the following can be proved,

- The nasty logarithms all go into the wave function, modifying it and giving it some dependence on Q^2, but modifying it in the same way for any processes. Hence,

$$\phi_\pi(x) \longrightarrow \phi_\pi(x, Q^2). \tag{41}$$

- Corrections to T_H are just perturbative corrections in the form of a power series in α_s. Hence T_H is still reliably calculated in perturbation theory.

Returning to ϕ_π and I_π, we have several pieces of firm knowledge, but not actually enough to be sure of the numerical value of I_π. A normalization of ϕ_π itself comes from the decay $\pi^+ \to \mu^+ + \nu$. The crucial matrix element for this decay[1],

$$\langle 0|\bar{d}\gamma_\mu(1 - \gamma_5)u|\pi^+(p)\rangle = -\sqrt{2}\, f_\pi p_\mu, \tag{42}$$

is the defining relation for the pion decay constant f_π, which is numerically 93 MeV. The diagram for the matrix element is shown in Fig. 9. The decay depends upon the short distance overlap of the quark and antiquark, and one may calculate using techniques already outlined to relate the matrix element to an integral over ϕ_π, whence

$$\int_0^1 dx\, \phi_\pi(x) = \frac{f_\pi}{2\sqrt{3}}.$$ (43)

Figure 9. Hadronic part of the pion decay amplitude

Further knowledge of ϕ_π follows from another item which will be reported but not elaborated, namely the evolution equation for ϕ_π of Brodsky and Lepage[1]. For Q^2 already large, further changes in ϕ_π with changing Q^2 can be calculated using perturbation theory. Doing so leads to the evolution equation,

$$Q^2\frac{\partial}{\partial Q^2}\phi_\pi(x, Q^2) = \int_0^1 dy\, V(x,y)\phi_\pi(y, Q^2),$$ (44)

where the kernel $V(x,y)$ can be examined in the source. One outcome of the evolution equation is that

$$\lim_{Q^2\to\infty}\phi_\pi(x, Q^2) = const. \times x_1 x_2.$$ (45)

For this formula to be valid actually requires not just $Q^2 \to \infty$ but even requires $\ln Q^2 \to \infty$. I.e., it is not just an asymptotic formula, but a "superasymptotic" formula, and it may not apply at experimental Q^2 even if the rest of the formalism does. However, taking ϕ_π proportional to just $x_1 x_2$ gives what is called the "asymptotic" distribution amplitude and is at least a good first guess for the distribution amplitude at experimental Q^2.

The normalization requirement then yields,

$$\phi_\pi^{ASY} = 6x_1 x_2 \cdot \frac{f_\pi}{2\sqrt{3}},$$ (46)

and

$$I_\pi^{ASY} = \int_0^1 \frac{dx_1}{x_1}\phi_\pi^{ASY} = 3 \cdot \frac{f_\pi}{2\sqrt{3}},$$ (47)

and

$$Q^2 F_\pi^{ASY} = (4\pi f_\pi)^2 \cdot \frac{\alpha_s(Q^2)}{\pi} \approx 0.15 \text{ GeV}^2.$$ (48)

The numerical value uses $\alpha_s = 0.3$.

Data for $Q^2 F_\pi$ can be found[11] and is clearly larger, by nearly a factor of 3, than the pQCD calculation with the asymptotic distribution amplitude and the stated α_s. One may suppose that another distribution amplitude is needed; that another α_s is pertinent; that other information on the integral I_π or on ϕ_π itself should be sought; or that the data should be remeasured. CEBAF will undertake the last; and we shall consider some of the other items.

Chernyak and Zhitnitsky have suggested a distribution amplitude[12]

$$\phi_\pi^{CZ} = 30 x_1 x_2 (x_1 - x_2)^2 \cdot \frac{f_\pi}{2\sqrt{3}}. \tag{49}$$

The work underlying this suggestion is based on QCD sum rules and in style is the same as the consideration on the $\Delta(1232)$ distribution amplitude that we will discuss at the end of these lectures. We will just consider the consequences of the Chernyak-Zhitnitsky distribution amplitude. The Chernyak-Zhitnitsky and asymptotic distribution amplitudes are compared in Fig. (10).

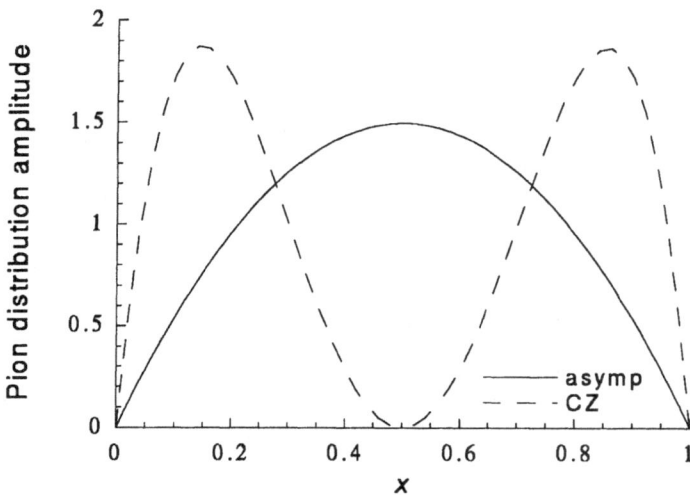

Figure 10. Two suggestions for the pion distribution amplitude. The units of the vertical axis are $f_\pi/2\sqrt{3}$.

There has been some discussion as to whether ϕ_π has a camel-like shape, but this of course displays some ignorance of camels. The asymptotic distribution amplitude is shaped like a dromedary camel, the Chernyak-Zhitnitsky like a bactrian camel. More importantly, the integral I_π is larger for the CZ distribution amplitude,

$$I_\pi^{CZ} = 5 \cdot \frac{f_\pi}{2\sqrt{3}} = \frac{5}{3} I_\pi^{ASY}, \tag{50}$$

and gives a good match to the the F_π data.

Clearly we should examine other ways to measure I_π. The $F_{\pi^0\gamma^*\gamma}$ form factor goes with the process virtual $\gamma +$ (almost) real $\gamma \rightarrow \pi^0$. A theorist might draw it in lowest order as the first of Fig. 11, and Cornell has measured it (using photons radiated from colliding electron beams to produced a final-state pion). Note that no powers of α_s are involved in the calculation. The Cornell result[13] agrees well with pQCD for Q^2, the virtuality of the virtual photon, above 1 GeV^2 if we use the *asymptotic* distribution amplitude.

We should like still another way to measure I_π, and we can mention the semi-exclusive process of photoproducing pions at high transverse momentum, $\gamma + p \rightarrow \pi +$ anything[14]. One lowest order diagram is shown Fig. 12, and the amplitude is proportional to $\alpha_s I_\pi$. The short distance production of the pion does have a serious background which is production of a high-transverse-momentum quark or gluon followed by its fragmentation into a pion plus other stuff, also illustrated[15] in Fig. 12. This process

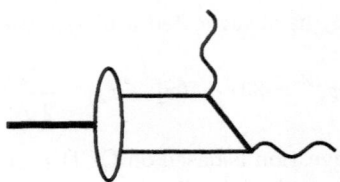

Figure 11. The $\pi^0\gamma\gamma$ form factor

in fact describes the pion production well at what we can call moderate transverse momentum, where there is currently data. In fact, explaining this data was a triumph in the 1980's, but one decade's success becomes another decade's background. At higher transverse momentum (above about 5 1/2 GeV for 100 GeV initial photons or above about 2 GeV for 20 GeV initial photons) the direct or short-range process dominates, at least in the calculations, and measuring pions at these transverse momenta could provide another piece of information on the pion distribution amplitude.

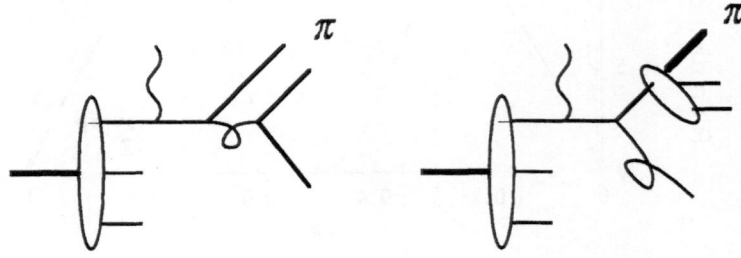

Figure 12. Photoproduction of pions at high transverse momentum

CAN SUDAKOV SAVE US?

The title of this section leads to another question: Save us from what? The answer to that is to save us from criticisms of using pQCD for exclusive reactions[16]. More precisely, the Sudakov form factors give us a way to separate what is reliable and non-reliable in a pQCD calculation and to eliminate without over-eliminating the parts that are not reliable. To continue, we need to learn what Sudakov did and to consider the effect of Sudakov form factors on composites.

What did Sudakov do?

Sudakov, in the middle 1950's, calculated the the form factor for free electrons[17]. He, of course, was working in QED rather than QCD, but his analysis carries over.

He worked to arbitrary order in perturbation theory. The higher orders cannot be evaluated exactly, but Sudakov did isolate the leading terms for very high Q^2 and did so for all orders of perturbation theory and found that they could be analytically summed. The result is the lowest order vertex modified by a form factor,

$$\Gamma_\mu(p,p') \equiv \gamma_\mu F(Q^2), \tag{51}$$

with

$$F(Q^2) = \exp\{-\frac{e^2}{2\pi}\ln\frac{Q^2}{|p^2|}\ln\frac{Q^2}{|p'^2|}\} \to \exp\{-\frac{e^2}{2\pi}\ln^2 Q^2\}. \tag{52}$$

The form factor is small as $Q^2 \to \infty$. Elastic scattering becomes unlikely because if the electron is hit hard, it will very likely radiate. "Processes in which a large number of real γ are simultaneously emitted will occur with much greater probability."

The same procedure also works for QCD for hypothetical isolated quarks. The main difference is the running coupling parameter, which leads to a change in the form of the logarithm-squared:

$$\ln Q^2 \times \ln Q^2 \to \ln Q^2 \times \ln(\ln Q^2). \tag{53}$$

The physical phenomenon is still the same. Isolated quarks that are hit hard tend to radiate.

Effect of Sudakov form factors on composites

Now consider gluons possibly radiated from a quark-antiquark pair close together when a photon is absorbed[18]. The pair, representing a real meson, is overall color neutral but we can take a case that is not electrically neutral. Then there is no problem absorbing a photon. We can radiate a gluon from either the first or second quark (see Fig. 13) since the quarks are connected by the need to share momentum in a (hoped for) elastic scattering. But now comes the consequence of having a color neutral object. If the quarks are near to each other, the amplitudes for gluonic emission will cancel. If they are far apart, the gluons will not interfere and we have normal bremsstrahlung.

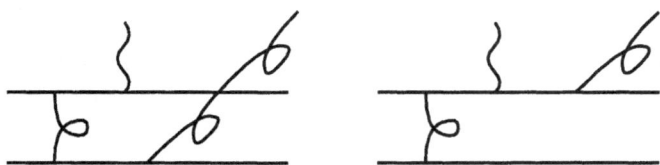

Figure 13. Gluon bremsstrahlung for composite system

To say it again, if the quarks are far apart in an elastic scattering process, the Sudakov form factors come into play for the individual quarks and there will be a suppression of the elastic scattering amplitude at high-momentum transfers. If the quarks are near each other, there is no Sudakov form factor effect and no suppression.

A few technical notes will help the discussion[18]. The expression for the form factor before k_T is integrated can be written down again and the transverse part of it put into coordinate space. This is sometimes called "impact parameter space." The result is

$$\begin{aligned} F_\pi &= \int dx\, d^2k_T\, dy\, d^2\ell_T\, \psi^*(y,\ell)T(x,k_T,y,\ell_T,Q)\psi(x,k_T) \\ &= \int dx\, dy \int d^2b_T\, \psi^*(y,b_T)T(x,y,b_T,Q)\psi(x,b_T), \end{aligned} \tag{54}$$

where

$$\psi(x, b_T) = \int d^2 k_T \, \psi(x, k_T) \, \exp(-i k_T \cdot b_T). \tag{55}$$

Li and Sterman show that the suppression at large b_T can be (approximately) put into the wave function in the form

$$\psi(x, b_T) \rightarrow \phi(x) \exp\{-S(x_1, b_T, Q) - S(x_2, b_T, Q)\} \tag{56}$$

where as usual x_1 and x_2 are the momentum fractions of the quarks inside the pion. The suppression of elastic scattering, or the chance of radiation, depends not only on the transverse separation but also on how hard the quark is struck, i.e., on x and on Q. To a decent approximation

$$\begin{aligned}
e^{-S(x,b,Q)} &\approx 0 \quad \text{if} \quad b > 1/\Lambda_{QCD}, \\
e^{-S(x,b,Q)} &\approx 1 \quad \text{if} \quad b < 1/xQ.
\end{aligned} \tag{57}$$

Since x can be either x_1 or $1 - x_1$, the condition for no suppression of the wave function is

$$1 - \frac{1}{bQ} < x_1 < \frac{1}{bQ}. \tag{58}$$

Thus for small separations b, there is no x_1 that is suppressed. For middle size b, the middle x_1 are unaffected, but contributions from near the end points of the integration region are suppressed. Incidentally, the physics of the suppression for $x_1 \rightarrow 0$ region is that x_2 is moving fast, and it is the fast quark that is susceptible to the bremsstrahlung that leads to suppression. The need for suppression of the endpoints is something one faced in the criticisms leveled at pQCD. There is indeed a suppression, and it is self consistent. There is no need for something external or *ad hoc* to obtain it. And it does not remove contributions that one would expect to be reliable, such as the short distance contributions where the gluon propagators happened to be nearly on-shell. One finds that the overall magnitude of the suppression is far from what one gets by putting in an *ad hoc* cutoff on the endpoint regions.

There is a plot based on the work of Li and Sterman[18] that shows $Q^2 F_\pi$ for varying Q between 10 and 50 times Λ_{QCD}. The result for F_π using the Chernyak-Zhitnitsky distribution amplitude is about 2/3 of its naive value obtained $\alpha_s = 0.3$. It is closer to constant than one might expect. As Q increases, the running coupling parameter decreases; but also the hard scattering kernel T is forcing a larger fraction of the result to come from small distance, where the Sudakov suppression is less, and the two effects roughly cancel.

BARYONS

We have spoken mostly about mesons so far. This has been for purposes of illustration and simplicity. The same methods work for baryons, and we now turn our attention to them. We will be mainly interested in the valence wave function,

$$|p\rangle = \psi_{qqq}|qqq\rangle + \ldots, \tag{59}$$

and we will often abbreviate still more via $\psi_N = \psi_{qqq}$.

Baryon distribution amplitudes

The baryons are more complicated merely because of having one more constituent, but we are now expert and a few technical comments should not be hard to follow. The proton state in more detail, albeit still neglecting the transverse-momentum degrees of freedom, is

$$|p\rangle = \int [dx_1\, dx_2\, dx_3\, \delta(1 - \sum x_i)] \times (color) \ \times$$

$$\{\phi_S \frac{1}{\sqrt{6}}|2udu - uud - duu\rangle_{\uparrow\downarrow\uparrow} + \phi_A \frac{1}{\sqrt{2}}|uud - duu\rangle_{\uparrow\downarrow\uparrow}\} \quad + \quad \text{perm.,} \qquad (60)$$

where the up and down quark combinations guarantee isospin $1/2$. The square bracketed combination after the integral sign is usually abbreviated $[dx]$. The functions ϕ_S and ϕ_A are symmetric and antisymmetric, respectively, when $x_1 \leftrightarrow x_3$. Commonly we let

$$\phi_N = \phi_S - \sqrt{3}\phi_A \qquad (61)$$

and it is easy to reconstruct the symmetric and antisymmetric pieces given ϕ_N.

The reader will be happy to know that the Δ resonance-state vector is simpler than the nucleon's. For the helicity $1/2$ projection,

$$|\Delta\rangle = \int [dx] \times (color) \times \phi_\Delta \frac{1}{\sqrt{3}}|uud + udu + duu\rangle_{\uparrow\downarrow\uparrow}. \qquad (62)$$

Evolution equation for baryons

The evolution equation for baryons[1]—the description of how the distribution amplitude changes with changes in scale Q^2 once Q^2 is large—is similar in form to the meson version,

$$Q^2 \frac{\partial}{\partial Q^2} \phi_N(x, Q^2) = \int [dy] V(x, y) \phi_N(y, q^2), \qquad (63)$$

where x stands for the whole set (x_1, x_2, x_3) and similarly for y. A general solution can be gotten using the separation of variables method, and reads

$$\phi_N(x, Q^2) = \sum_{n=0}^{\infty} g_n(Q^2) \cdot x_1 x_2 x_3 P_n(x), \qquad (64)$$

where P_N are the Appell polynomials[19]. There is one zeroth-order Appell polynomial

$$P_0 = 1, \qquad (65)$$

-two first order Appell polynomials,

$$\begin{aligned} P_1 &= x_1 - x_3, \\ P_2 &= 3x_2 - 1, \end{aligned} \qquad (66)$$

three second-order Appell polynomials, and so forth. They are orthogonal with weight $x_1 x_2 x_3$, meaning

$$\int [dx]\, x_1 x_2 x_3 P_m(x) P_n(x) = K_n \delta_{mn}, \qquad (67)$$

where the K_n are known constants. The Q^2 dependent functions can be shown to satisfy

$$g_n(Q^2) = B_n \left[\frac{\ln Q^2/\Lambda^2}{\ln Q_0^2/\Lambda^2} \right]^{-\gamma_n}, \qquad (68)$$

229

where Q_0 is some benchmark momentum transfer, λ is the same QCD parameter that appears in the running of the coupling parameter, and the γ_n are a set of monotonically increasing numbers beginning

$$\gamma_0 = 0, \qquad \gamma_1 = \frac{20}{9\beta_0}, \qquad \gamma_2 = \frac{24}{9\beta_0}, \qquad \dots, \tag{69}$$

with $\beta_0 = 11 - (2/3)n_f$ where n_f is the number of active quark flavors. The quantity in square brackets above is sometimes abbreviated as just $\ln Q^2$. The constants B_n are not yet determined. Then the proton distribution amplitude is

$$\phi_N = f_N \cdot 120 x_1 x_2 x_3 \left(1 + B_1 (\ln Q^2)^{-\gamma_1} P_1(x) + \dots \right), \tag{70}$$

and the factor 120 is inserted so that

$$\int [dx]\, \phi_N(x, Q^2) = f_N. \tag{71}$$

One should note that the P_1 and higher terms are squelched as Q^2 gets large, but further note that for this squelching to be significant requires very, very high Q^2. The word "superasymptotic" comes to mind once again. At experimentally accessible Q^2 many P_n could be important, and indeed some authors have questioned whether the expansion of ϕ_N in a series of Appell polynomials converges very quickly.

The nucleon form factor

The Dirac and Pauli form factors of the nucleon are defined by matrix elements of the electromagnetic current,

$$\langle p' | j_\mu^{em} | p \rangle = \bar{u}(p') \left[F_1 \gamma_\mu + \frac{i}{2m_N} F_2 \sigma_{\mu\nu} q^\nu \right] u(p), \tag{72}$$

and the magnetic and electric form factors are

$$G_M = F_1 + F_2, \qquad G_E = F_1 - \frac{Q^2}{4m_N^2} F_2. \tag{73}$$

The scaling at high-momentum transfer is that

$$F_1, G_M, G_E \sim \frac{1}{Q^4} \tag{74}$$

and

$$F_2 \sim \frac{1}{Q^6} \tag{75}$$

because of helicity selection rules.

Given the definition, one may calculate either the Dirac or the magnetic form factor—which have the same high Q^2 limit—to be

$$F_1 = \int [dx]\,[dy]\, \phi_N^*(y) T_H(x, y, Q) \phi_N(x), \tag{76}$$

where the hard scattering amplitude T_H is calculated in perturbation theory from diagrams like the second one of Fig. 4.

230

Baryon distribution and transition amplitudes

What is the distribution amplitude for the proton? One may start with the asymptotic distribution amplitude (all $B_i = 0$ in the expansion above, for $i \geq 1$). After that, there is a plethora of suggested distribution amplitudes, some of which are identified with the names Chernyak-Zhitnitsky, King-Sachrajda, Gari-Stefanis, Chernyak-Oglublin-Zhitnitsky, ..., and Eckardt-Hansper-Gari[20]. Calculations have also been made for the $\Delta(1232)$ distribution amplitude by Bonekamp & Pfeil[21], Farrar, Oglublin, Zhang, & Zhitnitsky[8], and Poor & me[7]. The graphs in Fig. 14 show the distribution amplitudes of King & Sachrajda for the proton, and of Poor and myself for the Δ. Distribution amplitudes for the $S_{11}(1535)$ have also been calculated[7], and the general methods for getting or modeling distribution amplitudes are discussed in the next section. The distribution amplitude for the proton is interesting in that it does not favor a configuration where the proton's momentum is evenly divided among the quarks. Rather it has large amplitude for a configuration where a single quark carries a large share of the proton's momentum. The Δ distribution amplitude is more regular, and in fact not too far in appearance from the asymptotic distribution amplitude.

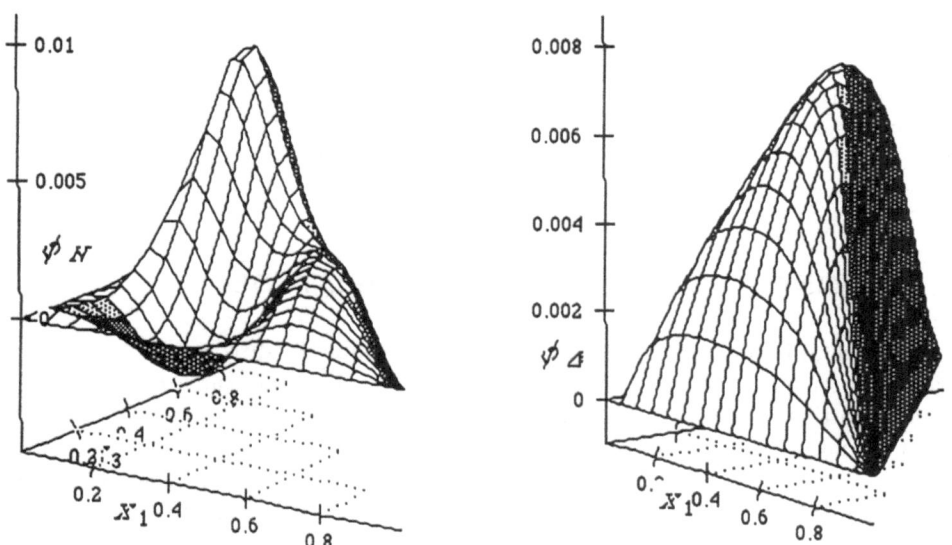

Figure 14. Distribution amplitudes for the proton and delta

Using the King-Sachrajda distribution amplitude for the proton and those of Poor & me for the Δ or S_{11} leads to results

$$Q^4 G_M(p \to p) \simeq \text{Data,}$$
$$Q^4 \text{``}G_M(p \to S_{11})\text{''} \simeq \left(\frac{1}{2} \text{ to } \frac{2}{3}\right) \times \text{Data,}$$
$$Q^4 \text{``}G_M(p \to \Delta)\text{''} \ll \text{Data.} \tag{77}$$

(Numerically, the right-hand sides above are about 1.1 GeV^4, 0.8 GeV^4, and 0.1 GeV^4, in order. The quantities called "G_M" are in all cases $Q/(m_N\sqrt{2})$ times the helicity amplitude.) The $p \to \Delta$ form factor comes out small because of cancellations in the integral between the ϕ_S and ϕ_A parts of the proton distribution amplitude. It needs to be stressed that nothing has been done to force a cancellation. It emerges from the forms of the proton and Delta distribution amplitudes and the form of the hard scattering amplitude, all of which are straightforwardly calculated.

More discrimination

We are interested in learning the nucleon distribution amplitude experimentally. The elastic electron-nucleon scattering is depends on only one integral involving the distribution amplitude, so thinking of what we learn experimentally, we get only one piece of information. This is clearly insufficient to determine a whole curve, so we need to seek further.

One process that has been suggested for additional study is proton Compton scattering[22], $\gamma + p \rightarrow \gamma + p$. At high enough t and s, it should be susceptible of a pQCD calculation. The result for the function with which the polarization matrices should be multiplied to get the scattering amplitude is of the form

$$M_{\mu\nu} = \int [dx]\,[dy]\,\phi_N^*(y) T_{\mu\nu}^H(x, y, s, t) \phi_N(x). \tag{78}$$

The hard scattering kernel has been calculated in [22]. The point of studying Compton scattering is that each different scattering angle, each different ratio of t/s, involves a differently weighted integral over the distribution amplitudes. Studies has been made of the invertability of a given set of (not yet existing) data to obtain the distribution amplitudes, and the results are encouraging. Of course, as yet the study has been made neglecting higher-twist corrections, meaning things like Sudakov form factors which are in principle corrections of the order of $\langle k_T^2 \rangle / s$.

DETERMINING THE DISTRIBUTION AMPLITUDE

Although today what we would like is to determine the distribution amplitude or wave function empirically, it seems reasonable to see what theory can predict. There are ways of getting information about the distribution amplitude theoretically, and using the information that can be gotten to constrain models for the distribution amplitude, like the ones that have already been presented and used in the previous section. The methods involved can be lattice gauge theory or QCD sum rules, and the things that can be learned are certain moments of the distribution amplitude. We will here give the reader an idea of how the QCD sum rule results are gotten.

Basically, QCD sum rules proceed by writing down some expression involving the product of two (or more, in the general case) operators, and evaluating the expression in two different ways, one involving things we know, and the other involving things we want to learn.

What we want to learn is the distribution amplitude of the hadrons, and we will present the calculation using notation pertinent for the $\Delta(1232)$ [7, 8, 21]. We cannot actually get the whole thing, but only moments of the distribution amplitude, defined by

$$f_\Delta \phi_\Delta^{klm} = \int [dx]\, x_1^k x_2^l x_3^m \phi_\Delta(x_1, x_2, x_3). \tag{79}$$

The normalization is $\int [dx] \phi_\Delta = f_\Delta$, so

$$\phi_\Delta^{000} = 1. \tag{80}$$

We will soon need to connect to the coordinate space wave function, which is as always related to the fourier transform of the momentum-space wave function,

$$\psi_{cs}(y_1, y_2, y_3) = \int [dx]\,[d^2 k_T]\, e^{-i\sum_1^3 (y_i x_i p^+ - y_{iT} \cdot k_{iT})} \psi(x, k_T). \tag{81}$$

Here, y_i are spatial coordinates and $x_i p^+$ are quark longitudinal momenta. The wave function at the origin in coordinate space is

$$\psi_{cs}|_0 = f_\Delta \psi_\Delta^{000} = f_\Delta, \tag{82}$$

and derivatives at the origin are related to the moments,

$$\frac{1}{(p^+)^{k+l+m}} (i\partial_1^+)^k (i\partial_2^+)^l (i\partial_3^+)^m \psi_{cs}|_0 = f_\Delta \phi_\Delta^{klm} \tag{83}$$

where $\partial_i^+ \equiv \partial/\partial y_i^-$.

The crucial expression to be evaluated is called the "correlator", and is

$$I \equiv i \int d^4y\, e^{ipy} \langle 0|T\, V^{klm}(y) J(0)|0\rangle \tag{84}$$

where T is the time-ordering operator, and the operators V and J are to be specified. One evaluation of the correlator proceeds by putting hadronic intermediate states between V and J, so we should choose them such that the Δ is the lowest contributing state. A choice for J is

$$J(0) = \left(\bar{u}(0)\gamma^+ C\bar{u}^T(0) \right) \bar{d}(0) \quad [+\ perm], \tag{85}$$

where $\bar{u}$ and $\bar{d}$ are quark fields, $\gamma^+ = \gamma^0 + \gamma^3$, and the superscript T denotes transpose. The charge-conjugation matrix C ensures nice lorentz transformation properties for J. The crucial thing about this operator is that it has isospin 3/2 and positive parity, so that there is no vacuum-to-nucleon matrix element, and writing out the Δ-state in terms of its wave function gives

$$\langle \Delta|J(0)|0\rangle \propto \psi_{cs}|_0 = f_\Delta, \tag{86}$$

where the proportionality constant is known. Similarly,

$$V^{klm} \equiv \left[((\partial^+)^k u)^T C\gamma^+ ((\partial^+)^l u) \right] (\partial^+)^m d, \tag{87}$$

and

$$\langle 0|V^{klm}|\Delta\rangle \propto f_\Delta \phi_\Delta^{klm}. \tag{88}$$

Thus we can complete the hadronic evaluation of the correlator and obtain

$$\Im I = (const) f_\Delta^2 \phi^{klm} \delta(p^2 - m_\Delta^2) + HR + cont., \tag{89}$$

where the constant is known, HR represents contributions of higher resonances, $cont.$ represents contributions from the $N\pi$, etc., continuum, and $\Im$ stands for the imaginary part.

Next we need a purely quarkic evaluation of the correlator. In terms of time to carry out, though not to describe here, this is the most time-consuming part of the calculation. We will consider three types of quarkic or QCD contributions, illustrated in Fig. 15. First is the purely perturbative contribution, which for each V^{klm} we can work out as

$$I_1^{klm} = \frac{c_1^{klm}}{\pi^4} (p^+)^{N+3} p^2 \ln(-p^2), \tag{90}$$

233

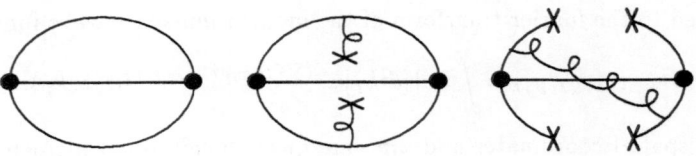

Figure 15. Diagrams for quarkic evaluation of QCD sum rules

where $N = k + l + m$ and the constants c_1^{klm} are known.

Next come some more interesting contributions. The QCD vacuum, the ground state of nature, is complicated. It contains gluon pairs, quark pairs, higher correlations among them, etc. The second diagram Fig. 15 shows a situation where a quark line emits a gluon that disappears into the vacuum, and the other quark line (conserving color overall) absorbs a gluon from the vacuum. The probability of such a thing happening depends upon the gluon content of the vacuum, which is measured by

$$\langle 0| \frac{\alpha_s}{\pi} G_{\mu\nu}^a G_a^{\mu\nu} |0\rangle \equiv \langle 0| \frac{\alpha_s}{\pi} G^2 |0\rangle, \tag{91}$$

where a is a color index. The value of this quantity is obtained by seeing what value is needed for one sum rule, in another context and long since done, to give the correct experimental answer for whatever that sum rule was calculating. We then get,

$$I_2^{klm} = \frac{c_2^{klm}}{\pi^2} (p^+)^{N+3} \frac{1}{p^2} \langle \frac{\alpha_s}{\pi} G^2 \rangle. \tag{92}$$

The dimensions work, since the dimension of G^2 is [mass^4].

Along the same lines, we can have quarks going into and coming out of the vacuum, as in the third diagram in Fig. 15. It happens that breaking just one quark line gives a contribution proportional to the quark mass, so such diagrams are ignored. We then have,

$$I_3^{klm} = \frac{c_3^{klm}}{\pi} (p^+)^{N+3} \frac{1}{p^4} \langle 0| \sqrt{\alpha_s} \bar{q} q |0\rangle^2. \tag{93}$$

Now we have two evaluations of the same object. We can try to set the result of the hadronic evaluation equal to the sum $I_1 + I_2 + I_3$. Each evaluation is an approximation, and of course, the answers do not match. In particular, the hadronic evaluation involved δ-functions and the quarkic evaluation does not. However, we may hope that they match if averaged, and a way to do a weighted average emphasizing the region near threshold is to do a Borel transformation, which involves the parameter M. This means replacing,

$$\Im I \longrightarrow \frac{1}{\pi} \int_0^\infty dp^2 \, e^{-p^2/M^2} \Im I. \tag{94}$$

Doing the Borel transformation to both evaluations of the correlator, equating them, and doing some rearranging leads to

$$(const.) f_\Delta^2 \phi^{klm} = e^{m_\Delta^2/M^2} \left\{ -\frac{c_1^{klm}}{\pi^4} M^4 \left[1 - \left(1 + \frac{s_0}{M^2} \right) e^{-s_0/M^2} \right] \right.$$
$$\left. - \frac{c_2^{klm}}{\pi^2} \langle \frac{\alpha_s}{\pi} G^2 \rangle + \frac{c_3^{klm}}{\pi^2} \frac{\langle 0| \sqrt{\alpha_s} \bar{q} q |0\rangle^2}{M^2} \right\} \tag{95}$$

where the constant is known. Actually something else was done along the way. The terms $HR + cont.$ in the hadronic evaluation were replaced by the simplest possible non-trivial substitution, which is a step function $\Theta(p^2 > s_0)$ times the purely-perturbative term. This accounts for the terms containing s_0 in the above formula. The quantity s_0 should reasonably be some number around or perhaps a bit below the mass-squared of the second Δ resonance.

The left hand-side above is independent of M^2. The right-hand side should then be independent of M^2 also. It is a simple matter to plot it and see if this is so. A typical good case is illustrated in Fig. 16. The curve is flat to within $\pm 1\%$ for M^2 in the range 1 to 2.5 GeV^2! When the results are so good, the averaging is working well (one can easily think what is going wrong for large and small M), and the value of the moment for the klm in question can be read off the plot.

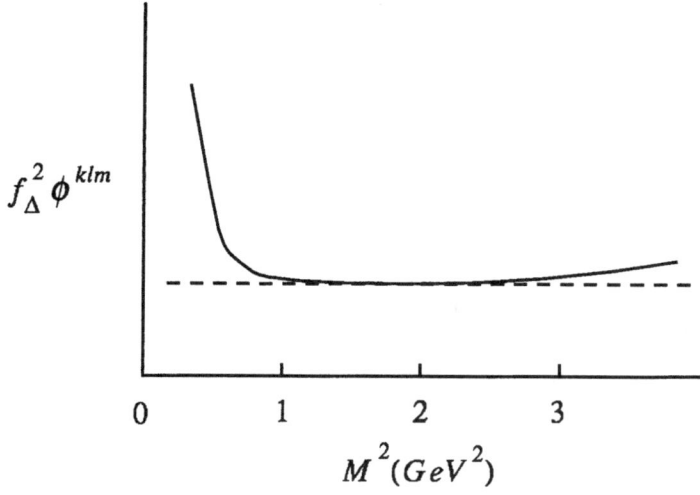

Figure 16. Stability plot for QCD sum rule. The value of the right-hand side in its constant region is proportional to the moment in question.

We can in this way get a number of moments, and if we could get all moments, we could get the distribution amplitude. We cannot. The coefficients c_3^{klm} and c_2^{klm} increase relative to c_1^{klm} as $N = K + l + m$ increases. This makes one worry about the size of further terms that we have not calculated and thus about the reliability of higher moments.

We do seem to have good moments up to second order, i.e., for $N \leq 2$. For example, we have[7]

$$
\begin{aligned}
\phi_\Delta^{001} &= 0.35 \pm 0.02, \\
\phi_\Delta^{002} &= 0.16 \pm 0.02, \\
\phi_\Delta^{101} &= 0.09 \pm 0.02.
\end{aligned}
\tag{96}
$$

One way to turn this into a model for the distribution amplitude is to expand in Appell polynomials to the same order as we have moments. There is a 1 to 1 match between the number of coefficients of the Appell polynomial expansion to a given order and the number of independent moments to that order. For the Δ we get

$$
\phi_\Delta = 0.24 \, \text{GeV}^2 \, x_1 x_2 x_3 \left\{ 1 - 0.35 P_2 + 0.10 P_3 + 0.04 P_5 \right\}.
\tag{97}
$$

As we have already both shown a plot of this function (Fig. 14) and have implicitly used it in the calculation of the $N \to \Delta$ transition, we can stop the present discussion here.

CONCLUSION

Our goal is to obtain a deeper understanding of hadron structure. A starting point is to learn the valence quark wave functions or distribution amplitudes. A tool for doing this involves using perturbative QCD to calculate how mesureable quantities depend upon the inherently not-perturbatively-calculable wave functions.

A crucial first question regards the inherent accuracy of pQCD at experimental momentum transfers. The neatest results depend not only upon pQCD but also upon factorization, which follows from the idea that transverse momenta may be neglected in the hard scattering process, allowing calculated quantities to depend upon integrands that are simple products of distribution amplitudes and hard scattering amplitudes. The results that follow from factorization and pQCD appear to be working for the $\pi^0\gamma\gamma$ form factor, so the field apparently has one definite success. Whether we can really neglect the transverse momenta in the hard scattering amplitudes in other processes is still under debate; the rôle, however, of the Sudakov form factors in naturally protecting us from situations where the transverse momenta are important does give us hope.

Scaling is an important result of the pQCD analysis of scattering amplitudes, and data often confirms it. Indeed, one instance where scaling fails (in the $N\Delta$ transition) can also be claimed as a success of pQCD in that the coefficient of the leading term is calculable in pQCD and turns out to be anomalously small. Scaling does not prove QCD. But scaling, short distance interactions, and pQCD go hand in hand, so in the regime where scaling is working—if the observed scaling is not just luck—then perturbation theory calculations in QCD should be valid.

Finally, many of the processes we look at depend on just one integral over the distribution amplitude, which is not enough for our purposes. We want the whole distribution amplitude. This motivates studying other processes in the future, such as Compton scattering from a nucleon, where each different scattering angle involves a different integral over the nucleon distribution amplitude and gives the possibility of inverting to find the distribution amplitude itself, or such as electro-production of pions at high transverse momentum, where each different Q^2 (say measured as Q^2/s) similarly gives a different integral over the pion distribution amplitude.

REFERENCES

1. G. P. Lepage and S. J. Brodsky, Phys. Rev. D 22, 2157 (1980).

2. F. Coester, Progress in Particle and Nuclear Physics 29, 1 (1992).

3. T. Gousset and B. Pire, hep-th/9511274 (1995).

4. G. R. Farrar and D. R. Jackson, Phys. Rev. Lett. 43, 246 (1979); G. P. Lepage and S. J. Brodksy, Phys. Lett. 87B, 359 (1979); A.V. Efremov and A. V. Radyushkin, Theor. Math. Phys. 42, 97 (1980).

5. C. E. Carlson, in *New Vistas in Electro-Nuclear Physics*, Proceedings of a NATO Advanced Study Institute in Banff, Canada, August 1985, ed. E. L. Tomusiak, H. S. Kaplan, and E. T. Dressler, p. 101 (Plenum, New York, 1986) or in *New Aspects of Nuclear Dynamics*, Proceedings of a NATO Advanced Study Institute in Dronten, the Netherlands, August 1988, ed. by J. H. Koch and P. K. A. de Witt Huberts (Plenum, New York, 1989).

6. P. Stoler, Phys. Rev. D 41, 73 (1991); Phys. Reports 226, 103 (1993).

7. C. E. Carlson and J. L. Poor, Phys. Rev. D 38, 2758 (1988)

8. G. Farrar, A. A. Oglublin, H. Zhang, and I. R. Zhitnitsky, Nucl. Phys. B311, 585 (1989).

9. D. Menze, W. Pfeil, and R. Wilcke, *Physics Data: Compilation of Pion Photoproduction Data* (Zentralstelle für Atomkernenergie-Dokumentation, Eggenstein-Leopoldshafen, 1977). I thank Jim Napolitano for discussions about this plot.

10. See ref [3] and A. H. Mueller, Phys. Reports 73 237 (1981).

11. C. J. Bebek *et al.*, Phys. Rev. D 17, 1693 (1978).

12. V. L. Chernyak and A. R. Zhitnitsky, Phys. Reports 112, 173 (1984).

13. V. Savinov for the CLEO collaboration, Contribuiton to the PHOTON95 conference, Sheffield (1995), hep-ex/9507005.

14. C. E. Carlson and A. B. Wakley, Phys. Rev. D 48, 2000 (1983) and T. Hyer, Phys. Rev. D 48, 147 (1993) and *ibid.* D 50, 4382 (1994).

15. P. Auranche *et al.*, Phys. Lett. 135B, 164 (1984).

16. E.g., N. Isgur and C. Llewellyn-Smith, Nucl. Phys. B317, 526 (1989); J. Bolz and P. Kroll, Wuppertal Report WU B 95–35 (hep-ph/9603289).

17. V. V. Sudakov, JETP 3, 65 (1956).

18. H.-N. Li and G. Sterman, Nucl. Phys. B381, 129 (1992).

19. E. T. Whitaker and G. N. Watson, *A Course of Modern Analysis*, fourth edition (Cambridge, 1969), p. 300, problem 22.

20. V. L. Chernyak and A. R. Zhitnitsky, Nucl. Phys. B201, 492 (1982); I. R. Zhitnitsky, A. A. Oglubin, and V. L. Chernyak, Sov. J. Nucl. Phys. 48, 536 (1988); M. Gari and N. G. Stefanis, Phys. Lett. B 175, 462 (1986) & Phys. Rev. D 35, 1074 (1986); I. D. King and C. T. Sachrajda, Nucl. Phys. B279, 785 (1987); R. Eckhardt, J. Hansper, and M. F. Gari, hep-th/9607380.

21. J. Bonekamp, Thesis, Bonn Report No. BONN-IR-89-43, 1989.

22. Glennys R. Farrar, Hua-yi Zhang, Phys. Rev. D 41, 3348 (1990), E, *ibid.* D 42, 2413 (1990); A. S. Kronfeld and B. Nizic, Phys. Rev. D 44, 3445 (1991), E, *ibid.* D 46, 2272(1992).

PARTICLES, SPACE AND TIME

Vincent Icke

Leiden Observatory
P.O. Box 9513, 2300 RA Leiden
The Netherlands
and
Universiteit van Amsterdam
Dept. for Astrophysics
Kruislaan 403, 1098 SJ Amsterdam
The Netherlands

CLASH OF TITANS

When an extraterrestrial or a child asks me what we have *really* learned about the nature of the Universe, I reply: *our Universe consists of particles, space and time.* Even for adult terrestrials, this does not seem to be too bad a summary, so I will use it as a starting point for a brief overview. I will try to show that this seemingly innocuous statement harbours a grave problem, in the following sense: the theory of particles (called somewhat preemptively the *Standard Model*) and the theory of space and time (General Relativity, GRT) are extremely good in their own realm of application, the world of the very small and the very large, respectively. Taken together, however, we get a theory that does not conform in the least to the world as we know it.

This shows up most clearly in the fact that the Standard Model predicts that the energy density of the vacuum is not zero, in the sense that vacuum fluctuations are always present in it. Oversimplifying somewhat we may say that in relativistic quantum mechanics the uncertainty relations imply that one can never be quite sure that a cubic metre of spacetime is empty. Particle number is no longer a constant of the motion.

These vacuum fluctuations contribute to the effective mass density of the Universe. Suppose that we have a field with quanta of mass m. Then integration over momentum space shows that the energy density e_V of the vacuum is

$$e_V = (2\pi)^{-3}\hbar^{-3} \int_0^K 4\pi k^2 \frac{1}{2}\sqrt{k^2c^2 + m^2c^4}\ dk \simeq \frac{cK^4}{16\pi^2\hbar^3}\quad \mathrm{J\,m^{-3}}\ ,\qquad (1)$$

in which $K \gg m$ is the upper cutoff of the wave number. By itself this need not be a meaningful number; we could transform it away by a renormalization procedure. However, in GRT this cannot be done and we must consider the gravitational effects of this vacuum-energy density. In the case of GRT we expect that K is inversely proportional to the coupling constant $8\pi G/c^2$, and so

$$K \approx \sqrt{\frac{\hbar c^3}{8\pi G}} , \tag{2}$$

$$e_V = \frac{c^7}{1024\pi^3 \hbar G^2} = 1.46 \times 10^{109} \quad \mathrm{J\,m^{-3}} . \tag{3}$$

Considering that the energy-density equivalent in an Einstein-De Sitter universe would be

$$e_0 = \rho_0 c^2 = \frac{3H_0^2 c^2}{8\pi G} = 4.24 \times 10^{-10} \quad \mathrm{J\,m^{-3}} , \tag{4}$$

for a Hubble parameter of $H_0 = 50 \ \mathrm{km\,s^{-1}\,Mpc^{-1}}$ (which is $1.62 \times 10^{-18} \ \mathrm{s^{-1}}$) we conclude that the clash between Einstein on the one hand and Dirac and Feynman and their cohorts on the other, is a clash of titans indeed: they differ by a factor of 10^{118}! Even if we only took K from quantum electrodynamics we would expect that it would be roughly equal to $m_e c$, and we would still get

$$e_V = \frac{m_e^4 c^5}{16\pi^2 \hbar^3} = 9.00 \times 10^{21} \quad \mathrm{J\,m^{-3}} , \tag{5}$$

still at least a factor 10^{31} larger than observations appear to allow. It is manifestly false that $\Lambda = 10^{118}$. If it isn't anywhere near, it might as well be zero. *Why doesn't space weigh anything?*

Such a massive pile of problems is a cairn along the winding path to a region where good new physics can be found. Nobody has yet returned alive from that land; let us see what difficulties we expect to find there.

SPACETIME AS REAL STUFF

First, consider space and time. Ever since Descartes we have known that true emptiness cannot exist; he conjectured that a force comes about through the direct physical contact between objects. In the second part of his *Principes de la Philosophie*, Article 16, Descartes wrote:

> *Concerning emptiness, in the sense given to that word by the philosophers, namely a space containing no substance, it is obvious that there is no such space in the universe, because the extent of space around or enclosed by an object is not different from the extent of that object. And even as from the sole fact that an object is extended in length, height and depth, we have reason to conclude that it is a substance, because we suppose that it is not possible for anything to have no extent, we must conclude the same about space which is supposedly empty: namely that, because it possesses spatial extent, it also has substance.*

In other words: space has physical attributes, namely its three dimensions, so it must be regarded as real stuff. This powerful notion lay hidden for three hundred years, until it was rediscovered independently by Einstein.

240

FROM GLOBAL TO LOCAL LORENTZ SYMMETRY

Ever since Einstein we have known that space and time are part of the stuff of our world, and not invisible graph paper. Thus, efforts to determine the structure of particles go in parallel with the search for the structure of spacetime. Einstein gave us a geometrical answer for the latter: a distance recipe (Lorentz-Minkowski) suffices. The theory boils down to the patching together of local Lorentz frames into a global whole, which gives it the form of a gauge-field theory based on local Lorentz symmetry.

Special relativity starts from the invariance of the speed of light. The equation of motion for light is the equation for a sphere in 3-space: $s^2 = c^2 t^2 - r^2$ where $s = 0$ for a light ray. Now we note that the above leads to a maximum value $v \leq c$ for all speeds v. That means that the above global Lorentz symmetry cannot be maintained. To stay consistent, we may require *local* symmetry only. After all, if signals travel with a finite speed, how would our colleagues at Arcturus know that we have just performed some Lorentz transformation here? We must restrict ourselves to the infinitesimal patch around the origin of our arbitrarily chosen standpoint:

$$ds^2 = c^2 dt^2 - dr^2 \,, \tag{6}$$

Local Lorentz symmetry means that, wherever you are, you can always find coordinates such that the above holds ('freely falling coordinates'). Thus, your neighbours in spacetime will also be able to do this. However, it isn't guaranteed that you will agree with the neighbours that their coordinate system $\{x^\mu\}$ is the same as yours. The best you can hope for is a patch-up between the two of you, by means of a bilinear form in the infinitesimal coordinates of some common coordinate system $\{\xi^\mu\}$ which you've both agreed to use, and for which local Lorentz symmetry holds too:

$$ds^2 = g_{\mu\nu} d\xi^\mu d\xi^\nu \,. \tag{7}$$

The equation of motion corresponding to the spacetime structure encoded in $g_{\mu\nu}$ can be found by an argument similar to what one uses in classical mechanics. First, consider free motion only, the kinematics of motion without force. Classical motion is subject to certain symmetries, namely Galilei symmetry and time-reversal invariance. These symmetries generate conserved quantities: momentum and energy. That is to say, the state of rest is equivalent to motion with constant energy and momentum: $dv^i/dt = d^2 x^i/dt^2 = 0$. Deviations from inertial motion are attributed to an external force, $dv^i/dt = F/m$. This was basically the method that Galilei and Huygens used in their quest for the equations of motion of classical mechanics.

In special relativity we use Lorentz symmetry instead of Galilei; in GRT we extend this to local Lorentz symmetry and find that the structure of spacetime is given by $g_{\mu\nu}$. Thus the equivalent of motion under the influence of a force is force-free motion in curved spacetime. Loosely speaking: "curved spacetime gives curved paths".

Now let us proceed to the algebraical expression of that statement. A classical free particle moves according to the law of inertia: $dv^i/dt = d^2 x^i/dt^2 = 0$. If an external force is present, say due to a gravitational potential Φ, we have $d^2 x^i/dt^2 = -\partial\Phi/\partial x_i$. In GRT the analogue of classical free motion must be written as $du^\mu/ds = 0$, where $u^\mu \equiv dx^\mu/ds$ is the four-velocity, having changed from using the time derivative to a derivation with respect to the interval s. If we arrange with 'the neighbours' elsewhere in the Universe to refer all descriptions to a global but otherwise arbitrary coordinate system $\{\xi^\mu\}$ (much as at a conference one usually agrees to speak English), we find

$$u^\mu = \frac{dx^\mu}{ds} = \frac{\partial x^\mu}{\partial \xi^\alpha} \frac{d\xi^\alpha}{ds} = v^\alpha \frac{\partial x^\mu}{\partial \xi^\alpha} \,. \tag{8}$$

Here v^α is the *local* four-velocity, the 'free-fall motion'. This produces an expression for the way in which the four-velocity changes with respect to interval:

$$0 = \frac{du^\alpha}{ds} = \frac{d^2\xi^\alpha}{ds^2} = \frac{d}{ds}\left(\frac{\partial\xi^\alpha}{\partial x^\mu}\frac{dx^\mu}{ds}\right) \tag{9}$$

and, after working out the differentiations, one uses the metric tensor to obtain

$$\frac{d^2 x^\lambda}{ds^2} + \Gamma^\lambda_{\mu\nu}\frac{dx^\mu}{ds}\frac{dx^\nu}{ds} = 0 \tag{10}$$

$$\Gamma^\lambda_{\mu\nu} \equiv \frac{\partial x^\lambda}{\partial \xi^\alpha}\frac{\partial^2\xi^\alpha}{\partial x^\mu \partial x^\nu} . \tag{11}$$

The above equation of motion can be interpreted as "curved spacetime gives curved orbits": the four-acceleration is no longer zero but proportional to Γ, in which we have dumped all the garbage due to the 'mismatch' in spacetime caused by the introduction of local Lorentz symmetry. Thus, although we can always *locally* transform away the curvature of spacetime ("freely falling coordinates"), we cannot do so globally.

SPACE, TIME AND PARTICLES CONNECTED

Einstein made a conjecture that encompasses the slogan 'Nature is made of particles, space and time' in a particular way. He did this by using, on the left hand side of his famous field equation, non-Euclidean geometry to describe the structure of spacetime via a local generalization of the global Minkowski distance recipe. On the right hand side he used a description of matter based on two approximations. First, he used a continuum representation of matter, averaging over the individual particles to obtain densities. Second, he connected the presence of matter to the structure of spacetime by demanding correspondence with the Newtonian equations in the limit for weak fields and small velocities. Accordingly, his equation reads something like

$$\left\{\begin{array}{c}\text{structure of}\\\text{space and time}\end{array}\right\} = \left\{\begin{array}{c}\text{distribution of}\\\text{particles and fields}\end{array}\right\} .$$

If we read the equals sign as implying an interaction, this says 'particles interact with spacetime'. Of course this interpretation immediately shows that the equation is incomplete, because the back-reaction of the structure of spacetime on the properties of the particles is not included, other than by their trajectories as pointlike test particles in the field $g_{\mu\nu}$.

How are matter and spacetime curvature connected algebraically? In other words, what connects matter and the distance recipe in spacetime? When we compare the Newtonian and the Einsteinian equations for the trajectory of a particle, we notice immediately that Φ and $\Gamma^\lambda_{\mu\nu}$ are apparently related. Of course it cannot be that $\Gamma = \Phi$ or something as simple as that, because then Newtonian theory would be relativistic already! We must incorporate a physical ingredient in our theory. The one that Einstein used is: in the limit for small velocities and small curvatures in a static spacetime we must recover the classical equations.

In that case, the only components that remain are those related to the time-indices: everything is zero except Γ^μ_{00}. Using the above calculations one may see that this is closely related to the fact that the limit of the four-velocity u^μ for small three-velocities is $u^\mu = (c, 0, 0, 0)$. The zero-component does not vanish, just like the 0-component of the energy-momentum does not vanish for $v \to 0$ (this is the famous mc^2-term).

Furthermore, from the definition of the metric we know that $ds^2 \simeq g_{00}dt^2$ (the other components of $g_{\mu\nu}$ vanish in the limit for small curvature), and therefore the low-velocity limit of Eq.(3.5) is

$$\frac{d^2x^\mu}{dt^2} + \Gamma_{00}^\mu = 0 \;, \tag{12}$$

from which we finally conclude that the desired correspondence with classical gravitation can be obtained if we relate Γ to the gradient of the gravitational potential:

$$\Gamma_{00}^\mu \Longleftrightarrow \frac{\partial \Phi}{\partial x_\mu} \;. \tag{13}$$

In order to connect with the classical Newtonian case, we felt obliged to equate Γ with a coordinate derivative of the classical potential. Then Γ is related to a linear combination of first derivatives of $g_{\mu\nu}$:

$$\Gamma_{\lambda\mu}^\kappa = \frac{1}{2}g^{\nu\kappa}\left\{ \frac{\partial g_{\mu\nu}}{\partial x^\lambda} + \frac{\partial g_{\lambda\nu}}{\partial x^\mu} - \frac{\partial g_{\mu\lambda}}{\partial x^\nu} \right\} \;, \tag{14}$$

and we finally conclude that *the role which the potential Φ plays in classical Newtonian gravity is taken over by the metric tensor $g_{\mu\nu}$ in general relativity.*

The classical potential is related to the presence of matter by means of the Poisson equation

$$\Delta\Phi = 4\pi G\rho \;, \tag{15}$$

which should give us a hint about how the presence of matter can be connected to the structure of spacetime. The point here is that the density field ρ could not possibly be used, because it is in no way Lorentz invariant. In fact, we can see immediately that a Lorentz transformation of ρ should go as

$$\rho' \propto \gamma^2\rho \;. \tag{16}$$

One Lorentz factor γ comes from the change of the effective mass via $E = \gamma mc^2$; the other one comes from the fact that a volume seen in motion decreases by one factor γ because of its Lorentz-FitzGerald contraction. Accordingly we suspect that ρ should be part of a tensor because a Lorentz scalar transforms with γ^0, a vector with γ^1 and a tensor with γ^2. Since ρ is classically a scalar field we also suspect that it is the 00-component of a tensor.

In the limit of small velocities without external forces we may take that tensor, which we'll call $T_{\mu\nu}$, to be diagonal. Then ρ is placed in the top-left corner. What will we have on the remaining three places of the diagonal? Since ρ is a mass density, and since in relativity we have to take mass and energy as equivalent, it seems natural to use an energy density. The mass density is the mass of a collection of particles per unit volume. The corresponding energy per unit volume we know as the *pressure P* of the collection of particles. The entries elsewhere in the tensor can be found by Lorentz transformation of T; the complete form is the *energy-momentum tensor*

$$T_{\mu\nu} = P\eta_{\mu\nu} + (\rho c^2 + P)u_\mu u_\nu \;. \tag{17}$$

The final order of business now is to construct a tensor $G_{\mu\nu}$ from $g_{\mu\nu}$ and its derivatives that has the same transformation properties as $T_{\mu\nu}$. The most obvious choice, taking T simply proportional to g, is not sufficient because it would not include Newtonian gravity; in order to obtain that, as we had seen above, we must include the derivatives of g. In particular, in the Newtonian limit we retain the 00-components only, namely

$$\frac{\partial^2 g_{00}}{\partial x^\alpha \partial x_\alpha} \propto T_{00} . \tag{18}$$

Apparently, the desired tensor must contain at least second derivatives of g. This implies that *a fourth-rank tensor must be involved!* Einstein showed that the correct expression is related to the monstrous *Riemann-Christoffel curvature tensor*

$$R^\lambda_{\mu\nu\kappa} \equiv \frac{\partial \Gamma^\lambda_{\mu\nu}}{\partial x^\kappa} - \frac{\partial \Gamma^\lambda_{\mu\kappa}}{\partial x^\nu} + \Gamma^\alpha_{\mu\nu}\Gamma^\lambda_{\kappa\alpha} - \Gamma^\beta_{\mu\kappa}\Gamma^\lambda_{\nu\beta} . \tag{19}$$

This animal must be reduced to second rank before we can equate it to $T_{\mu\nu}$. This is done by contracting it over one index. The most general expression for the required tensor is then

$$R^\lambda_{\mu\lambda\nu} - \frac{1}{2}g_{\mu\nu}g^{\alpha\beta}R_{\alpha\beta} + \Lambda g_{\mu\nu} = R_{\mu\nu} - \frac{1}{2}g_{\mu\nu}R + \Lambda g_{\mu\nu} , \tag{20}$$

and in units where the gravitational constant G is retained explicitly, correspondence with the Poisson equation shows that

$$R_{\mu\nu} - \frac{1}{2}Rg_{\mu\nu} + \Lambda g_{\mu\nu} = -\frac{8\pi G}{c^2}T_{\mu\nu} . \tag{21}$$

This is a physical choice; it does not have the mathematical necessity of a gauge theory because of the way in which $T_{\mu\nu}$ was put in by hand. It is Einstein's guess, based on correspondence with Newtonian mechanics, and it works very well on large scales: black holes, relativistic stars, the Universe. But it works not at all on small, atomic, quantum-mechanical scales. The hassle is that we are obliged to include the vacuum zero-point energy in $T_{\mu\nu}$. As we saw in Sec.1, it shows up in the form of a finite value $\Lambda \approx 10^{118}\rho_0 c^2$ of the cosmological constant which is totally excluded by cosmological observations.

Possibly there is another guess one could make by searching for an expression that corresponds with (say) Schrödinger's Equation instead of Poisson's Equation, but nobody has yet succeeded in doing so.

Note that the rather peculiar and counter-intuitive behaviour of Λ, and the related possibility of inflation ("you get something out of nothing") is due to our initial assumption for the connection between $g_{\mu\nu}$ and the potential Φ, and via Φ and the Poisson equation to the (thermo)dynamic mass- and energy densities ρ and P. The connection between matter and spacetime curvature is still a conjecture, since we do not have a quantum gravity theory.

One immediate cause for worry is that it seems like 'double dipping' to introduce spacetime in $g_{\mu\nu}$ as well as in Λ. Ought we not to include the Planck-scale fluctuations in some (possibly extended) form of $g_{\mu\nu}$ rather than in the potential term that produces Λ? After all, a potential is a classical beast that would be wiped out by second quantization.

TWISTS AND WRINKLES

Having described the structure and behaviour of space and time as they appear in our summary expression 'the Universe is made of particles, space and time', let us consider the particles. The trick of getting a field $g_{\mu\nu}$ that corrects for the consequences of using a *local* Lorentz symmetry instead of a global one is common to all current theories of interaction. First let me try to explain in somewhat pedestrian terms what happens here.

The similarity between the symmetries of Nature and simple rotations enables us to understand how a symmetry can produce a force field. Take before you, on a smooth table, a small tablecloth, or something similar (e.g. a large piece of aluminium foil). The material must be a uniform colour, without any patterns. Make sure it is quite smooth. We are not looking so closely that the individual fibres are visible, and we will pretend that the material extends to infinity: our tablecloth is a small piece of an unbounded model universe. Now rotate the whole cloth through an arbitrary angle. Any piece of the surface, when inspected individually, appears the same as before: the cloth is *invariant under global rotations*.

But it would be impossible, even in principle, to do something like this with the real Universe. Imagine that we want to perform a global symmetry transformation. Then we would have to let the symmetry act in all of space at exactly the same time. But this is impossible to do in reality because no signal can propagate faster than the speed of light. We must *accept only local symmetry rotations*, that is, a symmetry where the amount of rotation *differs* from event to event in spacetime.

Return to the tablecloth before you. Put your finger on a point near the centre and give the cloth an arbitrary twist, keeping the edges of the cloth in place. When you remove your finger, you notice that the piece of the surface you have just rotated still appears the same; at that one point, the cloth is invariant under local rotations. But in the vicinity of the twisted point, something has happened: *a spray of wrinkles radiates outward from it.* The local twist cannot be connected smoothly with the undisturbed cloth at large distances: the difference must be patched up. Because of relativity, all symmetries must be local, and *any local symmetry creates a field.* The wrinkles are related to the 'field lines'.

GAUGE TWISTS AND VELPONS

Elementary particles belong to certain families. Within such a family the particles treat each other as equals, at least in the ideal case. That is to say that they are, in some sense, interchangeable: because of the perfect equality one would not notice such a swap. This operation is a symmetry. If one were to subject the whole Universe at once to such a *global* symmetry, nothing would change at all.

If we pick a fundamental fermion multiplet with N members we expect that the symmetry group that acts on this N-plet should behave as a rotation in some abstract N-dimensional space. The rotation of a particle over the *mixing angle* θ literally makes the particle 'turn into' a different one! Some symmetries are actually connected with rotation in space (which creates the angular momentum of a particle) or rotation in spacetime (Lorentz symmetry). Other symmetries behave like rotations too, but not in ordinary space; rather, these symmetries are rotations about other directions than the axes of space and time. Apparently, the vacuum possesses more possible directions than those of spacetime.

A symmetry can be responsible for generating a field. Now in the quantum picture a field is built up from field quanta; and the exchange of a quantum produces a force. Thus, *any local symmetry creates a force.* The quanta of such a field are gauge quanta, or, more precisely, *gauge bosons*, because Lorentz invariance and exact gauge symmetry combined demand that the particle have mass zero and integral spin. The force that corresponds to the exchange of gauge bosons can be considered as a binding agent, a kind of glue between the particles to which the bosons are coupled.

Each force has its own set of glue quanta. A generic name (other than the insipid 'gauge boson') for these does not exist in the professional literature. Therefore, I will succumb to the temptation to name something, and use the generic name *velpon*, after one of the most common brands of glue in my home country.

All known forces are due to gauge symmetries. A local gauge twist causes wrinkles in the vacuum, because the mismatch between the twisted space and the unperturbed vacuum in the distance must be patched up somewhere in between. In the non-quantum picture, of which our tablecloth is a model, the difference is made good by wrinkles, 'field lines' that radiate from the twisted spot. But in a quantum world, where interaction is all-or-nothing, *the twist must be taken away by a single quantum.* The exchanged quantum, the velpon, then becomes the carrier of the vacuum wrinkle caused by the local gauge twist.

THE LAGRANGIAN IN QED AND IN GRAVITY

The above explanation is only an analogy and needs to be made more precise for practical purposes. This I will do by briefly showing how the 'wrinkle' picture of fields can be cast in algebraic form. It all begins by guessing a proper symmetry from some notion of similarity between particles, in particular fermions. For example, one may note that a proton is really rather like a neutron, with the exception of a small difference in mass and a difference in electric charge. It takes a little faith to overlook these differences, but one may with some justification surmise that the mass difference will ultimately appear to be due to the charge or something like that. Or one may note that in a sense an electron is rather like a neutrino, in that they always appear together in weak decays.

Start with such a guessed global symmetry. Then one notes that using this symmetry globally is contrary to the spirit of relativity. But using a *local* symmetry is not possible unless one inserts a new field to counteract the mismatch caused by locality. The quanta of this mismatch field transmit a force. In this way, a local symmetry of a basic multiplet of fermions produces bosons that couple to the fermions in a way that is prescribed by the symmetry.

Suppose that our mechanical system is described by a Lagrangian $\mathcal{L}$, which is a function over spacetime $\{x_\mu\}$ of a generalized coordinate vector q and its corresponding momentum $q_{,\mu}$ (we use the abbreviation $q_{,\mu} \equiv \partial q / \partial x^\mu$). The action corresponding to this is found by integrating the Lagrangian density over an arbitrary four-volume Ω:

$$S = \int_\Omega \mathcal{L}(q, q_{,\mu}) dx_\nu \,, \tag{22}$$

where the dynamical variables q and $q_{,\mu}$ are to be seen as functions of x_μ. Because Ω is arbitrary, the requirement $\delta S = 0$ implies

$$\int_\Omega \frac{\partial \mathcal{L}}{\partial q} \delta q + \frac{\partial \mathcal{L}}{\partial q_{,\mu}} \delta q_{,\mu} \, dx_\nu = \int_\Omega \left(\frac{\partial \mathcal{L}}{\partial q} - \frac{\partial}{\partial x^\mu} \frac{\partial \mathcal{L}}{\partial q_{,\mu}} \right) \delta q \, dx_\nu + \oint_\Omega \frac{\partial \mathcal{L}}{\partial q_{,\mu}} \delta q \, dx_\mu = 0 \,, \tag{23}$$

from which the Lagrangian equations of motion follow directly because the surface integral is zero. Note that we are allowed to subject $\mathcal{L}$ to the same symmetry under which q is supposed to be symmetric, because $\mathcal{L} = f(q, q_{,\mu})$.

Now let us change q infinitesimally by some symmetry $\mathbf{L}$. A global symmetry does not change the equations of motion, because $\mathbf{L}$ commutes with δ. However, if $\mathbf{L}$ is a *local* symmetry, then $\mathbf{L} = \mathbf{L}(x_\mu)$, and therefore

$$q \to q + \delta q \quad \text{and} \quad \delta q = \epsilon(x_\mu) q . \tag{24}$$

It follows immediately that if $q_{,\mu} \to q_{,\mu} + \delta q_{,\mu}$, we get

$$\delta q_{,\mu} = \epsilon_{,\mu} q + \epsilon(x_\mu) q_{,\mu} \tag{25}$$

so that the integrand of $\delta \mathcal{S}$ becomes

$$\delta \mathcal{L} = \left(\frac{\partial \mathcal{L}}{\partial q} q + \frac{\partial \mathcal{L}}{\partial q_{,\mu}} q_{,\mu} \right) \epsilon + \frac{\partial \mathcal{L}}{\partial q_{,\mu}} \epsilon_{,\mu} q . \tag{26}$$

The term in brackets drops out because of the equations of motion, and we conclude that $\delta \mathcal{L} \neq 0$ because of the derivative $\epsilon_{,\mu}$: the *local* character of the transformation $\mathbf{L}$ spoils the proper extremum behaviour of $\mathcal{L}$, and no good equations of motion result!

In other words, the fact that $\mathbf{L}$ changes from event to event in spacetime produces a *mismatch* between $\mathbf{L}\mathcal{L}$ at one event and the $\mathbf{L}\mathcal{L}$ elsewhere. The key idea now is, to patch this up by adding extra terms to the Lagrangian to correct the mismatch. It is by no means obvious that this can be done successfully!

Because the culprit is a vector $\epsilon_{,\mu}$, we try to patch up $\mathcal{L}$ by adding a vector field to it. For the moment, let us call this field A', and the corresponding Lagrangian is

$$\mathcal{L}' = \mathcal{L}'(q, q_{,\mu}, A') , \tag{27}$$

of which we will now rigorously require that $\delta \mathcal{L}' = 0$. This requirement prescribes a functional dependence of $\mathcal{L}'$ on its arguments, as follows. The δq and $\delta q_{,\mu}$ are found as before; the variation $\delta A'$ is, of course, a linear combination of ϵ and $\epsilon_{,\mu}$ (for infinitesimal transformations). The most general form for $\delta A'$ is then

$$\delta A' = U A' \epsilon + C^\mu \epsilon_{,\mu} , \tag{28}$$

with constant scalar U and vector C^μ, to be determined afterwards. To get a proper equation of motion from the patched-up Lagrangian, we require

$$\delta \mathcal{L}' = \frac{\partial \mathcal{L}'}{\partial q} \delta q + \frac{\partial \mathcal{L}'}{\partial q_{,\mu}} \delta q_{,\mu} + \frac{\partial \mathcal{L}'}{\partial A'} \delta A' = 0 . \tag{29}$$

Inserting the expressions for δq and so forth yields a linear equation in ϵ and $\epsilon_{,\mu}$. Because the magnitude of ϵ is arbitrary (provided it is infinitesimal), each coefficient of ϵ and $\epsilon_{,\mu}$ must vanish independently. This gives

$$\frac{\partial \mathcal{L}'}{\partial q} q + \frac{\partial \mathcal{L}'}{\partial q_{,\mu}} q_{,\mu} + \frac{\partial \mathcal{L}'}{\partial A'} U A' = 0 \tag{30}$$

$$\frac{\partial \mathcal{L}'}{\partial q_{,\mu}} q + \frac{\partial \mathcal{L}'}{\partial A'} C^\mu = 0 \tag{31}$$

with the consistency requirement that

$$C^\mu C_\mu = 1 . \tag{32}$$

The latter means that C_μ has an inverse; if it did not, then some of the above equations would be linearly dependent and the system could not be solved. Now define the vector field A_μ as

$$A_\mu \equiv C_\mu A' \tag{33}$$

to find that

$$\frac{\partial \mathcal{L}'}{\partial q_{,\mu}} q + \frac{\partial \mathcal{L}'}{\partial A_\mu} = 0 \ . \tag{34}$$

The remarkable thing is, that *this equation is in fact a prescription for the way in which the Lagrangian must depend on its arguments*. We see directly that it requires that the vector field A_μ, which was introduced to patch up the mismatch created by the locality of $\mathbf{L}$ (i.e. the dependency $\mathbf{L} = \mathbf{L}(x_\mu)$) occurs in $\mathcal{L}'$ *only* through the combination

$$q_{;\mu} \equiv q_{,\mu} - q A_\mu \ , \tag{35}$$

the *covariant derivative* of q. The form $\mathcal{L}' = \mathcal{L}'(q, q_{,\mu}, A')$ allows us only one way to insert $q_{;\mu}$ into $\mathcal{L}'$, namely in exactly the same way as $\mathcal{L}$ depends on $q_{,\mu}$. This must be so because $q_{;\mu}$ contains a term that is linear in $q_{,\mu}$, and another term that can be made zero by letting $\mathbf{L}$ equal the identity. Thus we get

$$\mathcal{L}' = \mathcal{L}(q, q_{;\mu}) \tag{36}$$

and from now on we use this form.

Note that in the covariant derivative the local symmetry prescribes that q and A_μ couple by means of the product $q A_\mu$; in quantum electrodynamics this appears in the form where the dynamical variables q and $q_{,\mu}$ are replaced by the derivative ∂ and a constant factor ie:

$$q_{,\mu} - q A_\mu \to \partial - ieA \to i\psi^*(\gamma \cdot \partial - ie\gamma \cdot A)\psi \ , \tag{37}$$

which is the famous 'minimal coupling' term in the Dirac equation (the γ's are Dirac matrices).

Having now found that there is only one functional form of the Lagrangian which allows us to patch up the mismatch due to the local symmetry, it remains to determine the constants U and C^μ. First, we note that

$$\delta A_\mu = C_\mu \delta A' = C_\mu U \epsilon(x_\alpha) A' + \epsilon_{,\mu} = C_\mu C^\nu U \epsilon A_\nu + \epsilon_{,\mu} \ . \tag{38}$$

Second, we recall the expressions for the variations $\delta\mathcal{L}$ and $\delta\mathcal{L}'$, which lead directly to

$$\frac{\partial \mathcal{L}'}{\partial q} = \left.\frac{\partial \mathcal{L}}{\partial q}\right|_{q_{;\mu}} - \left.\frac{\partial \mathcal{L}}{\partial q_{;\mu}}\right|_q A_\mu \ , \tag{39}$$

$$\frac{\partial \mathcal{L}'}{\partial q_{,\mu}} = \left.\frac{\partial \mathcal{L}}{\partial q_{;\mu}}\right|_q \ , \tag{40}$$

$$\frac{\partial \mathcal{L}'}{\partial A'} = -\left.\frac{\partial \mathcal{L}}{\partial q_{;\nu}}\right|_q C_\nu q \ . \tag{41}$$

Inserting these into the equation resulting from $\delta\mathcal{L}' = 0$, we find

$$\left(\frac{\partial \mathcal{L}}{\partial q} q + \frac{\partial \mathcal{L}}{\partial q_{;\mu}} q_{;\mu}\right) - \frac{\partial \mathcal{L}}{\partial q_{;\nu}} q U A_\nu = 0 \ . \tag{42}$$

The term in brackets vanishes because of the equation of motion for $\mathcal{L}$, and because we had $\mathcal{L}' = \mathcal{L}(q, q_{;\mu})$. It follows immediately that $U = 0$. The definition of A_μ then gives, by means of the expression for $\delta A'$, that

$$\delta A_\mu = \epsilon_{,\mu} \ . \tag{43}$$

This demonstrates quite clearly how the vector field A_μ comes in because of the *local* character of the symmetry: if **L** were independent of x_α, we would have $\epsilon_{,\mu} = 0$!

One further point remains to be settled. By patching up the Lagrangian, we have let a genie out of a bottle, namely the field A_μ. We are now obliged to take this field seriously, and to identify it with an actual particle. In that case, we must of course allow A_μ to occur in the Lagrangian as a free field (i.e. as more than just an entity which couples to the q-field by qA_μ). It may be a trifle much to ask, but can the locality of **L** prescribe the form of the occurrence of this free field too?

The patch-up vector field A_μ may occur itself in the Lagrangian as a dynamical variable, together with its spacetime derivative $A_{\mu,\nu}$, in the same way that we had a dependence on the dynamical variables of the q-field. Because $\mathcal{L}$ is linear, we can insert extra terms by simple addition, so we can restrict ourselves to finding the sub-part $\mathcal{L}''$ that depends on the A's only, and then add it to what we had already (note that the coupling term has already been disposed of!) We use the same variational form:

$$\delta\mathcal{L}'' = \frac{\partial\mathcal{L}''}{\partial A_\mu}\delta A_\mu + \frac{\partial\mathcal{L}''}{\partial A_{\mu,\nu}}\delta A_{\mu,\nu} . \tag{44}$$

As usual, we insert δA_μ, and require that each coefficient of ϵ and $\epsilon_{,\mu}$ vanish independently. This yields the equations

$$\frac{\partial\mathcal{L}''}{\partial A_\mu} = 0 \quad , \tag{45}$$

$$\frac{\partial\mathcal{L}''}{\partial A_{\mu,\nu}} + \frac{\partial\mathcal{L}''}{\partial A_{\nu,\mu}} = 0 \quad . \tag{46}$$

Accordingly, we find that the patch-up field *cannot itself occur in the Lagrangian*. Consequently, the field A_μ *is not an observable;* but it *can* couple to the dynamical variable q by means of the term qA_μ. In Feynman terms: the A_μ can only occur between vertices, it is an intermediary, a *virtual particle*. The above shows that the new field can occur in $\mathcal{L}$ only through the combination

$$F_{\mu\nu} = A_{\nu,\mu} - A_{\mu,\nu} . \tag{47}$$

That is to say, *the curl of the field is an observable!* This should of course look very familiar to aficionados of Maxwell's Equations.

This completes the demonstration that the requirement of local symmetry of the Lagrangian is so severe, that not only the way in which the patch-up field A_μ couples to the q-field, but also the way in which it must occur in the Lagrangian is prescribed entirely. This almost total lack of arbitrariness is what makes the local symmetry concept so compelling.

It can be shown that the same kind of construction works for vector fields q^a (in fact, this is what the original Yang-Mills paper was all about). In that case, it can be shown that the strictness and cleanness with which the form of the Lagrangian is prescribed is due to the group structure of the symmetry.

We have four such cases in Nature:

(1) the case of a phase-rotation symmetry $U(1)$, (i.e. the multiplication with a complex scalar function as treated above), which produces electromagnetism;

(2) the case of the "isospin symmetry" $SU(2)$ (i.e. multiplication with a factor derived from a 2×2 symmetry via $\exp(\frac{1}{2}ig\tau \cdot \omega)$, where τ are Pauli matrices and ω is an arbitrary smooth function over spacetime), which produces the weak interaction;

(3) the group $SU(3)$, leading to the colour interaction;

(4) Lorentz symmetry, which gives rise to the gravitational interaction (General Relativity).

In the Yang-Mills case, the group has non-zero structure constants f_{bc}^a, and following exactly the same line of resoning one may show that the "wrinkle" or "patch-up" field A^a can occur in the Lagrangian only through the combination

$$F_{\mu\nu}^a = A_{\nu,\mu}^a - A_{\mu,\nu}^a - \frac{1}{2} f_{bc}^a \left(A_\mu^b A_\nu^c - A_\nu^b A_\mu^c \right) , \tag{48}$$

which clearly shows the occurrence of non-linear terms due to the non-Abelian character of the group. The range of the index a depends on the group dimension; for $SU(N)$, it is $N^2 - 1$.

If the local symmetry is Lorentz symmetry, one may show in precisely the same way – though with much more effort – that the patch-up fields (which in gravity theory are traditionally called Γ instead if A) can occur in the Lagrangian only through the combination

$$R_{\mu\nu\kappa}^\lambda \equiv \Gamma_{\mu\nu,\kappa}^\lambda - \Gamma_{\mu\kappa,\nu}^\lambda + \Gamma_{\mu\nu}^\eta \Gamma_{\kappa\eta}^\lambda - \Gamma_{\mu\kappa}^\eta \Gamma_{\nu\eta}^\lambda , \tag{49}$$

which is the Riemann-Christoffel tensor.

CLASH

The remarkable fact is that both GRT and the Standard Model describe fundamental interactions by means of a gauge field. On small scales, these are quantum fields due to the symmetries $U(1)$, $SU(2)$ and $SU(3)$. On large scales, the gauge field is the Christoffel object $\Gamma_{\mu\nu}^\lambda$ due to local Lorentz symmetry.

It would be great if these similarities allowed us to bring all known forces together in one formalism. Then the expression 'the Universe is made of particles, space and time' would get a truly compelling uniformity, so that the field equations would read something like

$$\left\{ \begin{array}{c} \text{structure of} \\ \text{space, time and} \\ \text{particles} \end{array} \right\} = 0 .$$

Alas, we are disappointed. The structure of spacetime cannot be included with the density of particles if we use the Standard Model in the matter tensor. In field theory a potential is no longer something that can be freely changed by adding an arbitrary scalar term; due to the local (as opposed to global) character of the fields, a potential becomes an entity in itself, witness for example the occurrence of A_μ. In electrodynamics this is merely the vector potential, in quantum electrodynamics it stands for a real particle, the photon, which cannot be transformed away. Einstein's conjecture runs into profound trouble because the reality of potentials implies that the zero-point energy of the vacuum must be included in the Einstein equation.

PIGS IN SPACE

So the theory of forces and matter tells us that in the vacuum, this space 'supposed to be empty', spontaneous particle-antiparticle pairs arise. And that means that this apparently empty vacuum plays an active role in Nature: it has properties that have a profound influence on the behaviour of particles and their interaction. This is apparent, among other things, in the anomalous magnetic moment of the electron and in the Casimir-Polder force, in which the zero-point energy corresponding to the spontaneous pairs exerts a measurable influence on the force between conductors.

If one introduces matter into space in this manner, things simply won't fit. We cannot have pigs in space, because their attendant vacuum fluctuations would ruin the Universe. Einstein built his theory of spacetime expressly in such a way that Newtonian mechanics was recovered in the limit for small speeds and weak fields. But Newtonian theory is all wrong on a small scale, so it would be a stupendous marvel if the Einstein equation gave the right result all the way. If only he had believed in quanta, maybe he could have forged a correspondence with the Schrödinger or Dirac equation!

There are several ways out. People have tried to cancel the vacuum fluctuations against each other. That is in itself not so bizarre: a force is the net result of a quantum sum over *all* Feynman diagrams, and counterdiagrams might be dreamed up, as in the case when the c-quark was predicted from the absence of the decay of the kaon into a pair of muons. But the cancellation would have to be so extraordinarily perfect that it is contrived in the extreme. Only unbroken supersymmetry seems to help. In these theories, there is a symmetry that connects fermions and bosons. Quite a desperate move, because bosons and fermions are as un-alike as possible! However, such un-kosher combinations can be made, using generators Q_α, four-momentum p^μ, Pauli matrices τ_μ and the anticommutation rule

$$\left\{Q_\alpha, Q_\beta^\dagger\right\} = (\tau_\mu)_{\alpha\beta}\, p^\mu \ . \tag{50}$$

The vacuum is the state which has

$$Q_\alpha|0\rangle = Q_\alpha^\dagger|0\rangle = 0 \ , \tag{51}$$

so that the anticommutator {} immediately produces

$$\langle 0|p^\mu|0\rangle = 0 \ , \tag{52}$$

that is to say, the energy-momentum density of the vacuum is zero: $\Lambda = 0$. For this mechanism to work, each fermion should have a bosonic counterpart and vice versa. Our world doesn't look like that in the least, so we are not much further along.

Who's gotta yield, the left- or the right-hand side of the Einstein equation? Both, in a sense, but the left more than the right. The right-hand side, $T_{\mu\nu}$, is a continuum average and doesn't contain individual particles. That must modified: no more spacetime averages, no densities, because in a quantum formulation we should not expect space and time to be continuous in the conventional sense. Averages and derivatives $\partial/\partial x^\mu$ would lose their meaning.

The left-hand side, the Einstein tensor, at first appears to be the strongest fortress because it is purely mathematical. However, its description of the structure of spacetime in terms of a distance recipe (via $g_{\mu\nu}$) can probably not be quantized. After all, if one were to interpret g in terms of a collection of particles instead of a classical field, we get a paradox: *if spacetime is made of particles, how could such a particle move through space and time?* Stated somewhat differently, the usual particle symmetries (U(1), SU(2), SU(3) and relatives) connect fermions and bosons, whereas the Lorentz symmetry says something about spacetime and not about particles. Furthermore, it is my prejudice that quantum behaviour is a much more strongly established physical effect than the Einsteinian variant of gravity. Anyone can see the Balmer series with minimal equipment. Compared with that loud-and-clear demonstration of quantization, the usual relativistic tests (Mercury, gravitational lensing) seem weak and indirect. A black hole would be a convincing thing, analogous to the hydrogen atom, but its properties are still inferred only indirectly.

So how can we make particles out of g? When matching $g_{\mu\nu}$ to quantum degrees of freedom we need not assign every component to a dynamical field. In QED, the classical vector potential A_μ reappears as the photon; in GRT we have a correspondence between the Newtonian potential Φ and $g_{\mu\nu}$. But part of the covariance with respect to g could be purely a coordinate effect, i.e. there might be one specific 'nature-given' set of coordinates where some fields vanish. The large-scale homogeneity and isotropy of the Universe is a case in point: spherical coordinates might be a preferred reference frame, in which case Lorentz covariance is no longer guaranteed. Having decided which components of g should appear as quantum degrees of freedom, we must decide how to assign these to observable particles. Quantum superposition allows us to construct states by linear combination, as in the case of the $U(1) \otimes SU(2)$ unification: the photon γ and the neutral vector boson Z^0 are superpositions of the $U(1)$ and $SU(2)$ velpons. Compare also the Englert-Brout-Higgs mechanism for making particles massive: the degrees of freedom of the scalar particles are used to generate two extra degrees of freedom for the W and the Z, which – if they were massless – would have only two helicity states each, instead of the four of a massive particle.

One could make $\Lambda = 0$ at one specific point in time, but the expansion of the Universe would shift us away from that point and we'd be just as badly off. Similarly, if we inflate the Universe by using some sort of value K derived from grand unification or a similar theory, we should expect that today we'd still be not too far away from an $n = 1$ state in the bottom of the potential of the GUT, and thus have a substantial fraction of the GUT potential still around.

Currently I am trying to do away with the problem more radically by just stating that, by fiat, gravitons do not interact with vacuum fluctuations. The attractiveness is that this removes the need for renormalization of gravity: there'd be no more loops in the graviton propagator. But taken literally that should also exclude one-loop diagrams from the interaction between gravitons; gravity would no longer be non-linear, against all the evidence.

At this point, I'm talking pie-in-the-sky. It is like Bohr's treatment of the hydrogen atom. Bohr knew perfectly well that an accelerated electron in an atom ought to radiate like crazy. But he pretended it doesn't, just to see what happens. But I'm not Bohr, and I don't know what the equivalent of the hydrogen atom is. The Schwarzschild black hole may fit the bill. But the effects I've calculated are nowhere near observable, and probably all false. Nobody, so far, has found an escape route.

ENVOI

What I have discussed here is known and unknown. So well known that I'm quite embarrassed to talk about it, and so utterly unknown that I feel similarly embarrassed. In between we may be lucky to find fertile ground where we can practice what Medawar called *the art of the soluble*.

REFERENCES

1. Icke, V., *The Force of Symmetry*, Cambridge University Press, Cambridge (1995).

2. Utiyama, R., *Invariant theoretical interpretation of interaction, Phys. Review* **101** (1956) 1597-1607.

3. Weinberg, S., *The cosmological constant problem, Rev. Mod. Physics* **61** (1989) 1-23.

CONTRIBUTORS

C. Carlson
Physics Department
College of William and Mary
Williamsburg, VA 23187
carlson@spiffy.physics.wm.edu

H. Clement
Physikalisches Institut
Universität Tübingen
Auf der Morgenstelle 14
D-72076 Tübingen
Germany
clement@pit.physik.uni-tuebingen.de

P. Grabmayr
Physikalisches Institut
Universität Tübingen
Auf der Morgenstelle 14
D-72076 Tübingen
Germany
grabmayr@pit.physik.uni-tuebingen.de

U. Heinz
Institut für Theoretische Physik II
Universität Regensburg
D-8400 Regensburg
Germany
ulrich.heinz@physik.uni-regensburg.de

H. Horiuchi
Department of Physics
Kyoto University
Kyoto 606-01
Japan
horiuchi@scphys.kyoto-u.ac.jp

V. Icke
Afdeling Sterrenkunde en Natuurkunde
Rijksuniversiteit Leiden
2300 RA Leiden
The Netherlands
icke@strw.leidenuniv.nl

V. Koch
Lawrence Berkely Laboratory
Berkely, CA 94720
vkoch@nsdssd.lbl.gov

R. Malfliet
KVI
9747 AA Groningen
The Netherlands
malfliet@kvi.nl

V. R. Pandharipande
Physics Department
University of Illinois at Urbana-
    Champaign
Champaign, IL 61820
maryo@uiuc.edu

J. Schukraft
CERN
CH-1211 Geneva 23
Switzerland
jurgen.schukraft@cern.ch

W. Trautmann
GSI Darmstadt
D-64220 Darmstadt
Germany
w.trautmann@gsi.de

PARTICIPANTS

H. Akimune Research Center for Nuclear
  Physics
Osaka University
Osaka 567
Japan
akimune@rcnp.osaka-u.acjp

H. Arslan
Department of Physics
Zonguldak Karaelmas University
67100 Zonguldak
Turkey
arslan @ elmas.bim.karaelmas.edu.tr

M.F. van Batenburg
NIKHEF
1009 DB Amsterdam
The Netherlands
marcelvb@nikhefk.nikhef.nl

Th. S. Bauer
Physics Laboratory
Free University
1081 HV Amsterdam
The Netherlands
bauer@fys.ruu.nl

H. P. Blok
Faculteit Natuurkunde en Sterrenkunde
Vrije Universiteit de Boelelaan
1081 HV Amsterdam
The Netherlands
henkb@nikhefk.nikhef.nl

D. J. Boersma
R. v.d. Graaff Laboratorium
Rijksuniversiteit Utrecht
3508 TA Utrecht
The Netherlands
davidb@nikhefk.nikhef.nl

D. Bonekaemper
Institut für Kernphysik
Universität Münster
D-48149 Münster
Germany
bonekam@uni-muenster.de

A. Boukour
Physics Department
Univ. Libre de Bruxelles
B-1050 Bruxelles
Belgium
aboukou@orca.vub.ac.be

W. Brodowski
Physikalisches Institut
Universität Tübingen
Auf der Morgenstelle 14
D-72076 Tübingen
Germany
brodowski@pit.physik.uni-tuebingen.de

C. Carlson
Physics Department
College of William and Mary
Williamsburg, VA 23187
carlson @ spiffy.physics.wm.edu

F. Ceretto
Max Planck Institut für Kernphysik
D-69117 Heidelberg
Germany
federica@fossy2.mpi-hd.mpg.de

H. Clement
Physikalisches Institut
Universität Tübingen
Auf der Morgenstelle 14
D-72076 Tübingen
Germany
clement@pit.physik.uni-tuebingen.de

L. van Daele
Department of Subatomic and Radiation
  Physics
University of Gent
B-9000 Gent
Belgium
lieven@inwfaxp2.rug.ac.be

R. D. Dahl
Niels Bohr Institute
DK-2100 Copenhagen
Denmark
dahl@alf.nbi.dk

B. Decroix
Department of Subatomic and Radiation
  Physics
University of Gent
B-9000 Gent
Belgium
bruno@inwfaxpl.rug.ac.be

B. D. Diederich
Nuclear Physics Building
Old Dominion University
Norfolk, VA 23529
diededc@cebaf.gov

S. S. Dimitrova
Nuclear Physics Laboratory
Keble Road
Oxford OX1 3RH
England
s.dimitrova1@physics.oxford.ac.uk

D. Dutta
CEBAF, MS 90
Newport News, VA 23606
ddutta@cebaf.gov

A. Dvoredsky
106-38 Kellogg Lab.
CALTECH
Pasadena, CA 91125
andread@hermes.desy.de

R. Eckhardt
Institut für Theoretische Physik II
Arbeitsgruppe Mittelenergiephysik
Ruhr-Universität Bochum
D-44780 Bochum
Germany
roberte@mep.ruhr-uni-bochum.de

A. Fokin
Department of Physics
University of Lund
Sölvegatan 14
Lund
Sweden
kosu_fokin@garbo.lucas.lu.se

M. Geurts
NIKHEF
1009 DB Amsterdam
The Netherlands
m.geurts@math.utwente.nl

T. Gharib
Department of Physics
Faculty of Science
Ain Shams University
Cairo
Egypt
gharib@asunet.shams.eun.eg

S. Gieseke
Institut für Theoretische Physik
Universität Bremen
D-28359 Bremen
Germany
gieseke@physik.uni-bremen.de

M. J. van Goethem
KVI
9747 AA Groningen
The Netherlands
vangoethem@kvi.nl

P. Grabmayr
Physikalisches Institut
Universität Tübingen
Auf der Morgenstelle 14
D-72076 Tübingen
Germany
grabmayr@pit physik.uni-tuebingen.de

D. Groep
NIKHEF
1009 DB Amsterdam
The Netherlands
davidg@nikhefk.nikhef.nl

V. M. Hannen
Institut für Kernphysik
Universität Münster
D-48149 Münster
Germany
hannen@ikpuni-muenster.de

J. Hansper
Institut für Theoretische Physik II
Arbeitsgruppe Mittelenergiephysik
Ruhr-Universität Bochum
D-44780 Bochum
Germany
joergh@mep.ruhr-uni-bochum.de

M. N. Harakeh
KVI
9747 AA Groningen
The Netherlands
harakeh@kvi.nl

U. Heinz
Institut für Theoretische Physik II
Universität Regensburg
D-8400 Regensburg
Germany
ulrich.heinz@physik.uni-regensburg.de

M. Hoefman
KVI
9747 AA Groningen
The Netherlands
hoefman@kvi.nl

G. 't Hooft
Inst. voor Theor. Physica
3508 TA Utrecht
The Netherlands
thooft@fys.ruu.nl

H. Horiuchi
Department of Physics
Kyoto University
Kyoto 606-01
Japan
horiuchi@scphys.kyoto-u.ac jp

M.-T. Hütt
II. Physikalisches Institut
Universität Göttingen
D-37073 Göttingen
Germany
huett@up200.dnet.gwdg.de

H. Huisman
KVI
9747 AA Groningen
The Netherlands
huisman@kvi.nl

V. Icke
Afdeling Sterrenkunde en Natuurkunde
Rijksuniversiteit Leiden
2300 RA Leiden
The Netherlands
icke@strw.leidenuniv.nl

B. Jacak
Los Alamos National Laboratory
Los Alamos, NM 87545
jacak@p2hp2.1anl.gov

N. Jachowiz
Department of Subatomic and Radiation
Physics
University of Gent
B-9000 Gent
Belgium
natalie@inwfaxp2.rug.ac.be

M. D. Kadi-Hanifi
Alger 16000
Algeria
bouayed@ist.cerist.dz

D.N. Kadrev
Institute for Nuclear Research and
  Nuclear Energy
Sofia 1784
Bulgaria
kadrev@inrne.acad.bg

V. Kastens
Institut für Theoretische Physik
Universität Bremen
D-28359 Bremen
Germany
kastens@physik.uni-bremen.de

S. Kocaoba
Faculty of Art and Science
Yildiz Technical University
80270 Sisli - Istanbul
Turkey
(no e-mail)

J. H. Koch
NIKHEF
1009 DB Amsterdam
The Netherlands
justus@nikhefk.nikhef.nl

V. Koch
Lawrence Berkeley Laboratory
Berkeley, CA 94720
vkoch@nsdssd.lbl.gov

J. Konijn
NIKHEF
1009 DB Amsterdam
The Netherlands
joop@nikhef.nl

F. Laue
GSI
D-64291 Darmstadt
Germany
f.laue@gsi.de

T. E. Leth
Institute of Physics and Astronomy
University of Aarhus
DK-8000 Aarhus
Denmark
(no e-mail)

N. A. Loktionova
Lebedev Physical Institute
117 333 Moscow
Russia
loktion@sgi.lpi.msk.su

R. Malfliet
KVI
9747 AA Groningen
The Netherlands
malfliet@kvi.nl

M. Mang
GSI
D-64291 Darmstadt
Germany
m.mang@gsi.de

G. Martens
Institut für Theoretische Physik
Universität Bremen
D-28359 Bremen
Germany
gmartens@physik.uni-bremen.de

G. H. Martinus
KVI
9747 AA Groningen
The Netherlands
martinus@kvi.nl

R. Medaglia
DAPHNIA/SPHN, bat.
703 CEA-Saclay
Orme des Merisiers
F-91191 Gif-sur-Yvette
France
rosella@phnx7.saclay.cea.fr

J. G. Messchendorp
KVI
9747 AA Groningen
The Netherlands
messchendorp@kvi.nl

H. K. T. van der Molen
KVI
9747 AA Groningen
The Netherlands
vandermolen@kvi.nl

F. A. Natter
Physikalisches Institut
Universität Tübingen
Auf der Morgenstelle 14
D-72076 Tübingen
Germany
natter@pit.physik.uni-tuebingen.de

J. Negele
Center for Theoretical Physics
MIT
Cambridge, MA 02139
negele@mitlns.mit edu

R. Ogul
Physics Department
Selcuk University
42031 Kampus-Konya
Turkey
rogul @ mevlana.cc. selcuk.edu.tr

R. Ostendorf
GANIL
F-14021 Caen Cedex
France
ostendorf@frcpn11.in2p3.fr

V. R. Pandharipande
Physics Department
University of Illinois at
  Urbana-Champaign
Champaign, IL 61820
maryo@uiuc.edu

P. Papaconstantinou
Division of Nuclear and Particle Physics
University of Athens
157 71 Athens
Greece
ppapakom@atlas.uoa.gr

V. Pascalutsa
Institute for Theoretical Physics
University of Utrecht
3508 TA Utrecht
The Netherlands
v.pascalutsa@ fys.ruu.nl

V. R. Pomeroy
Physics Department
University of New Hampshire
Durham, NH 03824
vrp@curie.sr.unh.edu

D. Ridikas
Department of Physics
University of Bergen
N-5007 Bergen
Norway
danas.ridikas@fi.uib.no

D. Rondeshagen
Institut für Kernphysik
Universität Münster
D-48149 Münster
Germany
rondesh@ikp.uni-muenster.de

P. Rosinsky
Department of Nuclear Physics
Comenius University
842 25 Bratislava
Slovakia
rosinsky@fmph.unibask

J. Schmelzer
Fachbereich Physik
Universität Rostock
D-18051 Rostock
Germany
juern@cv.jinr.dubna.su

O. Scholten
KVI
9747 AA Groningen
The Netherlands
scholten@kvi.nl

J. Schukraft
CERN
CH-1211 Geneva 23
Switzerland
jurgen.schukraft@cern.ch

M. Seip
KVI
9747 AA Groningen
The Netherlands
seip@kvi.nl

A. Shafi
Department of Physics
The George Washington University
Washington, D.C. 20052
azizs@cebaf.gov

Y.-H. Shin
GSI
D-64291 Darmstadt
Germany
shin@rzhp9a.gsi.de

J. K. Spasova
Department of Theoretical Physics
Konstantin Preslavsky University
Shoumen 9712
Bulgaria
spasova@uni-shoumen.bg

R. Starink
NIKHEF
1009 DB Amsterdam
The Netherlands
ronalds@nikhefk.nikhef.nl

M. F. M. Steenbakkers
NIKHEF
1009 DB Amsterdam
The Netherlands
martijn@nikhefk.nikhef.nl

W. Trautmann
GSI Darmstadt
D-64220 Darmstadt
Germany
w.trautmann@gsi.de

P. Vogt
KVI
9747 AA Groningen
The Netherlands
vogt@kvi.nl

J. Volmer
NIKHEF
1009 DB Amsterdam
The Netherlands
volmer@nikhefk.nikhef.nl

J. G. Wang
Falkiner High Energy Physics
   Department
School of Physics
University of Sydney
Sydney N.S.W. 2006
Australia
wangjg@physics.usyd.edu.au

R.G.T. Zegers
KVI
9747 AA Groningen
The Netherlands
zegers@kvi.nl

INDEX